Ute Christina Herzfeld

Atlas of Antarctica

Topographic Maps from
Geostatistical Analysis of
Satellite Radar Altimeter Data

With 169 Figures

 Springer

Professor Dr. Ute Christina Herzfeld
Cooperative Institute for Research in the Environmental Sciences
National Snow and Ice Data Center
University of Colorado Boulder
Boulder, CO 80309-0449
USA

Geomathematics
University of Trier
54286 Trier
Germany

email: *herzfeld@iceberg.colorado.edu*

ISBN 3-540-43457-7   Springer Berlin Heidelberg New York

Images on the front cover: Sea ice and ice shelf with German Research Vessel Polarstern in Atka Bay, Weddell Sea; Iceberg offshore of New Schwabenland; *Images on the back cover:* Penguin Colony; Ice edge and sea ice in the polynia in Halley Bay, Antarctica; Author at Jakobshavn Isbrae, West Greenland.

Library of Congress Control Number: 2004107463

Springer is a part of Springer Science+Business Media
springeronline.com
© Springer-Verlag Berlin Heidelberg 2004
Printed in Germany

Cover design: Erich Kirchner, Heidelberg
Production: Luisa Tonarelli
Typesetting and layout: Ute C. Herzfeld
Printing and binding: Druckerei Stürtz, Würzburg

Printed on acid free paper   30/3141/LT – 5 4 3 2 1 0

The Antarctic Atlas
is dedicated to
Helmut, Almut and Matthias Urs

Preface

Although it is generally understood that the Antarctic Ice Sheet plays a critical role in the changing global system, there is to date still a lack of generally available information on the subject. Climatic change and the role of the polar areas are often discussed in the media. Much of our knowledge depends on models, and basic things such as maps are missing. The generally used geophysical atlas of Antarctica is still the one published by Drewry twenty years ago, with literally huge "white areas" on the maps (Drewry, D.J. (editor), 1983, Antarctic Glaciological and Geophysical Folio, Cambridge, Scott Polar Research Institute). While there are maps of specific areas of Antarctica available (usually where field research is undertaken), there is not one comprehensive work.

Antarctica is a remote continent. — Despite about 200 years of Antarctic exploration, many places in the Antarctic have not been visited by mankind, because Antarctica is hard to reach from inhabited places, a foreboding place to travel because of glaciated terrain and harsh climate, and the fact that the continent is surrounded by a frozen ocean for a large part of the year. Information on the remote continent is available today from satellite data.

Antarctica is a frozen continent. — It contains over 90% of the Earth's ice mass, and most of its area is covered by ice.

Antarctica is a fascinating continent. — The same geographic properties that make Antarctica difficult to visit draw our imagination to this wild and different place. In 1984, I interrupted my dissertation project in mathematics for the chance to participate in an Antarctic expedition, with the German Research Vessel POLARSTERN, willing to work on more or less anything that would be asked of me to get down south to the Antarctic. As it turned out, I got to work on least-squares prediction and collocation, a method used in geodesy for interpolation, which led me into a new field of science — geostatistics — which would become the centerpiece of my mathematical research; but the expedition did not lead me to Antarctica, ending in South America after crossing the Atlantic. In 1987/88 I "finally" set foot on the icy continent, during a $3\frac{1}{2}$ month expedition to Antarctica, also with the POLARSTERN and the Alfred-Wegener-Institute for Polar Research (Bremerhaven, Germany), to the Weddell Sea sector of Antarctica. In those days, expeditions were planned and staffed in a way that helping hands were always needed, scientists helped to establish stations, transport loads and clean up fuel barrels; marine biologists needed help in counting species in huge piles of biomass dumped on the deck; geophysicists needed help for around-the-clock watch at instruments; trace-chemical stations needed to be established far away from the main stations, and glaciologic data needed to be processed. This provided a great way to learn about many aspects of the Antarctic environment as well as a fun way to interact and collaborate. Quickly I noticed that who volunteered first got to set up remote sampling locations, approached by helicopter or snowcat tracks, while who volunteered last may end up pumping Arctic diesel ... Whatever the tasks, the Antarctic holds a grip on almost anyone who has ever visited it. Expeditions to places without the ICE are just not the same, as I came to discover in later seagoing expeditions.

Our society's knowledge on Antarctica is still far less than knowledge about any other continent, despite decades of exploration of Earth from space and publications in scientific journals. A school teacher who does not know what distinguishes Antarctica from the Arctic is not an exception. Books on Antarctica that are available to a general readership are commonly coffee-table picture books, compiled by photographers on voyages to the Antarctic, often with a focus on wildlife. In contrast, the Atlas is focused on snow and ice research, relationship of ice and climate, geography, cartography and geodesy, geomathematics and satellite geophysics — on revealing the seventh continent from space. As such, the Atlas of Antarctica will fill an information gap.

Today's educators want to teach children exciting facts about remote places, and tourism increases in an adventure and exploration component. Together with this there is a growing interest in Antarctica, and, often related to the cost of travel to remote places, the knowledge-desiring public has a high level of education. An increasing public interest in Global Change and Climate-related questions also creates a need for books that convey facts at a level that lies between that of picture books and science treatises.

The Atlas of Antarctica is written with state-of-the-art scientific accuracy, so that it is hoped that it may be useful for students and researchers of glaciology, geophysics, remote sensing, cartography and Antarctic research, and at the same time introduce to the complex of questions and facts outlined above in a way that is informative and fun for the general reader interested in Antarctica.

Information about the frozen continent, Antarctica, is not only accessible by expeditions, without the availability of satellite data, there would still be large "white areas" on the map.

The "Atlas of Antarctica" utilizes satellite data — more precisely, radar altimeter data from SEASAT, GEOSAT, ERS-1 and ERS-2. Other than the better known satellite images, altimeter measurements provide information on elevation and, as such, may be the basis of topographic maps. The Atlas of Antarctica covers the entire Antarctic continent in a collection of accurate topographic maps, each with 3 km by 3 km grid resolution, surface elevation in grey shades and contours, and place names of glaciers, ice streams, and ice shelves, as well as of other major geographic features. Other than satellite images, the

Atlas maps provide elevation which is geophysically useful and accurate information. As such the Atlas maps are the highest-resolution elevation maps available today, derived from satellite radar altimeter data by application of a geostatistical method, which was specifically designed by the author for the problem of ice-sheet mapping. Part II (chapter (D)) forms the center piece of the Atlas.

In part I "Motivation and Methods", introductory chapters concern the following topics: (A) The Antarctic Ice Sheet and its role in the Global System, (B) Satellite Remote Sensing, and (C) Data analysis methods applied in the Antarctic Atlas production and map construction. These sections may serve as introductions to the realm of ice and climate, Antarctic geography, to satellite observations of the cryosphere, principles of satellite radar altimeter observations and data processing, and geostatistical estimation or interpolation applied in cartography.

Applications presented in part III include monitoring changes in Antarctic glaciers, ice streams and ice shelves (chapter (E)) and detailed regional studies of outlet glaciers of the inland ice that are particularly exciting (chapter (F)). Combination and integration of digital elevation models from radar altimeter data and Synthetic Aperture Radar (SAR) data, which show surface structures and flow features of glaciers and ice streams, is given as an example of employing data from two sources in glaciologic research (chapter (G)). The Atlas contains a total of 145 maps, 25 figures, index maps and diagrams, 5 tables, and furthermore references on related subjects in glaciology, geomathematics, remote sensing and geodesy (chapter (H)), a glaciologic glossary, an index of the Antarctic place names shown, and a list of Antarctic expeditions (chapter (I)).

The Atlas is for educators, glaciologists, researchers, students, tourists, anyone interested in Antarctica. It is my hope that the reader may share some of my enthusiasm for Antarctica and enjoy "discovering" some of its many fascinating geographic and glaciologic features.

Boulder, Colorado, Christmas 2003

Ute Christina Herzfeld

Acknowledgements

Many people have helped throughout the course of the Antarctic Atlas project. Shortcomings are, of course, the author's, and comments are appreciated. The Atlas is not intended to be a comprehensive treatment of Antarctica and Antarctic science — there are too many disciplines and too many publications, but to fill a gap in knowledge.

The original idea to apply geostatistics to satellite radar altimetry was suggested to me at a glaciology meeting in Seattle in 1988 by colleagues at NASA. Over the years, the Atlas project took on its own dynamics and was continued with data from ERS-1 and various groups of students at the Universities of California San Diego (UCSD), Colorado (CU) Boulder, and Trier, Germany. Funding was provided by NASA Office of Polar Programs grants NAGW-3790 and NAG5-6114. Dr. Robert Thomas, at the time program manager at NASA's Office of Polar Programs, asked whether it would be possible to make maps of all of Antarctica. Data were provided by Dr. H. Jay Zwally and co-workers Dr. John DiMarzio and Dr. Anita Brenner at NASA Goddard Space Flight Center. Dr. Craig Lingle and Li-her Lee (University of Alaska Fairbanks) collaborated on Lambert Glacier studies. All this help and support is gratefully acknowledged.

Many thanks are due to my undergraduate and graduate students who helped with various parts of the Antarctic Mapping project, including coding, data processing, map production and labeling of geographic features: to Michael Matassa and Chris Higginson at CU Boulder; to Michael Lambert and Cecily Freeman at UCSD; to members of the "map team" at the Universität Trier, Michael Matassa, Ralf Stosius, Marcus Schneider, Marion Stellmes and Birgit Kausch. Of those, Michael Matassa has been instrumental in helping establish the "map team" at Geomathe-Uni Trier, while Ralf Stosius has dedicated a lot of energy to the Atlas Book project in recent years helping me train younger map-team members, keeping lists of files and maps and, at times, corresponding with the publisher. Matthias Mimler assisted in programming the TRANSVIEW software.

Much of my time as a Visiting Fellow at the Cooperative Institute for Research in Environmental Sciences (CIRES) and the National Snow and Ice Data Center (NSIDC)/World Data Center A for Glaciology has been devoted to the Atlas Book project. I wish to thank CIRES for the Fellowship, and CIRES and especially NSIDC for their hospitality. I am indebted to Dr. Konrad Steffen (Acting Director, CIRES) and Dr. Roger Barry (Director, NSIDC) for their support and understanding.

Thanks are also due to Dr. Robert L. Parker (Institute of Geophysics and Planetary Physics, UCSD) for the contouring and coloring programs CONTOUR and COLOR used in the Atlas maps and the plot program PLOTXY, to Dr. Bruce Fast (Department of Applied Mathematics, CU Boulder) for help with integration of postscript and LaTeX, to Allyn McAuley and Melanie Percy for help with typesetting with support provided by K. Steffen, to Bruce Raup, Peter Gibbons, Matthew Applegate, Daryl Kohlerschmidt and Jennifer Gerull for help with figure production, and last not least to Scott Williams (all CIRES / NSIDC, CU Boulder) for help with LaTeX scripting.

My thanks go to the publishing editor Dr. Christian Witschel and collaborators in his team at Springer-Verlag, Heidelberg.

Special thanks are due to Dr. Helmut Mayer, without whose help the Atlas would have been impossible, for his extensive knowledge of the literature, for helping me with the real work the book created, and for many times putting other tasks behind the Atlas. I also wish to thank my daughter Almut for all the understanding of a five-year old. After all — should I admit that I may have never actually started *writing*, had it not been for Matthias Urs, my little son, who stepped into the world and required me to stay home more than usual. So, when he fell asleep in his crib in my home office, I started writing.

Ute Christina Herzfeld

Contents

(II) The Atlas

(III) Applications

(IV) References and Appendix

Part I

Motivation and Methods

(A) The Antarctic Ice Sheet and its Role in the Global System

(A.1) Main Geographic and Glaciologic Provinces of Antarctica

The most conspicuous property of Antarctica is its ice mass — most of the Antarctic continent's landmass is covered by ice — ice sheets, glaciers, ice streams and ice shelves. All these features consist of ice and may thus be summerized as "glaciers", but have different properties. An ice sheet covers large land areas and is flat and wide, the ice in the ice sheet flows at a low velocity, following gravitational forces, generally towards the ice sheet's margin. An ice stream is an area of ice flowing at a higher velocity than the surrounding inland ice, large ice streams drain the inland into the Circum-Antarctic ocean. The lower part of an ice stream draining into the ocean becomes afloat (at the line termed the grounding line), it forms a floating tongue. Many Antarctic ice streams and glaciers end in ice shelves, areas of ocean water covered by thick ice (more glaciologic terms are explained in the glaciologic glossary, section (I.1)). Often several glaciers and ice streams end in the same ice shelf, but there are also coastal areas of Antarctica that are not bordered by ice shelves. For instance, the Atlantic Ocean sector is fringed by ice shelves, with large ice streams flowing into those, including Slessor Glacier, Jutulstraumen, Stancomb-Wills Glacier and Recovery Glacier. Examples of coastline without ice shelves are Mawson Coast (W of Lambert Glacier), Knox Coast and Sabrina Coast in Wilkes Land.

Antarctica is commonly "divided" into East Antarctica and West Antarctica (see Figure A.1-1). While the divider between these two distinctly different parts of Antarctica is well-defined as the Transantarctic Mountains, running from east of the Weddell Sea across the continent to the west of the Ross Ice Shelf, the terminology constitutes a misnomer, as the center is to the south — hence some parts of "East Antarctica" are west, some are east of the Transantarctic Mountains. It is generally accepted that "East Antarctica" is that part including Queen Maud Land, Enderby Land, MacRobertson Land, American Highland and Wilkes Land. (In maps plotted with the Antarctic Peninsula pointing to the top of the map, East Antarctica is to the right of the Transantarctic Mountains, West Antarctica is to their left.) East Antarctica contains the geographic South Pole. The coast of East Antarctica borders the eastern part of the South Atlantic sector, the Indian Ocean sector, and the western part of the South Pacific Ocean sector of the Circum-Antarctic Ocean. The geologic shield of East Antarctica is covered by the Earth's largest ice sheet, the East Antarctic Ice Sheet. As we shall see in the map part of the Atlas, East Antarctica has a fairly simple geography, except for the marginal (coastal) areas, with huge basins and separated by broad ridges.

Correspondingly, the continent-scale flow systems are simple. The only large ice stream-ice shelf system in East Antarctica is the Lambert Glacier/Amery Ice Shelf system, this is also the most northerly of the large Antarctic ice-stream/ice-shelf systems (except for ice shelves on the Peninsula). Queen Maud Land is the large part of Antarctica between the Weddell Sea and Lambert Glacier/Amery Ice Shelf, closer to Amery Ice Shelf is Enderby Land. MacRobertson Land and American Highland are smaller areas bordering Amery Ice Shelf (on the W and E, respectively). To the east is Wilkes Land, the largest and least

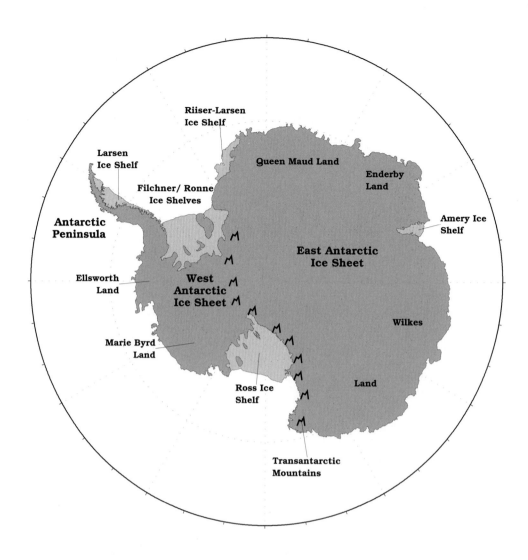

Figure A.1-1. Antarctica — Overview map.

explored part of Antarctica, extending from Lambert Glacier/Amery Ice Shelf to the Ross Sea sector and the Ross Ice Shelf. Surface gradients in East Antarctica are generally very low. Ice drains generally towards the ice sheet margin, and for 10% of the area to the Lambert Glacier/Amery Ice Shelf system. Three subglacial domes influence the flow direction over hundreds and thousands of kilometers: Valkyrie Dome near $(40° E/77° S)$ in Southern Queen Maud Land, Dome Argus ($78°$ $E/81° S$), and south of American Highland Dome Charlie ($125° E/75° S$) in Southern Wilkes Land. The Transantarctic Mountains include the Shack-

leton Range, the Pensacola Mountains (east of the Filchner Ice Shelf), continue in subglacial mountains, Queen Maud Land Mountains (bounding the southern Ross Ice Shelf), Queen Elizabeth Range, Churchill Mountains, Prince Albert Mountains and many other smaller ranges bordering the Ross Sea.

West Antarctica is geographically much more divided and smaller chambered than East Antarctica. Two large ice shelves, the Filchner-Ronne Ice Shelf (separated by Berkner Island but connected south of that island) and the Ross Ice Shelf. The

Filchner-Ronne Ice Shelf extends to 82.5° South (where Support Force Glacier enters) and to 83° South (where Foundation Ice Stream enters). The Ellsworth Mountains border Ronne Ice Shelf in the west and contain Antarctica's highest mountain, Vinson Massif (5140 m).

From west of the Ross Ice Shelf, the Antarctic Peninsula extends north to 63° S, with Clarence Island and Elephant Island near 61° S and the King George Islands at 62° S across the Bransfield Strait. The islands are considered part of Antarctica (and their presence contributes to the prestigious question of who first set foot on Antarctica, or sighted Antarctica). Contenders for having first set foot on the mainland include Nathaniel Palmer, an American whaler, Thaddeus Bellingshausen, captain for Russia, and Norwegian biologist Carsten Borchgrevink, who participated in H.J. Bull's expedition to Borchgrevink Coast, Victoria Land, in 1894-95 and led the British Antarctic expedition in 1898-1890.

The Peninsula has a mountainous spine and a small-chambered topography with many mountain ranges, local glaciers, ice shelves and islands, the largest of which is Alexander Island. The northern part is Graham Land, the southern part Palmer Land.

The Ross Ice Shelf extends to 85.5° S, it is bordered by the Transantarctic Mountains on the west, with many ice streams draining from the interior of Wilkes Land and the Polar Plateau. In contrast, the east coast (Siple Coast) is characterized by a much less relief-energetic landscape of large ice streams that drain from Marie Byrd Land. A large volcanic province extends from the Jones Mountains at 94° W to the Fosdick Mountains at 145° W in Marie Byrd land. The coast of Marie Byrd Land and Enderby Land contains many ice shelves.

(A.2) Climate Change, Sea-Level Rise, and Changes in the Cryosphere

The Antarctic Ice Sheet plays a major role in the Global System. Under warming climatic conditions, as have been observed in the past to present decades, the Antarctic ice is reacting sensitively.

As a result of warming, one would expect an ice sheet to loose mass and a glaciers to retreat. Indeed, some glaciers in Antarctica have been retreating, an example is Pine Island Glacier on Walgreen Coast, Ellsworth Land. Not all Antarctic ice streams are retreating — an advance of Lambert Glacier averaging 1 km/yr over 10 years has been observed (Herzfeld et al. 1994, 1997).

Increasing temperatures do not necessarily result in mass loss in the Antarctic. To the contrary, in a warming climate, precipitation in Antarctica may increase; precipitation at Antarctic latitudes means snowfall, which alters the low accumulation rate and adds mass.

The ice sheet mass balance is the difference between mass input and mass output (or mass loss). Mass input processes are snowfall, rainfall, and condensation. Mass loss from an ice sheet can occurr through evaporation, sublimation, surface and bottom melting, water runoff and iceberg discharge. At present, the average accumulation rate in Antarctica is only 162 mm/yr water equivalent, much lower than the average accumulation rate of Greenland, which is 273 mm/yr water equivalent. Antarctica is an arid continent, and most of the accumulation is by redepostion of drifting snow. The equilibrium line, the line between accumulation and ablation areas, is close to sea level, and the ablation area constitutes only 1% of the total area of Antarctica. In contrast, two thirds of the area of the Greenland Ice Sheet belong to the ablation area.

The mass balance of the Greenland and Antarctic Ice sheets is, however, only known with an accuracy of 25% (Warrick et al. 1996). Mass and energy exchange processes change on seasonal to interannual time scales, and the general patterns of these smaller time-scale changes vary under climatic change. The response of large glaciers to changes is on the order of 1000 to 10000 years.

In a warming climate, the sea level will increase due to several causes. First, melting of glacier ice and mass loss by calving increases sea level. Numbers and models vary widely. The total ice volume of Antarctica is 29.000.000 km^3. It does not make much sense to convert all that to water equivalent or ocean height equivalent, because it is unlikely that the entire Antarctic ice mass might melt, and other effects such as rebound of the continental plates would change the world's coasts, as a result of the redistribution of loads. Currently, the sea level rises about 2 mm/yr, half of that has been attributed to the melting of small glaciers (i.e. glaciers excluding Greenland and Antarctica) by Meier (1984). It is not known whether the ice sheets in total are presently contributing to sea-level rise. Second, warming climate causes warming of the World's oceans, and hence expansion of the volume of seawater, incurring further sea-level rise. Oceanographic (acoustic) experiments have been conducted that resulted in observations of an expanding ocean due to warming.

Changes in the ice mass, however, do rarely take place as a uniform surface lowering, but (1) thinning of the ice occurs predominantly in coastal areas (as established by Krabill et al. (1999) and Steffen and Box (2001) for Greenland), and (2) fast-moving ice streams and outlet glaciers and ice shelves react most rapidly to climatic changes. Consequently, the most important part of the Antarctic to investigate when looking for changes are the marginal areas of the Antarctic ice sheets. Mass flux occurs across the ice-ocean boundary. Ice is transported seaward by ice streams and outlet glaciers, ending in ice shelves or directly in the Circum-Antarctic Ocean, and iceberg calving occurs.

Two methodological aspects of the Atlas project facilitate investigation of the Antarctic margin: (1) Mapping of regional areas as part of the Atlas Scheme, and (2) exploitation of the neighbourhood structure through the geostatistical concept of the regionalized variable (these concepts will be explained in chapter C).

Ten percent of the entire area of Antarctica that is covered by ice is constituted by ice shelves. In recent years, the ice shelves in the northern latitudes, i.e. on the Antarctic Peninsula, have been disintegrating rapidly, as a consequence of climatic warming: One of the first such occurrences was the break-up of Wordie Ice Shelf (1984) (Vaughan 1993), followed by the break-up of Wilkins Ice Shelf (west of Alexander Island), parts of George VI Ice Shelf west of Palmer Land (between Palmer Land and Alexander Island), and most recently, Larsen Ice Shelf (2002) east of the Antarctic Peninsula (MacAyeal et al. 2002, Rott et al. 2002).

Müller Ice Shelf, a small ice shelf at (67° 15' S/ 66° 52' W) in Lallemand Fjord, Marguerite Bay, existed for a long time as the northernmost ice shelf on the western side of the Antarctic Peninsula, because it was protected by land around Lallemand Fjord, until it broke up. However, Wordie Ice Shelf, located farther south at 69° S, broke up earlier. Generally, ice shelves sheltered by bays and promontories are less susceptible to break-up than less protected ice shelves. Mechanisms for break-up are unknown and presently an object of scientific research, while the occurrence of meltwater on the surface of an ice shelf is clearly an indicator for warming and thus for possibly impending disintegration. Amery Ice Shelf is presently showing signs that a large iceberg will break off in the next years; however, that does not necessarily indicate break-up, as calving of icebergs, including huge ones of several tens of kilometers in diameter, occurs with a catastrophic nature.

Changes in ice streams occur in both a climatic and a dynamic context: Whereas the existence of climatic change is widely known and its influence discussed in public media as well as in the glaciologic community, the existence of dynamic properties of glaciers is much less known and hence a lesser objective of debate. All glaciers have their inherent dynamics and can, by virtue of their flow properties, be described as dynamical systems. Glaciers ending in fjords may have an internal dynamic cycle of advance and retreat on the order of 1000 years (a number that has been established for Alaskan fjord glaciers, Meier and Post 1969). A situation where one glacier is advancing while a neighboring glacier (in the same climate!) retreats, is not a rarity in Alaska, Greenland or Antarctica — this cannot be attributed to climatic effects. Reaction time of large Antarctic ice streams to climatic change is on the order of several thousand to tens of thousands of years, e.g. Lambert Glacier. An example of neighboring Antarctic glaciers that advance and retreat respectively are Mertz and Ninnis Glaciers in Wilkes Land.

The dynamical properties of glaciers deserve more attention. Under a condition of a warming climate, the fast-moving glaciers play a key role in any scenario of a break-up of the ice sheet.

An interesting dynamic phenomenon is that of glacier surges: During a surge, a glacier accelerates to many times (up to 100 times) its normal velocity, its surface breaking up into characteristic crevasses. After the surge, which may last one to several seasons, the glacier returns to its usual speed. Some glaciers are surge-type glaciers, while others do not surge. Surges occur quasi-periodically. Some side glaciers of Lambert Glacier surge (most likely). Many unanswered questions are associated with glacier surges — the surge mechanism is unknown, why some glaciers surge and others do not surge, and why surge glaciers are limited to a subset of the earth's mountain ranges (see Meier and Post 1969, Raymond 1987, Kamb 1987, Herzfeld 1998, Herzfeld and Mayer 1997).

Types of fast-moving ice include (Clarke 1987) continuously fast-moving glaciers, surge-type glaciers, tidewater glaciers, and the West Antarctic ice streams, which form their own class, moving absolutely rather slowly but otherwise sharing properties of fast-moving ice. Fast-moving glaciers are relatively rare but important.

Climatic changes have been occurring for millions of years, they are part of the astrophysical cyclicity of the Earth in the solar system, as discovered by Milankovitch (Milankovitch cycles of c. 20 kyr, 40 kyr, 100 kyr and 400 kyr). These cause ice ages and warm ages. The last ice age ended $\approx$ 8000 years ago. On the other hand, anthropogenically induced changes contribute to global warming — increased CO_2 emission, destruction of the ozone layer, and other forms of pollution — it is this contribution to global change that is the reason for policy discussions, and the reason to make us think about effects of human behavior on the environment. The relative magnitude of human-induced changes is a matter of heated debate, and the answer to this problem depends on the assessment of "natural" changes. The latter may in part be inferred from the geologic record (e.g., marine sediment cores and stratigraphic studies of marine environments from previous geologic times).

(A.3) Modeling Versus Measuring

Studies of the effects of global warming on the cryosphere usually involve a modeling approach to processes in the global system. Any model is based on a physical model, commonly a dynamical system. Most models also contain observations on some variables. To simulate a scenario of change in the modeled system, the model is run, typically by keeping certain variables fixed while changing others (for instance, increasing temperature by 1 K), or by letting the model step through time. In both cases, the behaviour of the model and the outcome are noted and reported. As an example, in model runs the West Antarctic Ice Sheet is more susceptible to disintegration than the East Antarctic Ice Sheet. Difficulties exist in modeling more than one Earth system, for instance, in combining ice and ocean models, or only ice sheet models and ice shelf models (Ralf Greve, pers. comm., Huybrechts 1993).

Model results may be compared to observations for validation of the model, and, ultimately, understanding of the Earth system. The opposite approach to learning about the Earth starts from conduction of experiments, observations and measurements on the Earth's surface, from the air and, in the last decades, from satellites. Technical advance has brought new observational methods in all those fields, most dramatically in satellite remote sensing. New observational methods require new data analysis methods to utilize collected data and learn about processes on the Earth from those observations. The construction of the Antarctic Atlas from satellite radar altimeter data is part of the Earth-observation-and-data-analysis approach to Earth Sciences.

Collaboration between the modeling and the measuring/data analysis parts of the science communities is one of the most important tasks in the geosciences today. Calling for collaboration may sound trivial, but oftentimes real-world properties may not be formulated easily as model constraints; or model results may contradict glaciologic or geologic observations. The modeling community often holds that a model needs to be consistent in itself, or that adding another condition would make the model unnecessarily complicated and solution computationally prohibitive. Better models, and hence better understanding of our world, can only be achieved by taking better data into account; hence more and better observations along with improved mathematical data analysis methods are a prerequisite for learning about our planet.

(B) Satellite Remote Sensing

(B.1) An Overview of Ice Sheet Observations by Satellite

Antarctica is a remote and large continent, and hence, despite a long history of sea- and land-based expeditions, often combined with aerial observations, many areas have not been visited by mankind to this date. There are still "white spaces" in ground-based mapping of Antarctica. The best map of all of Antarctica compiled from expedition data is the widely used Antarctic glaciological and geophysical folio edited by Drewry (1983), it contains maps of a small scale only. Observations by satellite provide an alternative to expeditions, which is a less direct, literally "remote" form of observation — thus the term *satellite remote sensing* — and yields completely different types of data. The advantage of satellite observations is the coverage of large areas in a short time. Such coverage by remote-sensing data typically is of limited resolution or contains gaps, depending on the type of instrumentation installed as payload of the observing satellite, as shall be described in the sequel for some examples.

Two principally different types of satellite data are distinguished: (1) *image data* and (2) *geophysical line survey data*. Image data constitute raster images that consist of pixels in rows and columns, usually square, with a greyscale or color matching the intensity of the reflected signal. Image data may be recorded in one frequency channel or be composed of signals from several channels.

Geophysical line survey data from satellite are measurements of geophysical properties of the Earth's surface or the atmosphere, which follow the flight track of the satellite. These fall into two subclasses, (a) *single-track data* that consist of discrete measurements onlong the track line, and (b) *swath data* that consist of sets of several measurements, taken ideally normal to the track line. For swath data, the observations cover a stripe of a certain, instrument-dependent width. Satellite radar altimeter data are examples of geophysical line survey data collected by a satellite. The measured variable is elevation of the Earth's surface anywhere on continents and oceans, and on any land surface.

Image data also cover a stripe of a given width, located along or to the side of the satellite ground track, or sections of such a stripe. The best-known image data are LANDSAT data, which have been available since 1972 (LANDSAT 1,2,3 Multi-spectral Scanner (MSS)). LANDSAT data are collected in several frequency bands or channels. The latest generation of LANDSAT data are *LANDSAT 7 Enhanced Thematic Mapper Plus (ETM+) data*, available since 1999, which have a resolution of 15 m for Very Near Infrared (VNIR), 30 m for panchromatic and 60 m for Thermal Infrared (TIR) and absolute radiometric calibration.

The most complete source of LANDSAT images of Antarctica is the U.S. Geological Survey's Satellite Image Atlas of Glaciers of the World, volume B, Antarctica (Swithinbank 1988). Because LANDSAT images are relatively small and cloudfree scenes of polar areas are comparatively rare, many areas along the Antarctic margin are not covered in Swithinbank (1988). In those areas which are covered, the images in Swithinbank (1988) provide a great basis for comparison with satellite-altimetry-derived maps in this Atlas.

For selected Antarctic areas, maps based on image data have been compiled, for instance by the Australia Division of National Mapping for the Lambert Glacier/Amery Ice Shelf and surrounding areas and by the U.S. Geological Survey for the West-Antarctic Ice Streams. A topographic map of the Filchner-Ronne-Schelfeis based on satellite images and ground-based geodetic surveys was published by the (former) Institut für Angewandte Geodäsie, Frankfurt, Germany (Sievers et al. 1993).

Images of Antarctic ice streams have not only been compiled from LANDSAT data, but also from lower-resolution data. The AVHRR (Absolute Very High Resolution Radiometer) sensor aboard a NOAA (National Oceanic and Atmospheric Administration, USA) weather satellite was designed to observe clouds, but cloud-free images have provided a good source of images of Antarctic and Greenland ice streams and glaciers, albeit with only a 1-km resolution (Bindschadler and Scambos 1991).

A "Satellite Image Map of Antarctica" at scale 1:5.000.000 (at standard parallel 71° S), compiled from a mosaic of cloud-free AVHRR data from NOAA satellites 6, 7, 9, 10, 11, 12 and from years 1980-1994 has been published by the U.S. Geological Survey, Flagstaff, Arizona (Ferrigno et al. 1996). The fact that the scenes stem from 14 years of observations indicates how difficult it is to obtain cloud-free scenes for a complete coverage of Antarctica. The image composite is overlain with 500 m spaced elevation contours from a data base of the British Antarctic Survey. The map has a 1:3.000.000 inset showing the Antarctic Peninsula, also based on AVHRR data, and a 1:1.000.000 inset showing McMurdo Sound, based on LANDSAT MSS data. (Processing methods are described in the International Journal of Remote Sensing, 1989, v. 10, nos. 4, 5, p. 669-674.)

In March 2000, the TERRA Satellite was launched with the goal of monitoring the health of planet Earth. TERRA is equipped with a number of instruments for imaging and monitoring. The highest-resolution image data collected aboard the TERRA satellite are *Advanced Spaceborne Thermal Emission and Reflection Radiometer (ASTER) data*. ASTER data have 15 m resolution for VNIR bands, 30 m for SWIR bands and 90 m for TIR bands. Since TERRA was only launched in March 2000, and only few scenes are recorded per revolution around the Earth, not many areas have been covered so far.

Lower-resolution optical data collected by the TERRA satellite are MODIS and MISR data. Multi-angle SpectroRadiometer (MISR) data are, as the name indicates, not only collected in several frequency bands, but also in several look-directions from far-backward to backward to straight-down (nadir) to forward in angles to far-forward. The resolution varies with the frequency. More precisely, MISR data are collected in 4 bands (448 nm (blue), 558 nm (green), 672 nm (red), and 866 nm (near infrared)) and in 9 viewing angles (+/-70.5°, +/-60.0°, +/-45.6°, +/-26.1°, 0° , where +/- indicates forward/backward, and 0° nadir). The pixel size is 275 m x 275 m for all bands in the nadir camera and for the red band in all other cameras, and 1100 m x 1100 m for the blue, green and near-infrared bands in the forward and aft cameras. The swath width is 380 km. Consequently, the MISR instrument is best suited for studies at the regional scale. MISR scenes are so large that 16 consecutive days of orbital coverage already provide complete coverage. MODIS data have been utilized in the study of snow surfaces on land.

Backscatter data of higher resolution are obtained by the Synthetic Aperture Radar (SAR) instrument, which have become available to the scientific community through ERS-1/2, JERS-1 and RADARSAT (ESA 1992a,b, 1993; Canadian Space Agency et al. 1994). SAR data are particularly well suited for the study of polar areas, beacuse the SAR signal penetrates clouds, which often create a problem over glaciated areas rendering images useless for glaciologic studies. ERS-1 SAR data have a resolution of 12.5 m, but often are corrected for terrain effects (so then the resolution depends on the resolution of the terrain model used, e.g. 30 m for Alaska SAR data processed by the Alaska SAR Facility).

ENVISAT, recently launched by ESA, collects next-generation SAR data, named Advanced Synthetic Aperture Radar (ASAR) data.

Other than LANDSAT data, however, SAR data are not really images, but backscatter values which may be associated with grey values for display, and are collected in only one frequency. The fact that image analysis techniques depend heavily on multivariate statistical techniques designed for multispectral image data such as LANDSAT data requires development of new methods for the analysis of SAR data.

Different intensities of "grey" in a SAR image may be caused by surface or volume scattering of the signal, local aspect of the surface, or subscale geophysical variation. Returns from signals from ascending or descending satellite tracks differ also. These sources cannot be discriminated without additional information. Because an absolute reference is lacking, a characterization of surface patterns from SAR data needs to be based on relative differences in "grey" value, such as in the geostatisical classification methods (see e.g. Herzfeld 1999, Herzfeld et al. 2000a).

A major difficulty with the analysis of SAR data is that quantitative analysis is not directly possible. One promising avenue in that direction is the application of interferometry, a technique that exploits the phase differences of two images, but at the same location, possible in the rare situation of very close repeat of the ground tracks (Goldstein et al. 1993) and good correlation of the images to be compared (Zebker and Villasenor 1992). The best-known application is the extraction of the velocity of the ice (Goldstein et al. 1993). If no movement occurred and the environment did not change between the times of collection of the two images, it is possible to compute topography from pairs of SAR images using interferometry. There is ongoing work on construction of elevation maps from SAR stereo images, but that has yet to be completed. Examples of applications of interferometry are restricted to date to the study of smaller regions, and SAR images can only be collected for 10 minutes per revolution. The technique is not suitable for mapping large areas of the Antarctic ice, leaving ample necessity for altimetry-based mapping.

If ice movement is complex, then SAR interferometry is not possible any more, because two images from different times will not be coherent. The advantage of the geostatistical classification method is that only one image is needed for analysis, and complex ice movement manifests itself in surface patterns which can be classified (Herzfeld 1999).

The best data source for topographic mapping from satellite is radar altimetry. The first satellite carrying an altimeter became operational in 1978 (SEASAT). Together with data from the GEOSAT Geodetic Mission (1985-86) and the Exact Repeat Mission (1987-89) and data from ERS-1 (1992-96) and ERS-2, almost a 20-year record of altimeter data is available. This makes altimeter data the type of data most suited for the study of changes on a regional or continental scale for length of record. One disadvantage of studying Antarctica by satellite data is that the orbital coverage of the previously mentioned satellites does not extend to the poles.

Radar altimeter data may be analyzed geostatistically to construct maps of 3-km-by-3-km resolution, which have a high accuracy (Herzfeld et al. 1993, 1994). This data type and the geostatistical method are utilized in the derivation of the topographic maps in this Atlas. Therefore the principles of radar altimetry will be described in more detail in section (B.2). Bamber (1994) produced a map of Antarctica (north of 82°S) from ERS-1 altimeter data with 20-km grids. Limitations of this map are the lower resolution and the fact that the map is only reliable in areas with a slope of less than 0.65° (Bamber 1994). By total area most of Antarctica is flatter than 0.65°, but the steeper regions include the dynamically important ice streams and outlet glaciers. The geostatistical method (cf. Herzfeld et al. 1993) facilitates calculation of maps of higher accuracy and including steeper areas, but is computationally more intensive. The need for higher resolution is not well met if all of Antarctica is shown on one map sheet. An alternative is to construct an atlas, which in turn requires specific cartographic considerations. Observational and methodological aspects of data collection and analysis of radar altimeter data will be treated in sections (C.1) and (C.2), the geostatistical method applied here is the objective of section (C.3).

Complete coverage of Antarctica by SAR data was obtained during the two Antarctic missions of the RADARSAT, a satellite launched by the Canadian Space Agency. As its only instrument, RADARSAT carries a Synthetic Aperture Radar (SAR). During an "Antarctic mission", the SAR sensor aboard RADARSAT is turned so it "looks" at Antarctica and thus covers the notorious "hole" in the data coverage around the South Pole that is typical for most other satellite missions, including the altimetry missions used in the Atlas. As already noted, SAR yields image data and, in particular, no elevation data. A scatter image of Antarctica has been compiled by Jezek. Unfortunately, RADARSAT does not carry an altimeter.

As part of a project termed "RAMP" (Radarsat Antarctic Mapping Project), Jezek and coworkers have derived a DTM of Antarctica (Liu et al. 1999; Jezek et al. 1999). This is based on interpolation of ERS-1 radar altimeter data to 5 km, using mathematical algorithms (as RADARSAT has not been instrumented to collect elevation data, the "RAMP" grid is a bit of a misnomer). In selected small areas where higher-density data were available, for instance, from field work, the grid has been refined up to 200 m resolution. Unfortunately, the user cannot tell which data types are utilized in a given area (there is one grid, not overlays per data source). The grid has a nominal resolution of 200 m — possibly to make space for potential future improvements —, but the data support is that of the 5 km grid (except for the small study areas).

For contouring outlet glaciers and ice shelves along the margins of Antarctica, the geostatistical method utilized in the Atlas is better suited than mathematical algorithms. Tiling Antarctica into regions and studying individual maps affords a higher attention to detail.

The backscatter information contained in the SAR data provides information on ice surfaces that is complementary to elevation data, as it shows boundaries of exposed rock areas and ice-covered areas, flow features of glaciers, calving fronts of ice shelves, and some surface features of the inland ice. Hence, a good way of satellite data analysis lies in the combination of SAR and radar altimeter data. An example of such a combination, using RADARSAT SAR data and ERS-1 radar altimeter data, is given in section (G) for the Lambert Glacier/Amery Ice Shelf area.

Recent approaches in the study of the cryosphere concern mapping surface features from satellite and surface characteristics such as surface roughness, albedo, and snow and ice properties. The most commonly studied material property in snow-and-ice research is wetness of snow. Typically three classes are discerned, wet snow/ice, dry snow/ice, and an intermediate class, snow/ice in a zone where meltwater percolates through it, called the percolation zone.

In multisensor optical data (such as MISR data, collected by the TERRA satellite, launched in March 2000), the difference between signals from forward and backward looking beams relates to surface roughness. Because of its nine viewing angles, MISR data are suitable for the study of surface roughness. Rougher surfaces are backward scattering, even if the material itself may be forward scattering. For instance, ice is forward scattering at the particle scale, but microtopographic ice surface features such as sastrugi or crevasses are backward scattering. Hence, multiangular observations may be used for classification of snow and ice surface micromorphology and surface roughness, however, to date only preliminary work on ice surfaces has been done from MISR data (Nolin et al. 2002 ; Nolin and Herzfeld 2002a,b). MISR has a fairly low resolution, with pixel sizes of 275 m and 1100 m, depending on frequency and sensor, this of course limits the amount of detail that can be derived from a structural analysis of the surface data. More complex characteristics of surface roughness are detectable in SAR data, which have a higher resolution (12.5 m). Features include crevassed areas in ice streams and during glacier surges, and the quantitative characterization of more complex features naturally requires more advanced methods for characterization, such as the geostatistical surface classification method (Herzfeld 1998, Herzfeld et al. 2000a).

This is by no means a complete summary of ice observations by satellite and their analysis, but rather an introduction to this growing field with a focus on the most common types of satellite observations. There are other types of data that have been utilized in glaciological investigations, e.g. passive microwave data (for principles of passive-microwave remote sensing and deduction of geophysical parameters, see Steffen et al. (1992), for sea-ice-passive-microwave remote sensing, see Grenfell et al. (1992), Seelye et al (1992)).

The variety of sensors that can be carried as the payload of a single satellite may be exemplified by the instrumentation of ENVISAT, a large satellite that was launched into space in September 2002 by the European Space Agency. In contrast, there are small satellites that carry only one main instrument and a few others whose observations assist in correcting the data. ENVISAT follows ERS-1/2 type orbits. ENVISAT is designed by ESA to make the most complete set of observations of Earth ever collected by a satellite, including land, ocean, ice and atmosphere measurements.

The instruments of most interest in our context are the RA-2 altimeter, an instrument further developed from, but similar to ERS-1/2 altimeters, and the Advanced Synthetic Aperture Radar (ASAR) instrument. ASAR operates at C-Band (see section (B.2.3)) and hence ensures continuity with the image mode of ERS-SAR data and wave mode of ERS-1/2 AMI data. Improvements are facilitated by a full active array antenna with distributed transmit/receive modules which provides distinct transmit and receive beams, a digital waveform generation for pulse "chirp" generation and a ScanSAR mode for operation by beam scanning in elevation (ENVISAT website under www.esa.int). RA-2 utilizes Doppler Orbitography and Radiopositioning Integrated by Satellite (DORIS) as a tracking system. Data are processed at ground stations and provide centimeter accuracy for satellite orbits. A Laser Retroreflector (LRR) is mounted on a pillar attached to the nadir panel close to the RA-2 antenna and supports satellite ranging and RA-2 altitude calibration, it is not really an instrument of its own but a passive device that is used as a reflector by ground-based satellite ranging stations which send out high-pulsed laser signals (see also section (B.2.3)).

The Michelson Interferometer for Passive Atmospheric Sounding (MIPAS) is a Fourier transform spectrometer for the measurement of high-resolution gas emission. It operates in the near to mid infrared frequencies where many of the atmospheric trace gases exhibit important emission characteristics. MIPAS is geared at stratospheric chemistry (O_3, H_2O, CH_4, N_2O and HNO_3) and climatology (temperature, CH_4, N_2O, HNO_3) and thus of indirect relevance to snow and ice research.

The Medium Resolution Imaging Spectrometer (MERIS) instrument measures solar radiation reflected by the Earth with a spatial resolution of 300 m in 15 bands which are programmable in width and position, in the visible and near infrared frequencies (passive microwave). Global coverage can be achieved in 3 days. The primary objective is the measurement of colour in the oceans and in coastal areas, which is indicative of ocean production, chlorophyll concentration, aerosol loads and temperature. Cloud top height, water vapour and aerosol over land can also be observed.

The Advanced Along-track Scanning Radiometer (AATSR) aboard ENVISAT continues ATSR-1/2 data sets of sea-surface temperature (with 0.3 K accuracy), which are important for climate research. Together with ATSR data, a 10-year almost-gap-free data set will be available.

The main objective of the Microwave Radiometer (MWR) is the measurement of the integrated atmospheric water vapour column and cloud liquid water content, as corrections for the radar altimeter signal. In addition, MWR data are expected to be useful for the determination of land surface emissivity and soil moisture, for surface energy budget studies, and for ice characterization.

ENVISAT also carries the "Global Ozone Monitoring by Occultation of Stars" (GOMOS) instrument. Ozone concentration in the atmosphere is largely responsible for stratospheric heating through absorption of harmful UV radiation, it determines a significant part of the oxidative capacity of the troposphere, therefore ozone is more commonly known as a "greenhouse" gas. A so-called "ozone hole", a low in ozone concentration, has been discovered over Antarctica and raised public interest. GOMOS also measures NO_2, NO_3, $OClO$, temperature and water vapour during day and night and at altitudes between 100 km and the tropopause, with an altitude resolution of 1.7 km. The star occultation method is a new measurement principle, a predecessor is a sun occultation method used on an instrument called SAGE.

ENVISAT provided space for so-called payloads of opportunity, new instruments designed by individual research groups. One such instrument is the Scanning Imaging Absorption Spectrometer for Atmospheric Chartography (SCIAMACHY) for measurements in the troposphere and stratosphere.

For the next years, a significant increase in satellite technologies and in satellite missions is planned. However, it takes a number of years before data from any new mission or instrument yield results that are of interest to the science community or even the general public, because a lot of development and processing is necessary before so-called higher level products, that is, data that are meaningful to scientists, become available. Much later, products that can be communicated to the public can be expected. A few flashy sample products are usually publicized early for public-relations reasons.

For individual studies on satellite remote sensing of snow and ice areas the interested reader is referred to scientific journals in related areas, which include the International Journal of Remote Sensing, Zeitschrift für Gletscherkunde und Glazialgeologie, the Journal of Glaciology and the Annals of Glaciology, Antarctic Science, Polar Geography, Polarforschung, and Computers & Geosciences. Descriptions of data types are given on the website of the World Data Center for Glaciology A (http://www.nsidc.org), along with examples of their use.

(B.2) Satellite Radar Altimetry

Applications of altimetry include ocean circulation, ocean currents and eddies, surface waves and tides, sea-level change — an important indicator of global change — (see Nerem and Mitchum 2001), estimation of the gravity field of the Earth and determination of the geoid (Tapley and Kim 2001), marine geophysics and bathymetry (Sandwell and Smith 2001; see also Herzfeld and Brodscholl 1994, Brodscholl et al. 1992), and last not least, ice sheet observations, which are the topic here (for principles, see Zwally and Brenner 2001).

Several satellite missions have carried altimeters. Altimetry missions differ by instrumentation and by mission type (see Table B.2-1).

Table B.2-1. Summary of Past and Present Satellite Altimeter Measurement Precisions and Orbit Accuracies

Satellite	Mission period	Measurement precision (cm)	Orbit accuracy (cm)
GEOS-3	April 1975–December 1978	25	~500
Seasat	July 1978–October 1978	5	20 or 100
Geosat	March 1985–December 1989	4	10–50
ERS-1	July 1991–May 1996	3	8–15 (PRARE:5)
TOPEX/POSEIDON	October 1992–present	2	2–3
ERS-2	August 1995–present	3	7–8 (PRARE:5)

(B.2.1) Satellite Missions with Radar Altimeter Observations

(B.2.1.1) SEASAT

Satellite radar altimeters had predecessors on the SKYLAB satellite (launched 1973) and the GEOS-3 satellite (launched 1974), but the radar altimeter aboard SEASAT (launched 1978) was the first one that produced scientifically useful data. At first, satellite altimeter instruments were designed to survey the sea surface to determine the geoid, as the name SEASAT indicates. SEASAT was in operation only for a period of about three months (10 July to 9 October, 1978), this corresponds to late austral winter (late winter on the southern hemisphere).

Although SEASAT was only operational for such a short time, the existence of SEASAT altimeter data makes this data type the oldest satellite geophysical data type. Because of this relatively long history of observations, altimeter data may be utilized for monitoring recent changes in the Earth. This is particularly important for observations of ice surfaces where changes are happening in present times!

Although SEASAT was designed for oceanographic and ocean-geodetic observations, its data are of great value for ice research also. Data processing over flat ice areas works similarly to that over the ocean surface (Brown 1977; Partington et al. 1989), and initially routines for retracking were copied. Taking into account the specific and often complex surface properties of the ice lead to adaptation of processing methods and design of new ones (see section (B.3)).

SEASAT flew at an orbit height of about 785 km above the Earth's surface. Typical SEASAT groundtracks are shown in Figure B.2.1-1. The tracks have pseudo-rhombic gaps, formed by ascending and descending orbits, appear slightly curved in this projection, and converge towards the southern limit. SEASAT did not track very accurately, and data from areas with morphologic relief were not retracked (see "improved" pattern in the track figure for GEOSAT, Fig. B.2.1-2).

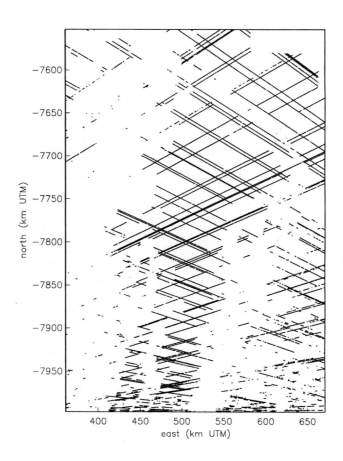

Figure B.2.1-1. Typical SEASAT ground track pattern. (All data collected during entire SEASAT Mission shown, but with a reduction in numbers to match capacity of plot software). Geographical area is Lambert Glacier/Amery Ice Shelf (see chapter (E)). Ground tracks are more continuous over the smooth and low-angle ice stream/ice shelf area, while large gaps occur over the mountaineous and rugged terrain west of the glacier, and smaller gaps occur over the mountaineous but less rugged terrain to the east of the glacier. Tracks of different revolutions do not repeat accurately.

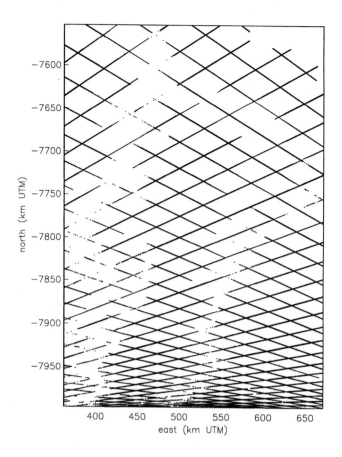

Figure B.2.1-2. Typical GEOSAT Exact Repeat Mission ground track pattern (All data collected during a time frame corresponding in season and length to SEASAT Mission shown, but with a reduction in numbers to match capacity of plot software). Geographical area is Lambert Glacier/Amery Ice Shelf (see chapter (E)). Ground tracks are more continuous than those of the SEASAT Mission, with some gaps occuring over the mountaineous regions along both margins of the ice stream/ice shelf area. Tracks of different revolutions repeat accurately. Ground tracks converge towards the northern and southern limits of GEOSAT radar-altimeter-data coverage at 72.1° N and 72.1° S, respectively, here visible for the southern limit.

(B.2.1.2) GEOSAT

After SEASAT, GEOSAT (U.S. Navy Geodetic Satellite) was launched into the same orbit (about 785 km above the Earth's surface) and with a similar altimetry instrument on board.

A major objective of the GEOSAT mission were measurements of the marine geoid. GEOSAT also provided sea-state and wind observations, which were useful for U.S. Navy operations. GEOSAT was operated in geodetic missions and exact repeat missions (see section (B.2.2)), the denser coverage Geodetic Mission data remained classified (unavailable to the scientific community) for several years. Exact Repeat Mission data provided the first long-term high-quality altimeter data available to the scientific community. Although GEOSAT was also designed primarily for marine studies, its data proved very useful for ice-sheet research, after methods had been adapted adequately for the processing and interpolation over ice. Both GEOSAT Geodetic Mission and Exact Repeat Mission data are evaluated in the Atlas.

GEOSAT collected data from March 1985 to December 1989, the first geodetic mission was 01 April 1985 to 30 September 1986.

GEOSAT flies around the Earth at an elevation of ca. 785 km, following a spiral track revolving around the globe, such that each track is at an oblique angle with the equator; the tracks do not reach all the way to the poles (orbital coverage of GEOSAT altimetry extends to 72.1° southern latitude) and the next track is offset to the previous one by a certain distance, which can be changed during the operation of the satellite (by signals that are sent to the satellite from computers at a ground control center / mission control center on the Earth). Data density increases towards the limit of coverage (72.1° southern latitude) because of the convergence of the groundtracks.

(B.2.1.3) ERS-1 and ERS-2

The ERS-1 and ERS-2 satellites were built by the European Space Agency (ESA), both satellites carry a suite of instruments including an altimeter. The ERS altimeters on both satellites are of the same type (but different in design from the GEOSAT altimeter). ERS satellites fly also at about 785 km above the Earth's surface, and they follow sun-synchronous orbits. ERS-1 collected altimeter data from July 1991 to May 1996, at which time the altimeter was switched off deliberately (unfortunately!), but other instruments continue to work (e.g., the SAR instrument, which has a much higher data volume). ERS-2 started in August 1995 and has been collecting data to the present (2003). So, for several months, both ERS-1 and ERS-2 were following each other with an 8-day or a 1-day separation and collecting data at the same time. ERS-1 and ERS-2 also carry a number of other instruments, including SAR instruments and instruments for precise orbit determination.

(B.2.1.4) Other Missions with Altimeters, and Related Missions

The heritage of satellite radar altimetry goes back to SKYLAB, launched in 1973, which was the first space-borne active microwave remote-sensing spacecraft. From the first satellite radar altimeter to today's altimeters, the technology has evolved considerably. SKYLAB carried a generalized active/passive microwave measurement system, S193, part of which was the altimeter. The altimeter was designed to measure the influence of ocean-state effects on the radar pulse characteristics. Flying at a relatively low altitude of 435 km and using a pulse-width of 0.1 microseconds, the SKYLAB altimeter obtained a resolution of 15 m, but operated only over short segments. Large features of the marine geoid such as anomalies induced by seafloor trenches could be resolved, which indicated that satellite altimetry works in principle.

In 1974, the NASA satellite GEOS-3 was launched into an 840 km orbit above the Earth's surface, it carried an altimeter that for the first time used pulse compression (see section (B.2.3)). GEOS-3 collected data from April 1975 to December 1978. The first satellite that yielded scientifically useful data was SEASAT (section (B.2.1.1)), launched in 1978. The SEASAT altimeter used improved technologies, notably a technique called full deramping (see section (B.2.3)).

Another satellite that carries an altimeter is the TOPEX/POSEIDON satellite of NASA and CNES (Centre National d'Etudes Spatiales, Toulouse, France) at a 1336 km orbit.

Recent and future missions are the following: JASON-1 (Ménard et al. 1999) is a joint US-French mission designed to measure sea-surface properties as TOPEX/POSEIDON, a follow-up mission JASON-2 is planned for 2004. A next-generation altimeter was launched recently (in September 2002) aboard ENVISAT, a large ESA satellite designed to observe the Earth — land, ocean, ice surfaces and the atmosphere — through a multitude of isntruments (as described in section (B.1)). ENVISAT follows the tendency to make many complementary observations from one satellite platform. Much of its payload continues instrumentation of ERS-1 and ERS-2, so observations can be used for monitoring. The radar altimeter, RA-2, is similar to the ERS-1/2 altimeters. DORIS is used for orbit determination (see

(B.2.3)). ENVISAT follows ERS-1/2 orbits. usually several years lie between the launch of a new satellite and the derivation of higher-level products that may be of interest to the scientific community, and, later, to the general public, because engineering, calibration and validation, hand in hand with derivation of adequate processing methods, needs to take place for any satellite data type.

GRACE (Gravity Recovery and Climate Experiment), a joint US-German venture, uses a pair of low earth-orbiting satellites whose relational distance is precisely determined (Wahr et al. 1998). Geoid accuracy from GRACE data will be better than 1 cm on spatial scales of greater than or equal to 200 km, corresponding to half wavelength. GRACE was launched in March 2002, the commissioning phase was completed in May 2003, and at present (Oct. 2003) the validation phase is started. GOCE (Gravity Field and Steady-State Ocean Circulation Explorer) is a planned high-resolution gravity mission of ESA (mid-2004?). Gravity gradiometry and GPS precision tracking in a very low orbit are expected to yield 1 cm accuracy of geoid undulations with a spatial resolution of 100 km (LeGrand and Minster 1999).

The fact that ERS-1, ERS-2 and ENVISAT are flying on a sun-synchronous orbit leads to aliasing errors of the solar constituents and thus limits accuracy. But information on that effect may be obtained from correcting non-sun-synchronous altimetry with upcoming GRACE/GOES geoids and apply that to older altimetry-derived sea surfaces. With 100 km or better accuracy, mean structures of boundary currents, the Antarctic circumpolar current, and mesoscale eddies can be mapped, which will be useful for ocean models, coupled ocean-atmosphere models and for simulation and prediction of interannual and decadal variability.

Altimeter measurements of higher accuracy than satellite radar altimeter measurements are expected from laser altimetry. A satellite laser altimeter instrument, the Geoscience Laser Altimetry System (GLAS) was recently launched aboard ICESAT on January 12, 2003. GLAS has a much smaller footprint than a radar altimeter, so GLAS data can be used to study features of higher resolution (subscale morphology, relative to radar altimetry), such as crevasses, surface roughness and micromorphology of glaciers. GLAS is designed especially for the study of glaciers and ice sheets. From simulations of typical GLAS and GRACE data, it is expected that a combined evaluation of both data types will yield information on Antarctic mass balance changes and postglacial rebound at scales of 250 km (Wahr et al., 2000; Velicogna and Wahr, 2002)

(B.2.2) Mission Types: Exact Repeat Missions and Geodetic Missions

There are two types of missions for altimetry mapping: (a) exact repeat missions, and (b) geodetic missions, which differ in their scientific objectives and their groundtrack pattern.

Exact Repeat Missions

During an *exact repeat mission*, the satellite orbits repeat exactly in location, in space as well as relative to the Earth, so the groundtracks repeat "ecactly" also. For instance, the GEOSAT repeat cycle is 35 days. Data from ERMs facilitate the study of changes of areas covered by the ground tracks, at regular time intervals. In the GEOSAT Exact Repeat Mission, the orbits were designed such that SEASAT orbits were repeated (hence here the term "exact repeat" may have a twofold meaning).

Since most satellites carry more than one instrument, the optimal mission design is influenced by the coverage and processing plans for several instruments. For instance, the Exact Repeat Mission track pattern facilitates comparison of two data sets from the same location (if the ground tracks repeat with sufficient accuracy such that footprints of the data sets overlap on the ground). A good example is a technique of deriving information from phase differences calculated for two SAR data sets from the same location, termed *SAR interferometry*. If the phase differences can be attributed to a single cause, or a combination of well-controlled causes, results on the dynamics of a glacier may be derived; this method is called *interferometry* (e.g. Goldstein et al. 1993, Feigl and Dupré 1999). Similarly, earthquakes over desert areas may be studied. For the interferometric method to work, groundtracks need to repeat very closely, for example, in the example of Gold-stein et al. (1993), the difference in groundtrack location was only 4 meters. Usually, the difference is much larger (typically tens or hundreds of meters or several kilometers).

It is important to notice that the orbit scheme is, of course, determined for the satellite under consideration of the potential applications of *all* the instruments aboard. So, the orbit type optimal for an application that may utilize interferometric evaluation of SAR data may not be optimal for mapping from altimeter data, and vice versa; or, two applications that both require altimeter data may be associated with two different requirements for an optimal orbit setup.

Ground reflectors are placed in several locations around the Earth to help the satellite maintain its orbital track within a certain limit of accuracy. When the satellite passes over the ground reflector, it sends a "CHIRP" to the reflector, and the reflector sends a "CHIRP" back. (A "CHIRP" is a signal that quickly goes through a range of frequencies.) Using the "CHIRP", the actual position of the satellite is calculated and related to the predetermined, theoretical orbit position, where the satellite should be at this time of the overpass over the ground reflector. If the difference exceeds a threshold, the satellite is "moved" back to the correct orbit location ("moved" — by a computer command). This is an automated procedure programmed to maintain orbit repeat accuracy. The disadvantage of ERMs are the large gaps between the resultant groundtracks — 40 km diamond-shaped gaps remain in the example of Lambert Glacier (and other areas of the same latitude) (see Figure B.2.1-2).

Geodetic Missions

To the contrary, during a *geodetic mission*, the satellite is allowed to drift. This results in much denser groundtracks (cf. Fig. B.2.2-1) than in the Exact Repeat Mission (Fig. B.2.1-2). Consequently, gaps between adjacent tracks are much smaller (only a few kilometers, with variable distance). For mapping larger areas, this is preferable. Changes in surface elevation can then be detected, if maps from two different time intervals are compared. Depending on the "repeat" of the orbit, it takes several months to achieve complete coverage of the Earth (or, of any given large area within the limits of orbital coverage).

GEOSAT GM data were unavailable to the scientific community for several years (the satellite was operated by the U.S. Navy), but were later released. The GEOSAT Atlas maps are based on 1985-86 GEOSAT GM data. We determined that 6 months of satellite altimeter data were sufficient to construct a map (see section (C.1), below).

(B.2.3) Radar Measurement Principles

The section on radar measurement principles follows largely Chelton et al. (2001). More detailed descriptions are found in Ulaby et al. (1981, 1986a,b).

The concept of satellite radar altimetry is to measure the distance from a satellite that orbits the Earth to points on the Earth's surface. The altimeter is an instrument carried aboard the satellite. The altimeter transmits a pulse of microwave radiation out toward the Earth's surface, the pulse is reflected at the surface and received back at the altimeter on the satellite. From the two-way travel time — the time it takes the signal to travel from the satellite to the surface and back —, the range satellite–surface is calculated, and from that the height of the satellite over the ground point, from which the signal was reflected, is determined. Knowing the position of the satellite at that time in its orbit around the Earth allows calculation of the elevation of the Earth's surface (in that point). Data recorded by the instrument aboard the satellite are communicated down to receiving and archiving facilities on the Earth. As simple as the principle of altimetry is, as complex is the actual accurate calculation, because many factors influence the travel time and the reflection of the energy. At the surface, part of the energy is lost due to interaction of the pulse with the rough surface (*surface scattering*) or due to partial penetration of the pulse into the upper part of the land, sea or ice (*volume scattering*). Both surface and volume scattering are related, because, depending on scale, scattering off a rough boundary surface

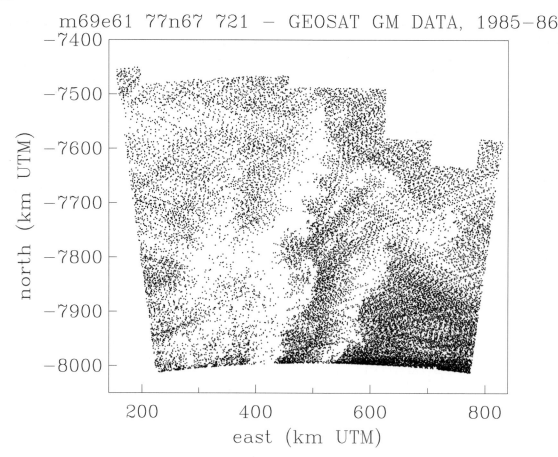

m69e61 77n67 721 — GEOSAT GM DATA, 1985–86

Figure B.2.2-1. Typical GEOSAT Geodetic Mission ground track pattern. (All data collected during GEOSAT Atlas time frame shown, but with a reduction in numbers to match capacity of plot software). Geographical area is that of Atlas map m69e61-77n67-721 Lambert Glacier. Retracking difficulties, and hence gaps in the ground-track pattern, occur over rough terrain (mountaineous areas) and breaks in slope (ice front, glacier margin).

at high resolution appears as scattering in the volume at lower resolution (imagine snow crystals at high resolution and a smooth snow surface at low resolution).

Many conditions influence the travel time — atmospheric conditions (clouds, water vapour, aerosols, dry gases, ionospheric electrons; see section (C.1) on corrections, and section (B.3) on specific problems related to the survey of ice surfaces).

The range $\hat{R}$ computed using two-way travel time t and speed of light c is

$$\hat{R} = \frac{ct}{2} \qquad (B.2-1)$$

(refraction is neglected).

Taking corrections $\Delta R_j, j = 1, \ldots n$ into account, the range from the satellite to the surface is

$$R = \hat{R} - \sum_{j=1}^{n} \Delta R_j \qquad (B.2-2)$$

where ΔR_j are corrections, e.g. atmospheric effects or scattering. Notice that all corrections are assumed to be positive ($\Delta R_j > 0$ for all $j = 1, \ldots, n$), consequently, the elevation would be overestimated without corrections. Determination of the ground location is termed "retracking" — there are many factors to consider in retracking, as we will see in the section on data processing later.

In addition to environmental factors requiring corrections, instrumental corrections are necessary.

The latter include corrections for tracker bias, waveform sampler gain, calibration biases, antenna gain patterns, Doppler shift, range acceleration, oscillator drift, and pointing angle.

Because the first satellite carrying an altimeter, SEASAT, was designed for oceanographic and ocean-geodetic observations, data processing followed the principles of processing over the oceans. SEASAT data are also valuable for ice research. Data processing over flat ice areas works similarly to that over the ocean surface (Brown 1977; Partington et al. 1989), and initially routines for retracking were copied. Taking into account the specific and often complex surface properties of the ice lead to adaptation of processing methods and design of new ones (see section (B.3)).

Frequency bands. Radar Equation.

Because of physical properties of the atmosphere and the sea surface, the optimal frequency range for satellite altimetry is 2-18 GHz, this is within the microwave frequency range.

Frequency band allocations (defined in Ulaby et al. 1981) are: S-band 1.55-4.20 GHz, C-band 4.20-5.75 GHz, X-band 5.75-10.9 GHz, and K_U-band 10.9-22.0 GHz.

Above 18 GHz, atmospheric attenuation increases rapidly; thus the signal would loose power on its way through the atmosphere. At frequencies below 2 GHz, Faraday rotations and refraction of electromagnetic radiation by the ionosphere increase, and interference from ground-based sources increases (navigation, communication, radar; civilian and military sources). Also the antennas would have to be too big.

The size of the antenna footprint on the Earth's surface is proportional to the wavelength of the electromagnetic signal and inversely proportional to antenna size. Use of smaller antennas is possible using *pulse compression*, the standard method is using *pulse-limited signals*.

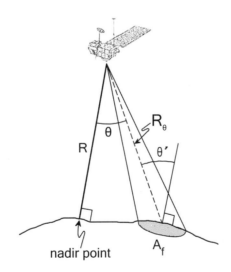

Figure B.2.3-1. Principle of satellite altimeter measurements. θ - antenna pointing angle. θ' - incidence angle. R -satellite altitude above nadir point. R_θ - slant range of radar measurement and pointing angle θ. A_f - antenna footprint area. The surface is an ellipsoid with large-scale geoid undulations, eg., an Antarctic ice surface.

The principle of satellite altimeter measurements is sketched in Figure B.2.3-1, where the following variables are introduced:

θ - antenna pointing angle

θ' - incidence angle

R -satellite altitude above nadir point

R_θ - slant range of radar measurement and pointing angle θ

A_f - antenna footprint area

The normalized radar cross section. Power of the transmitted versus power of the received signal. Because of the shape of the Earth and its topography — here one should not only think of topography of mountain ranges, but also topography of the sea surface, which relates to geoid undulations, and ice sheet topography — the antenna pointing angle θ is not equal to the incidence angle θ'. More explicitly, the antenna pointing angle is the angle between the boresight of the antenna, which corresponds to the peak of the antenna gain, and the line from the satellite's center of mass to the normal onto the Earth surface. The incidence angle is the angle between the line of propagation and the normal to the surface at the point of incidence. However, since geoid slopes rarely exceed 10^{-4} and incidence angles are $\approx 1\,°$ for altimetry (Brenner et al. 1990; Minster et al. 1993), $\theta \approx \theta'$, and also $R \approx R_\theta$ for altimetry.

Parameters of a radar system are:

λ – wavelength of transmitted and received electromagnetic radiation,

P_t – transmitted power [Watts],

G_t – transmitting antenna gain,

G_r – receiving antenna gain.

Since the same antenna is used for transmitting and receiving in altimeter systems, G is used for G_t and G_r.

The parameter dP_r, the backscatter power from a differential area dA within the antenna footprint (on the sea surface), is calculated using the

Radar Equation

$$dP_r = t_\lambda^2 \frac{G^2 \lambda^2 P_t}{(4\pi)^3 R^4} \sigma \qquad (B.2-3)$$

where σ is the proportionality factor, in $[m^2]$, and $t_\lambda(R,\theta)$ is the atmospheric transmittance (defined as the fraction of electromagnetic radiation at wavelength λ that is transmitted through the atmosphere from altitude R at off-nadir angle θ), and two-way transmittance is $t_\lambda^2(R,\theta)$.

The *radar equation* (eqn. B.2-3) characterizes the backscattered power from location dA as isotropically scattered by a sphere with cross-sectional area σ, hence σ is called *radar cross section* of the differential target area dA. In practice, backscattering is from a rough surface (not from spherical elements), and hence dP_r and σ can only be considered average quantities. Scattering properties of the surface (in the antenna footprint) are characterized by differential radar cross section per unit area, denoted σ_0, for which $\sigma = \sigma_0 dA$ holds. The term σ_0 is the *normalized radar cross section*, the unit is decibel; the value of σ_0 varies over the footprint area. Measurements of actual surface roughness are therefore important for calibration of new satellite sensors. A method to measure ice surface roughness on the ground has been developed by Herzfeld et al. (1999b), this may be used to understand the influence of surface roughness on the altimeter signal.

Total returned power is obtained by integrating over the illuminated area A_f:

$$P_r = t_\lambda^2 \frac{\lambda^2 P_t}{(4\pi)^3 R^4} \int_{A_f} G^2 \sigma_0 dA \qquad (B.2-4)$$

This equation assumes that the footprint is small enough such that R and t_λ are approximately constant over A_f; but G is under the integration because G depends on θ and (less) on azimuthal angle and therefore varies over A_f. Hence, σ_0 is the only unknown in the equation!

If one assumes that the surface is spatially homogeneous, or that an average σ_0 is known, then eqn. (B.2-4) can be rearranged to

$$\sigma_0 = \frac{(4\pi)^3 R^4}{t_\lambda^2 G_0^2 \lambda^2 A_{eff} P_t} P_r \qquad (B.2-5)$$

i.e., to calculate σ_0 from the other quantities.

(Here, G_0 is the boresight antenna gain, A_{eff} the effective footprint area, which is approximately

$$A_{eff} = \int_\theta \frac{G^2(\theta)}{G_0^2} \qquad (B.2-6)$$

for angles θ such that the signal hits A_f. Consequently, P_r, the returned power, and thus σ_0 depend only on surface roughness. The value of σ_0 is independent of the altitude of the measurement (if scattering properties are homogeneous in A_f).

More information on radar backscattering is found in Ulaby et al. (1986a,b). Radar backscatter theory is based on the principle of physical optics, alternatively, on geometrical optics. Commonly, a discretization of the surface into small facets is used in model calculations.

Atmospheric attenuation. Atmospheric attenuation (lack of transmittance of the signal) depends on latitude: The transmission is better over the polar latitudes than over the tropical latitudes, because the thickness of the troposhere is 18 km over the tropics, 10 km at mid-latitude, and least over the poles, and moisture is highest in the tropics also. In the troposphere, moderately strong water vapor absorption occurs at 22.235 Ghz, and there are strong oxygen absorption bands at 50-70 GHz and at 118.75 Ghz, and there is strong water vapor absorption at 183.31 Ghz (Chelton et al. 2001, p. 7).

Opacity (optical thickness) τ_λ is related to transmittance t_λ by

$$t_\lambda = e^{-\tau_\lambda} \qquad (B.2-7).$$

Attenuation through a cloud-free atmosphere follows Raleigh scattering (because atmospheric gas molecules are much smaller than radar wavelength).

Cloud attenuation. Clouds are opaque to infrared and visible radiation, but relatively transparent to microwave radiation. As a consequence, areas that are often clouded are difficult to observe in visible satellite imagery (eg. LANDSAT), but radar altimeter data and SAR data can all be used (even on cloudy days), because the radar signal penetrates clouds. This makes radar altimeter data, SAR data, and other microwave data particularly useful for the study of polar areas. Cloud attenuation does still affect microwave signals (this effect can be calculated), and rain has an even larger attenuation than non-raining clouds.

Two-way travel time. Pulse-limited altimetry. An important aspect of altimeter design is the area of the Earth's surface over which the range from the altimeter to the surface is measured, i.e. the area from which the signal is returned. The ideal footprint size depends on the science requirements or goals of the satellite mission. For sea-surface-oriented missions, the footprint size should be

large enough to resolve baroclinic mesoscale variability; clearly, an altimeter for ice-surface characterization would be designed according to different constraints. The latter has been realized in the GLAS instrument aboard ICESAT, which was launched on January 12, 2003 (see section (B.2.1.4)).

In addition, the footprint should be small enough that homogeneity of roughness characteristics of the surface inside the footprint may be assumed in order to simplify integration in the radar equation (see eqn. (B.2-4)). A footprint of 1-10 km satisfies the criteria for the sea surface. The footprint of an antenna is described in terms of the beam-limited footprint, defined to be the area on the surface within the field of view induced by the beam width (full width at half power) of the antenna gain pattern.

For a narrow-beam antenna, antenna beam-width γ, orbit height R, and footprint radius r are related by

$$\gamma = 2tan^{-1}(\frac{r}{R}) \approx \frac{2r}{R} \qquad (B.2-8)$$

The antenna beam width for a circularly symmetric antenna gain pattern is related to the antenna diameter d by

$$\gamma = k\frac{\lambda}{d} \qquad (B.2-9),$$

where λ is the wavelength and k a constant of the illumination pattern across antenna aperture. However, beam-limited antennas would often have to be impracticably large, and they are highly sensitive to antenna-pointing errors.

Hence, pulse-limited antennas are built: A very short pulse (duration a few nanoseconds) is transmitted from an antenna with a smaller diameter and correspondingly larger beamwidth. The footprint size over which the range to nadir mean surface level is determined is effectively defined by the pulse duration. Antenna diameters are $d = 1.6$ m for SEASAT, 2.1 m for ERS-1 and ERS-2, and 1.5 m for TOPEX/POSEIDON, since the pulse expands spherically, the travel time is independent of the antenna pointing angle θ (as long as the pointing angle θ does not exceed half beamwidth).

Average returned waveform. The pulse-limited footprint area represents the area on the surface where reflecting facets contribute to the returned signal measured by the altimeter. The radius of the outer perimeter of the expanding signal defines the maximum extent, but in a rough (or real!) surface, only a small fraction of the facets that approximate the surface are illuminated by the short pulse.

Pulse compression is a technical method to overcome the white noise internally generated in all radar transmitters (Chelton et al. 2001, p. 23). A high signal-to-noise ratio can be achieved by transmitting a long pulse — but that is undesirable, as just stated, because the pulse-limited footprint increases with pulse duration. The small footprint size of the short pulse can be preserved, however, by transmitting a frequency-modulated long pulse (a CHIRP) and analyzing the returned signal in a way that is equivalent to having transmitted a short pulse. The frequency modulation commonly used in altimetry radar signals consists of a transmitted signal (called CHIRP) that is swept linearly from a frequency f_1 to a lower frequency f_2 over a pulse duration τ'. This technique is used on the GEOS satellite, for example.

An onboard adaptive tracking unit (called ATU), consisting of a microprocessor, performs processing of the returned radar signals provided by the hardware parts of the instrument and generates information sent to ground receiving stations. The ATU also controls the instrument hardware via the "synchronizer" which generates the timing of transmitted and deramping chirps and other signals, such as signals sent to the satellite to keep it on the correct orbit.

Corrections include *instrument corrections* (corrections for Doppler-shift errror, acceleration error, oscillator drift error, pointing angle error) and *range estimation* (corrections for atmospheric refraction, dry tropospheric and wet tropospheric refraction, ionospheric refraction (dielectric properties of the upper atmosphere caused by the presence of free electrons)).

Precision orbit determination. Precision orbit determination is the procedure for determination of the three-dimensional location of the satellite's center of mass (called orbit ephemeris) at regularly spaced time intervals in a specified high-accuracy reference frame. Specific instruments in the payload of the satellite collect data necessary for orbit determination. The science and technology of precision orbit determination has advanced considerably with the development of satellites. At time of SEASAT launch in 1978, the orbit error was about 5 m (Lerch et al. 1982), errors were at about 1 m in the early 1980's (Schutz et al. 1985), while the orbit error for TOPEX/POSEIDON could be reduced to about 2 cm (see Table B.2-1).

More information on orbit determination is given in Chelton et al. (2001, p. 64-86). Briefly, orbit determination involves calculation of the satellite in motion, effects of gravitational forces, geographically correlated errors, effects of the geodetical reference frames used, forces of atmospheric drag and solar and terrestrial pressure. Precision satellite tracking systems are needed, examples of such systems are satellite laser ranging (SLR, used on all altimeter-carrying satellites except GEOSAT), DORIS (TOPEX/POSEIDON), PRARE (ERS-2), GPS (TOPEX/POSEIDON), TRANET/OPNET (SEASAT, GEOSAT), and TDRSS (TOPEX/POSEIDON).

TRANET, the initial satellite tracking system used for SEASAT in 1978, consisted of a global network of 40 ground receivers using the Doppler principle. SLR uses satellite laser ranging (thus the name), DORIS is again a Doppler system. PRARE was developed by the German Space Agency (DLR) and provides two-way dual-frequency range and rate data; it failed for ERS-1 but worked for ERS-2. The data are collected aboard the satellite and processed at Geoforschungszentrum Potsdam, Germany (Reigber et al. 1997).

The Global Positioning System (GPS) is a space-based radio navigation system, developed under the auspices of the U.S. Department of Defense. The space segment consists of a minimum of 24 satellites in circular orbits 20200 km above the Earth's surface. A minimum of 4 satellites is necessary to determine location, and by design of the constellation a GPS user anywhere on the Earth will almost always observe 6-11 satellites (Hofmann-Wellenhof et al. 1993).

(B.3) Analysis of Satellite Radar Altimeter Data over Ice Sheets and Glaciers

(B.3.1) Problems and Methods of Mapping Ice Surface Elevation

The first satellite radar altimeters were designed for mapping the ocean surface. Hence, processing methods were also oriented at satellite radar altimeter data collected over the oceans. While the ocean surface is by no means flat, but characterized by waves and influences on those by tides, wind, seastates and other environmental factors, the complex and different characteristics of the ice surface of Antarctica and Greenland require adapted and additional processing steps. The land surface is not flat on the large scale, no matter how flat is defined or the reference surface is defined, but has a locally varying slope. Over large areas in the interior of the Antarctic continent, the slope is fairly small (and thus processing is easier and more similar to those of ocean returns), but the areas that are of greatest interest in the study of changes in the ice sheets are the margins of the continent — it is here that outlet glaciers flow into ice shelves, calve into the ocean, and that most changes occur. The gradient is largest and most variable in the marginal areas of Antarctica. Consequently, the value of mapping methods increases with their ability to handle processing of data collected over complex terrain.

The ice surface has also numerous surface features that vary locally, such as crevasses, sastrugi, meltwater streams, and long-wave features induced by ice riding over a rough bedrock, the spacing of surface features varies from centimeters to kilometers.

In this section, methods developed specifically for processing of satellite radar altimeter data over ice surfaces are introduced. In section (C.1.1) "Corrections applied to satellite radar altimeter data for ice surface mapping", only those methods that are used in the data set underlying the Antarctic Atlas are mentioned, and also presented in a simpler context, for the reader who may be less familiar with or interested in remote-sensing techniques.

For accurate topographic mapping, the data need to be corrected for the effect that the radar signal that is received first at the satellite is the one returning from the point of closest approach (PCA) (= the point located most closely to the satellite at this time) rather than the signal returning from the point at nadir (= directly below the satellite, in a line from the satellite and normal to the Earth's surface). Only if the Earth's surface at nadir and in its neighbourhood is perfectly flat, the PCA and the nadir coincide; if the surface has a slope, the points differ. As a consequence, the first return signal appears to come from a point that is higher and at nadir. The correction is called the slope correction. Most slope correction methods proceed iteratively, using an aproximative surface calculated assuming a flat surface, and then reprocessing the data using the slope values from the first step. An example is the method by Brenner et al. (1983) used in the Atlas data.

The next step of complication in the analysis of waveforms returned from surfaces with non-trivial relief is to consider the situation of a return from an area that consists of two or more parts with different slope or roughness characteristics (see also B.3.2).

Martin et al. (1983) fitted a double-ramp-model waveform to returns from more complex surfaces. Where the waveform results from reflectors from two distinct surfaces, in particular, surfaces with two different slopes, this method works best. (This is the method used in the radar altimetry processing by the ice-sheet altimetry group at NASA Goddard Space Flight Center, and thus in the data utilized in this Atlas). Partington et al. (1987) suggest a method of retracking that uses the first return signal in the waveform as an indicator of the range. Brooks et al. (1983) retracked using the half-peak value in the waveform. Both the methods of Brooks et al. (1983) and of Partington et al. (1987) have been applied to study the topography of Amery Ice Shelf and Lambert Glacier — this ice-stream/ice-shelf system has played a considerable role in the development of satellite radar altimetry processing methods (see also chapters (E) and (F)). The first-return method results in a higher elevation value than other retracking methods, the difference is proportional to the width of the leading edge for ocean returns, but the method is particularly useful for retracking complex waveforms. Crevassed areas may be the cause for a large difference between the two values (Partington et al. 1987).

When crossing a break in slope, the altimeter has a tendency to lock onto a reflector, before it continues on track. This effect is called "snagging effect". Consequently, the ice surface appears too high or too low beyond a break-in-slope, depending on the travelling direction of the satellite. In two cases, the consequences of a snagging effect are particularly evident: (a) The margins of a glacier in a valley appear overdeepened. (b) Shelf-ice edges are often poorly tracked: When the satellite flies from the ice to the ocean side, the ice edge appears to be farther out, when it flies from the ocean up onto the shelf ice, there is a tendency for the ice edge location to be too far upstream. In the ice data records produced at NASA Goddard Space Flight Center, the ice edge is in the middle of both errors. Discussion of this effect and a correction method are given in Thomas et al. (1983): In a track where the satellite approaches the ice shelf edge, the range is determined from the earliest signal associated with the lower (the sea-ice) surface, and the tangent point of a parabolic trajectory of the lower surface away from the satellite marks the position of the ice edge.

In our Atlas, ice edges are often mapped surprisingly well. Details are discussed for specific glaciers in the map chapter (D).

(B.3.2) Derivation of Ice Surface Roughness and Morphology

The main and primary use of satellite radar altimetry is the mapping of elevations. The potential of using secondary information in satellite data records has been explored also. Secondary information is not the information that is planned to be recorded, but information that may also be derived from a given data set. For instance, radar altimeter data contain information on roughness or on other surface characteristics which is of interest in the study of ice sheets and glaciers. Specific methods may need to be developed for the extraction of secondary information from any data set, in other cases, the detection of secondary information may be a by-product of the primary data processing.

Partington et al. (1989) study the influence of snow grain size, surface roughness (amplitude of main features in the 0-3 m range), snow density, and snow temperature on the altimeter data waveforms and their derivatives, for model waveforms and averages of collected waveform data from Antarctica (Wilkes Land) and subareas of Greenland. As a result of their analysis, the authors suggest that (a) surface properties need to be taken into account to avoid retracking errors (the apparent surface is up to a few meters lower than the mean actual surface, if volume scattering dominates and thus the waveforms are positively skewed), (b) altimeter waveforms may indicate surface properties which may be related to wet/dry snow zones and percolation zones. Variations in snow grain size and snow density result in similar waveform variations, but surface roughness and temperature effects on the waveform may be discriminated. It should be noted that (a) the model is based on a number of simplifying assumptions on snow properties (any model needs to make assumptions), and (b) the concept of surface roughness is simplified a lot, as one type of feature of one height is assumed, rendering a one-parametric model. The surface topography of the ice is far more complex, however, as high-resolution measurements of the Greenland ice surface have shown (Herzfeld et al. 1999b; Herzfeld and Mayer, in press).

A related hypothesis is that crevassed areas may be detectable in satellite-radar-altimeter-waveform processing, put forward in Partington et al. (1987) in their study of Lambert Glacier/Amery Ice Shelf: There are different methods of calculating the range (the distance from the point of reflection on the ground to the point of receipt of the signal on the satellite, see sections (B.2) and (C.1)), two of those are the method of first return and the method of half-peaks (both are features discernable in waveforms). The two methods yield different elevation values, and the difference between the two resultant values is attributed to the complexity of the ice surface from which the return stems; in the studied case this is an ice-stream/ice-shelf system with crevassed areas. An important challenge in the study of an ice-stream/ice-shelf system is the detection of the grounding line (the line where the ice shelf becomes afloat on the water of the ocean). A point of the grounding line may be found in waveform analysis. This method has, however, not been used to construct a grounding line across Lambert Glacier, as the processing possibilities were only demonstrated for individual satellite tracks by Partington et al. (1987). *Mapping* of the grounding line requires construction of a map first, this is further described in the sections on Lambert Glacier in chapters (D), (E) and (F).

(C) Data Analysis Methods Applied in the Antarctic Atlas

(C.0) Introduction

Between the collection of raw altimeter data by the ERS-1 satellite and the presentation of maps showing surface elevation of the Antarctic Ice Sheet, there are many steps of data corrections and data processing, carried out by several groups of people at different institutions. Principles and methods of these corrections, processing and analysis algorithms are introduced in this chapter, in the order in which they are applied. Emphasis is placed on the context of each step and its necessity for creation of maps of the Antarctic ice surface, more than on presentation of the specifics of the algorithms.

First, radar altimeter data need to be corrected for various "errors" that occur during data collection (C.1), caused by geophysical and atmospheric conditions that affect the traveltime of the signal or the apparent traveltime. These steps are largely performed at the NASA Goddard Space Flight Center for the data used in this atlas. Data correction formats, algorithms and files have been developed by a geographically dispersed group of scientists and data analysts, organized, for ERS-1 data, by ESA. A few problems remained after correction, such as the bad-track problem, which was treated by the author's group. Corrections of radar altimeter data render data in tracks, which need to be interpolated. Geostatistical methods (section C.3) are the methods used for interpolation, esti-mation, or inversion of the trackline altimeter data onto a regular grid.

The question what constitutes a regular grid, however, has many answers when it comes to mapping the "ball" of the Earth to one or many sheets of paper. Hence, map projection questions need to be dealt with prior to embarking on a mapping mission of a region as large as the Antarctic (section C.2). A related question is that of the geodetical reference surface — the Earth may be approximated by a ball, or an ellipsoid (a ball "flattened at the poles", more precisely, compressed), or by the geoid, representing, simply speaking, the elevation of the ocean surface everywhere. These concepts and resulting effects on mapping the surface of a continent on Earth are introduced in section (C.4), and maps referenced to different surfaces are presented.

Processing methods presented in sections (C.2)–(C.4) have been developed or adapted to the problem at hand — mapping Antarctica from satellite radar altimeter data — by the author, and the resulting processing steps have been carried out by map teams of the author's Geomathematics group at the Institute of Arctic and Alpine Research at the University of Colorado Boulder, Colorado, USA, and of the Geomathematics Division, Universität Trier, Germany. Other or more specific references are given in the respective method sections.

(C.1) Corrections of Radar Altimeter Data

(C.1.1) Corrections Applied to Satellite Radar Altimeter Data for Ice Surface Mapping

Altimeter data from SEASAT, GEOSAT, and ERS-1/2 are used in mapping the Antarctic Ice Sheet. Altimeter data from the GEOSAT Geodetic Mission (GM 1995-96, 01 April 1985 – 30 Sept. 1986) were selected for the Antarctic Atlas project because of their higher density of groundtracks resultant from the fact that the satellite was allowed to drift rather than being forced to repeat wider-spaced orbits exactly (as in the GEOSAT Exact Repeat Mission of 1987–1989). Orbital coverage of GEOSAT altimeter data extends to 72.1° southern latitude. Data density increases toward the limit of coverage because of the convergence of the groundtracks. The GEOSAT Atlas is based on altimeter data from one year of GEOSAT GM data (15 June 1985 – 15 June 1986).

Orbital coverage of ERS-1 extends to 81.5° southern latitude, and hence extends the area that is mappable from altimeter data by 9.4° latitude compared to GEOSAT. In section (C.2.3) it is explained how the Atlas Mapping Scheme is designed to include map sheets of the same area where data from both satelites are available, and to extend to the southern limit of coverage for ERS-1 data.

For ERS-1 Atlas maps, the time interval of data selected was determined under the constraints that it should be of the same length as that of the GEOSAT Altas data, provide sufficient data coverage to construct a map, and minimize seasonal effects. The selected time interval is February 1st to August 1st, 1995, which overlaps with the 168-day-repeat-cycle mission (until April 10 – March 20, 1995) and with the 35-day-repeat-cycle mission. Orbit separation at the equator is 8.3 km for data from the two 168 day-repeat cycles, orbit separation at the equator is 39 km for a 35-day-repeat cycle; along-track spacing of ERS-1 radar altimeter data is 335 m in all missions. There was no ERS-1 mission of geodetic type with a data distribution as dense as for the GEOSAT Geodetic Mission. During the GEOSAT Geodetic Mission, the satellite was allowed to drift, whereas during the ERS-1 168-day-repeat-cycle mission the orbits were still repeated.

Both GEOSAT and ERS-1 data used in the Atlas project were processed by the Ice Sheet Altimetry Group at NASA Goddard Space Flight Center (H.J. Zwally, J. DiMarzio, A. Brenner, and coworkers), using the following correction methods:

(a) transformation of "waveform data records" (WDR) into "ice data records" (IDR),

(b) "retracking", correction for position (Martin et al. 1983); see also Partington et al. (1987), Thomas et al. (1983) on ice-edge detection,

(c) correction for atmospheric effects (Zwally et al. 1983),

(d) correction for solid earth tides (Zwally et al. 1983),

(e) slope correction of the ice surface (Brenner et al. 1983),

(f) correction for water vapor (wet atmospheric correction),

(g) referencing to Goddard Earth Model (GEM T2) satellite orbits (Marsh et al. 1989),

(h) resulting in elevations in meters above the WGS-84 ellipsoid.

A general discussion of methods for processing satellite radar altimeter data over ice surfaces is given in section (B.3), here the methods that are employed in the processing of the ice-record data sets that are utilized in the Antarctic Atlas are described briefly.

The process of retrieving the location of the point on the ground from which the satellite signal was returned is termed "retracking". A series of waveform data is received, and specifics of the waveform yield the information on the reflected energy. Many effects change the traveltime and the energy, thus the appearance of the waveform. The traveltime is affected by atmospheric effects (a wet atmosphere, for instance, is denser than a dry atmosphere, and the signal travels slower). Tides, caused by the gravitational effect between the moon and the Earth, as known for the oceans, also exist for the solid Earth. The tidal waves of the solid Earth are much smaller than the differences between low tide and high tide on the coast, but they still need to be corrected for.

Satellite altimetry works best over surfaces with a low slope angle and few surface features, and, in fact, altimeters were designed for surveying ocean surface topography ("SEASAT"). Therefore, retracking methods were first designed for ocean surfaces and their properties also. This is not appropriate for retracking data from ice surfaces, which may contain crevassed zones, grounding lines, and other breaks in slope.

Martin et al. (1983) therefore fitted a double-ramp-model waveform to returns from more complex surfaces. Where the waveform results from reflectors from two distinct surfaces, this method works best. Alternative methods for retracking of radar-altimeter data over ice surfaces include using the half-peak value in the waveform (Brooks et al. 1983) or the first-return signal as an indicator of the range (Partington et al. 1987) (see section (B.3)).

For accurate topographic mapping, the data need to be corrected for the effect that the radar signal that is received first at the satellite is the one returning from the point of closest approach (PCA) (= the point located most closely to the satellite at this time) rather than the signal returning from the point at nadir (= directly below the satellite, in a line from the satellite and normal to the Earth's surface; on the problem of the geodetical definition of the surface, see section (C.4)). Only if the Earth's surface at nadir and in its neighbourhood is perfectly flat, the PCA and the nadir coincide; if the surface has a slope, the points differ. As a consequence, the first return signal appears to come from a point that is higher and at nadir. The correction is called the slope correction.

At NASA GSFC, the slope correction is carried out according to a method described in Brenner et al. (1983). This slope-correction method proceeds iteratively, by first calculating a surface ignoring slope effects, then calculating slopes for that surface, and then reprocessing the data using the slope values from the first iteration. In using this method, the assumption is made that the slope in the first step is sufficiently correct. Of course, the slope of the surface of the second iteration step is

different. An alternative method for slope correction has been developed by Bamber (pers. comm.). For topographic mapping, we use slope-corrected data. For calculation of elevation changes, however, the slope correction is best not applied (see section (E) and Herzfeld et al. (1997)).

The satellite orbits are calculated with respect to Goddard Earth Model (GEM T2) (see section (B.2)).

Elevation is given with respect to WGS84, an ellipsoid model of the Earth. Advantages and disadvantages of mapping the ice surface above the ellipsoid or above the geoid are discussed in section (C.4).

Data processing for ERS-1 data and for GEOSAT data is similar (as far as the scope of this book is concerned), every new satellite does, of course, require a new set of correction algorithms, and it should not be assumed that a program written for the analysis of one data set might work for the next data set from a different instrument as well. The ERS-1 altimeter is different from the GEOSAT altimeter, but the GEOSAT altimeter is very close in design to the SEASAT altimeter, consequently each satellite has different data and accuracy characteristics. Even after corrections, there are remaining offsets between any two satellites. For SEASAT and GEOSAT, these were about 30 centimeters, for ERS-1 and GEOSAT, a systematic offset has not been calculated to our knowledge.

The geographically correlated, systematic orbit bias between SEASAT and GEOSAT ERM has been estimated as 0.083+/-0.131 m by Lingle et al. (1994) for the Lambert Glacier/Amery Ice Shelf region, i.e. GEOSAT ERM surface is systematically slightly higher.

Using the accuracy information in the table "Summary of Past and Present Satellite Altimeter Measurement Precisions and Orbit Accuracies" in section (B.2), we can estimate the offset between GEOSAT and ERS-1, or at least put a boundary on it:

$$\|elev_{GEOSAT} - elev_{ERS-1}\|$$
$$\leq prec_{GEOSAT} + prec_{ERS-1}$$
$$+ orbacc_{GEOSAT} + orbacc_{ERS-1}$$

$$(C.3 - 18)$$

32

which is 72 cm if the largest values for orbit accuracy are used, and 22 cm, if the lowest values for orbit accuarcy are used (Table B.2-1).

Our direct link to the Ice Sheet Altimetry Group was employed for transferring Ice-Data-Record (IDR) datasets. From each IDR dataset, those points with retracked and slope-corrected data were retained, prior to mapping. For each map sheet, a track plot is constructed to investigate coverage and ensure that coverage by retracked and slope-corrected data is sufficient. This was the case for all map sheets. A typical track map is given in Figure B.2.2-1 for GEOSAT data and in Figure C.1.1-1 for ERS-1 data. The satellite groundtracks appear slightly curved, due to the projection, and they converge toward the southern limit of the satellite coverage, hence data density increases towards the southern limit. The density of the GEOSAT GM data is much higher than for the ERS-1 data. The ERS-1 data stem from February 1, 1995–August, 01 1995, and hence partly from a 168-day-repeat-cycle mission (until March 20,.1995) and partly from a 35-day-repeat-cycle mission. The change of repeat cycles is visible in the track plots in that two track patterns are overlain. As stated previously, the time frame was selected to minimize seasonal effects, while processing only 6 months of data.

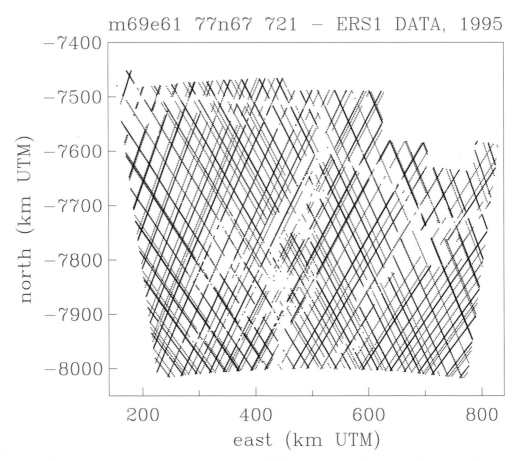

Figure C.1.1-1. Typical ground track pattern for ERS-1 data used in Atlas. (All data collected during ERS-1 Atlas time frame 01.02.1995-01.08.1995 shown, but with a reduction in numbers to match capacity of plot software). Geographical area is that of Atlas map m69e61-77n67-721 Lambert Glacier. Retracking difficulties, and hence gaps in the ground-track pattern, occur over rough terrain (mountaineous areas) and breaks in slope (ice front, glacier margin), but are generally relatively less frequent than for GEOSAT data.

(C.1.2) The Bad-Track Problem

Western Queen Maud Land (South)— ERS1 DATA, 1995

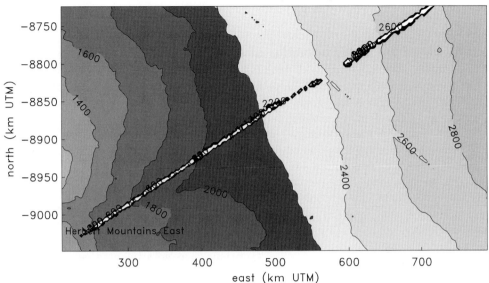

e339–15n78–815, WGS84, Gaussian variog., central mer. 357, slope corrected, scale 1:5000000, 990302

Figure C.1.2-1. Example of map with a bad track; ERS-1 data.

Western Queen Maud Land (South)— ERS1 DATA, 1995

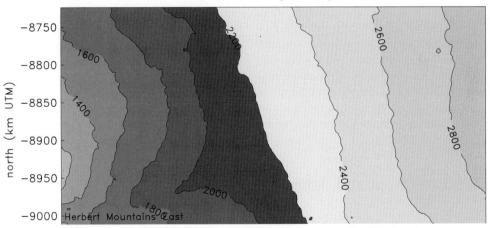

Figure C.1.2-1. Example of map with a bad track; ERS-1 data.

e339–15n78–815, WGS84, Gaussian variog., central mer. 357, slope corrected, scale 1:5000000, 990302

Figure C.1.2-2. Example of map after correction of bad-track problem; ERS-1 data.

After the processing carried out by NASA GSFC, as described in section (C.1.1) above, "bad" tracks with elevation (a) much lower than the surrounding area, or (b) of about constant small (50 m) difference to the surrounding area remained in several ERS-1 maps. An algorithm was developed to identify and remove these bad-track data (Herzfeld et al. 2000b). An example of a map with a bad track and corrected track using the algorithm by Stosius is given in Figures C.1.2-1 and C.1.2-2, respectively.

(C.1.3) The Need for Interpolation of Geophysical Line Survey Data

A disadvantage of radar altimeter data is that measurements are only taken by single radio beams along the groundtrack of the satellite (and evaluated for the center of the footprint — see section on corrections and retracking), so all data lie along groundtracks, and large gaps exist between tracks, i.e. satellite radar altimeter data are geophysical line survey data. The typical pattern of line survey data is evident from the track maps (see section (C.1.1)).

Radar altimeter data share these distribution properties with other geophysical or geological data from surveys that are carried out from vehicles that follow track lines: Aeromagnetics from planes, bathymetry from survey vessels, submarine magnetics or gravimetry from devices that are towed by ships, radio-echo soundings from sledges driven over a glacier. The resulting data distribution in any situation is characterized by dense information along the survey track and gaps of information in between. The data may be recorded continuously (radio-echo soundings, e.g. Jonsson et al. 1988) or discretely along a line (single-beam bathymetric surveys, satellite radar altimeter data), or on a stripe of varying width (multibeam deep sea sonar device, e.g. Herzfeld (1989a)).

Such data cannot be interpolated directly by isoline methods, but interpolation onto a regular grid is necessary beforehand. However, the spatial characteristics of geophysical track line data are not consistent with the properties expected of irregularly distributed data in mathematical interpolation routines, hence specific search algorithms for data points are required in interpolation algorithms to avoid artefacts that resemble the trackline pattern. The necessity of appropriate software to evaluate survey line data was outlined earlier (Briggs 1974), one such algorithm is described in Herzfeld (1990).

The interpolation method applied in the Antarctic Atlas project is kriging. Kriging methods are methods of geostatistical estimation. "Estimation" is probabilistic vocabulary, and corresponds to "interpolation and extrapolation" in the language of numerical analysis. For the analysis of altimeter data in the Atlas project, however, only interpolation and no extrapolation is needed, because the coverage of the Atlas maps matches the coverage of the Earth by satellite data. The geostatistical methods are explained in section (C.3).

(C.2) Map Projection and Atlas Mapping

One of the oldest problems in mapping the Earth is the definition of projections of the Earth's surface onto a two-dimensional map sheet. For mapping purposes, the Geoid is commonly approximated by a sphere or ellipsoid. Desirable properties of map projections are conservation of area (equal-area projection), of distances (equal-distance projection), of angles (equal-angle or conformal projection), which are mutually exclusive when mapping on a plane, and projection to a rectangular coordinate system (for examples see Hake 1982; Snyder 1987). Because it is not possible to satisfy all of these conditions, some projections have been defined that do not fulfill any conditions exactly, but a combination of them approximately (Hake 1982; Snyder 1987).

Orthogonal coordinate systems are advantageous for interpolation of irregularly distributed data and quantitative analysis, in particular if distance-dependent measures are involved. Geoscience data usually are collected with reference to geographic coordinates. Low distortion, orthogonal coordinates and distances given in meters are achieved by the Universal Transverse Mercator (UTM) projection. Mapping of large areas, in particular at high latitude, may require calculation in several adjacent map sheets.

(C.2.1) The UTM Projection

The distortion if using geographic coordinates is particularly severe for mapping at high latitude. A useful projection algorithm for mapping at all latitudes is the Transverse Mercator projection.

A Mercator projection is defined by a cylinder that is tangent to the Earth and a mapping to orthogonal coordinates. For the (common) Mercator projection, the tangent circle is the Equator, for the Transverse Mercator projection, the tangent circle is a meridian (called the central meridian of the projection). The advantage of the Transverse Mercator projection is that all latitudes are mapped with the same distortion. The disadvantage is that areas far away from the central meridian are strangely distorted. The solution provided by the UTM system is to rotate the cylinder around the Earth in steps of 6°. Zone 1 corresponds to central meridian 177° W. The standard central meridians are at 3° (for 0° – 6°), 9° (for 6° – 12°), 15° (for 12° – 18°) etc. The central meridian is projected with a factor of 0.9996, lines of true scale are approximately parallel and lie approximately 180 km east and west of the central meridian. The border meridians are projected slightly lengthened, for example at 50° latitude with a factor of 1.00015. The projection is defined everywhere except 90° away from the central meridian. In the UTM scheme, the projection is chosen such that the central meridian is mapped to East coordinate 500,000, units are in meters; the North coordinate along the central meridian is in meters from the equator (along the ellipsoid). Any meridian, however, may be selected as central meridian and utilized in the projection. Inversion of the projection back to geographic coordinates needs to use the same central meridian. UTM is conformal and close to an equal-area projection, it has orthogonal coordinates in meters.

(C.2.2) The Atlas Mapping Problem

For series of topographic maps in countries with a long tradition in mapping, algorithms have been designed to construct maps constituting an atlas: An atlas, in the sense of differential analysis (Hollmann and Rummler 1972, p. 63), is a set of maps that

(i) covers a given area completely (that is, each point in the area is contained in at least one map);

(ii) projections restricted to areas that appear on two (adjacent) maps (subsets of two maps) are identical on the intersection.

In cartography, only property (i) is required for an atlas, and the neighbourhood relationships need to be matched between sheets. For the user of a map series, an overlap of adjacent sheets is convenient for identification of regional relationships of location close to the map edge. One problem with most map series is that for high latitudes a different projection algorithm is used, which is not designed for studies at a resolution used for lower latitudes. Commonly, the Arctic and Antarctic are each mapped separately in one sheet, using the polar stereographic projection (e.g. US-UTM series; see Snyder 1987, p. 58). From the viewpoint of interpolation of irregularly distributed data onto a regular grid, an orthogonal coordinate system facilitates the algorithm and saves computation time. The latter is especially important if distance-dependent measures are used, such as in inverse-distance weighting or in geostatistical methods.

Distance may be calculated on the sphere or on the ellipsoid (cf. Moritz 1984; Torge 1980), but this requires transformations at each step of the interpolation algorithm which usually is dependent on the number of points squared. In comparison, the number of essential operations for coordinate transformation depends only linearly on the number of points. Methods involving the covariance function or the variogram (kriging, least-squares prediction; cf. Herzfeld 1992) would require estimation of the structure function over the ellipsoid which would be troublesome. It is apparent that coordinate systems with orthogonal coordinates in meter units are thus extremely convenient for interpolation purposes.

Common practice is not to change coordinate systems, but to simply use geographic coordinates, which is unproblematic for small areas. For large areas, neglection of the coordinate transformation results in a severe distortion of the spatial structure in the data. (Recall that 1° latitude is always about 111 km, but 1° longitude is cosine of latitude times 111 km; so, at 60° North/South it is only 0.5 times 111 km or 55.5 km.)

The distortion is particularly severe for mapping at high latitude. The Arctic and Antarctic are usually treated separately in one map using the polar stereographic projection. Typically, such maps of polar regions are at a small scale and do not show much detail. The importance of the polar system in the Earth's global systems and its role in 'global change' have become increasingly recognized, and detailed maps of the polar regions have become more important.

(C.2.3) The Solution: The Antarctic Atlas Mapping Scheme

The solution to these requirements lies in mapping the Antarctic continent by an atlas, which consists of many individual maps that have orthogonal coordinates, a sufficiently large scale and resolution, are organized in latitudinal rows and have sufficient overlap to show not only each point, but each feature with a neighbourhood on at least one map. The Antarctic Atlas Mapping Scheme has all these properties.

In the mapping scheme of the Antarctic Atlas, the UTM projection (see section (C.2.1)) is utilized for all maps, and standard central meridians are employed.

The UTM projection does not satisfy condition (ii) of an atlas, if two adjacent maps belong to two different central meridians; however, this defect may be compensated for by overlaps between neighboring map sheets. This is realized for the Antarctic Atlas Mapping Scheme, with a scale of 1:5.000.000 and 3 km grid nodes.

Sufficient overlap of adjacent sheets is convenient for the user of the atlas and necessary to ensure that each point of Antarctica is contained in at least one map despite of the distortion of the map edges introduced by the projection algorithm. With the help of the computer program TRANSVIEW especially developed to facilitate atlas design (Herzfeld et al. 1999a), we decided on the following setup for projection and map sheets:

Rows:

63° – 68° S,

67° – 72.1° S,

71° – 77° S,

75° – 80° S,

78° – 81.5° S

In rows 63° – 68° S and 67° – 72.1° S maps are of 16 degrees longitude nominal size and overlap 2 degrees on each side, so each map is offset against the next one by 12 degrees longitude.

Latitude 72.1° South marks the poleward limit of coverage of altimeter data from the SEASAT and GEOSAT satellites. Following the collection of satellite data, the GEOSAT Atlas was designed and calculated first. Map sheets of the ERS-1 Atlas are designed to match those of the GEOSAT Atlas, and to cover Antarctica to 81.5° S, the southern limit of ERS-1 coverage.

In rows 71° – 77° S, 75° – 80° S, and 78° – 81.5° S maps are of 36 degrees longitude nominal size, overlap is 6 degrees on each side, offset of two adjacent maps is 24 degrees longitude. The map names (e.g. m69e61-77n67-721) give central meridian (69°) and extent of the nominal map area (61° to 77° E, 67° to 72.1° S (the minus in the north coordinate was dropped in the file names for ease of reading)). The true map area is then the maximal area contained in the nominal map area with the

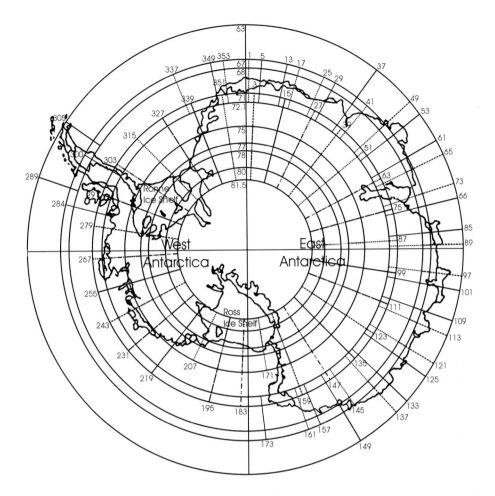

Figure C.2.3-1. Antarctic Atlas Mapping Scheme: Coverage of the Antarctic continent by maps in latitude rows 63°-68° S, 67°-72.1° S, 71°-77° S, 75°-80° S, 78°-81.5° S.

property that it has straight edges in the UTM coordinate system and the grid coordinates are multiples of 3000 (Herzfeld and Matassa 1999). All maps in one row have the same size and grid coordinates. The most extreme row (in that it features the largest differences between UTM and geographic coordinates) is the southernmost row.

Dimensions of map sheets are listed in Table C.2.3-1.

Grid Spacing. Maps are based on digital elevation models with 3 km grid nodes. The selection of the grid spacing has a glaciologic reason: The resolution needs to be high enough to allow for glaciologic modeling which includes longitudinal stress gradients, at least for larger glaciers and ice streams, but low enough to avoid too much complexity. More precisely, the horizontal resolution of the DTM should be sufficient for nu-

merical models that take into account longitudinal stress gradients, while being coarse enough to warrant neglecting the so-called T-term (a double integral over a coordinate perpendicular to the glacier surface of the second derivative of the shear stress, in the longitudinal direction) in the longitudinal stress equilibrium equation (Kamb and Echelmeyer 1986). The spacing of 3 km is approximately equal to three times the ice thickness in the vicinity of the grounding line (exactly for Lambert Glacier/Amery Ice Shelf and roughly for other large ice-stream/ice-shelf systems) which satisfies the numerical criterion for modeling glaciers stated above. On the other hand, the resolution of the altimeter data limits the grid resolution – 3 km grids satisfy the glaciologic criterion, but 2 km grids do not yield better maps. In this book, contoured maps have a scale of 1:5.000.000, which is not mandatory, as is demonstrated, for instance, in section (F).

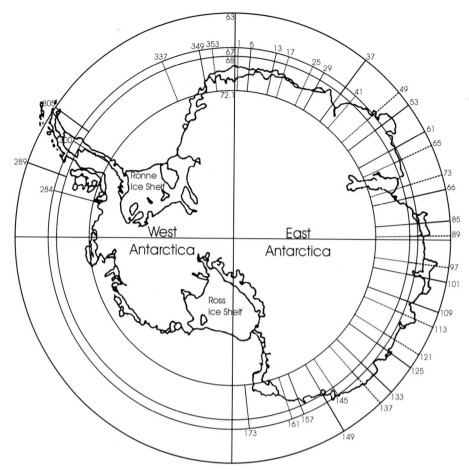

Figure C.2.3-2. Antarctic Atlas Mapping Scheme: Coverage of the Antarctic continent by maps in latitude rows 63°-68°S, 67°-72.1°S (GEOSAT and ERS-1 Atlas maps).

Table C.2.3-1. Antarctic Atlas Mapping Scheme

Latitude	E–W		N–S	
	Size [km]	no. gridnodes	Size [km]	no. gridnodes
63°–68°S	666	223	531	178
67°–72.1°S	546	183	543	182
71°–77°S	888	297	570	191
75°–80°S	684	229	477	160
78°–81.5°S	582	195	324	109

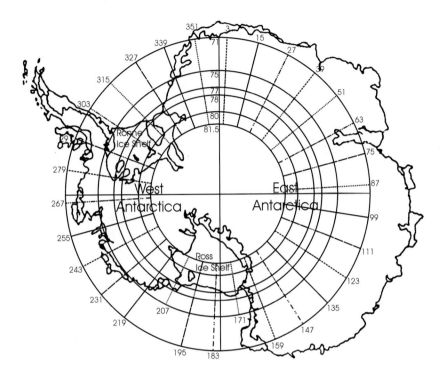

Figure C.2.3-3. Antarctic Atlas Mapping Scheme: Coverage of the Antarctic continent by maps in latitude rows 75°-80° S, 78°-81.5° S (ERS-1 maps).

(C.2.4) Map Sheet Calculation with TRANSVIEW

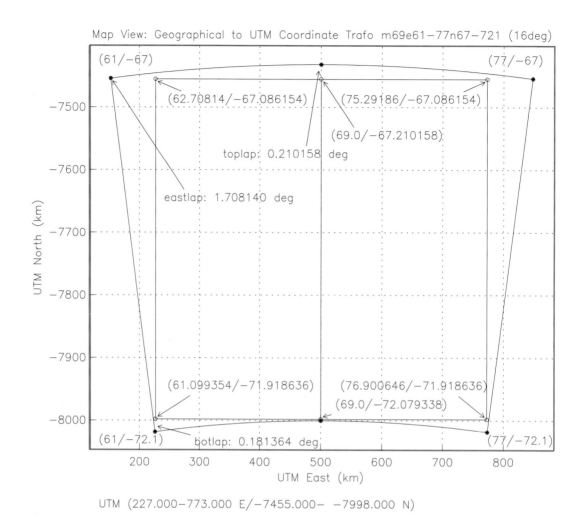

Figure C.2.4-1. Map sheet calculation with TRANSVIEW: Nominal area in geographic coordinates and actual area in UTM coordinates for latitude row 67°-72.1° S.

Calculation of map sheets is achieved with the program TRANSVIEW (Herzfeld et al. 1999a). The objective of our program TRANSVIEW is to provide a tool to calculate and visualize

(a) the shape of a map that is rectangular in geographic coordinates when transformed to UTM coordinates

(b) the amount of distortion for any map sheet on the Earth

(c) the largest map that is rectangular in UTM coordinates and inscribed in a given map that is rectangular in geographic coordinates

(c) the overlap necessary to map a large area in individual sheets using the UTM projection, and

(d) to provide a system that works also for Antarctica.

TRANSVIEW works for any rectangular area on the Earth. The only restriction is that the area needs to be located entirely on the Northern or on the Southern hemisphere. The UTM projection is defined relative to an appropriate central meridian (3° W, 9° W, 15° W, etc., uneven multiples of 3°). For map areas that do not contain a central meridian of the UTM projection, an appropriate

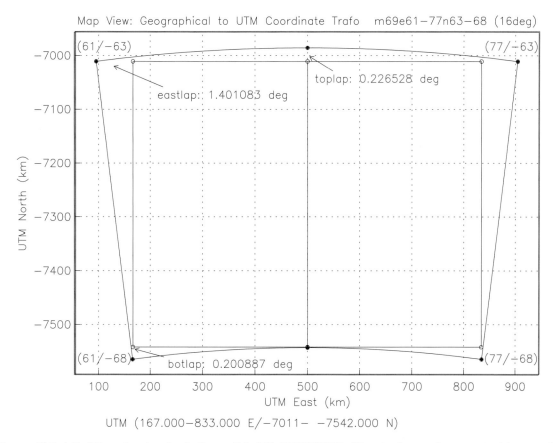

Figure C.2.4-2. Map sheet calculation with TRANSVIEW: Nominal area in geographic coordinates and actual area in UTM coordinates for latitude row 63°-68° S.

meridian needs to be determined by the user of the program. The latter is of particular importance for mapping small areas. The program discerns the location of the map relative to the central meridian and automatically selects the appropriate case for the transformation. TRANSVIEW is written in FORTRAN77, uses only FORTRAN77, and is directly portable to any computer with a compiler (main frame, work station, or personal computer).

For the examples of the maps for central meridian 69°, TRANSVIEW is applied to calculate and visualize the sheets in all rows to demonstrate the differences in shape and distortion (Figures C.2.4-1 to C.2.4-5).

42

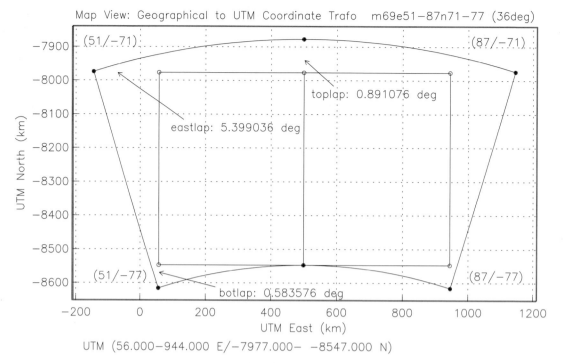

Figure C.2.4-3. Map sheet calculation with **TRANSVIEW:** Nominal area in geographic coordinates and actual area in UTM coordinates for latitude row 71°-77° S.

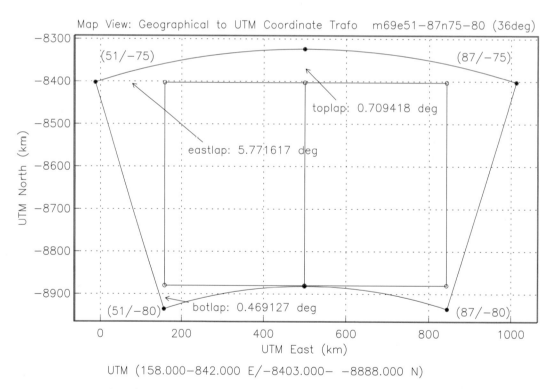

Figure C.2.4-4. Map sheet calculation with **TRANSVIEW:** Nominal area in geographic coordinates and actual area in UTM coordinates for latitude row 75°-80° S.

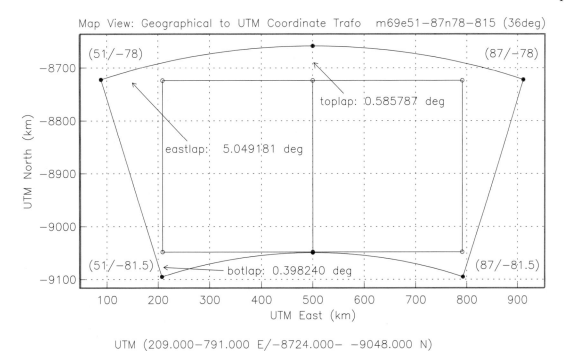

UTM (209.000−791.000 E/−8724.000− −9048.000 N)

Figure C.2.4-5. Map sheet calculation with TRANSVIEW: Nominal area in geographic coordinates and actual area in UTM coordinates for latitude row 78°-81.5° S.

(C.3) Geostatistical Estimation

The geostatistical method has allowed to push the limits of radar altimeter data evaluation. Mathematical interpolation methods apply the same algorithm to all data locations and do not take regional properties into account. Geostatistical methods, to the contrary, proceed in two steps: (1) a structure analysis of the spatial continuity properties (variability in surface structure in our case), termed *variography*, and (2) interpolation (and extrapolation) of the data using variogram properties of neighbouring data (see Fig. C.3-1). It is the first step that allows to take geophysical knowledge into account when interpolating. The variogram takes on the role of a modeling function. As an application, the variogram function most suitable for mapping ice surface topography with an emphasis on studying the Antarctic marginal areas is the variogram of an ice-stream/ice-shelf system near its grounding zone (calculated from radar altimeter data). In the central areas of the inland ice, spatial variability is low, and selection of a mathematical method is less crucial (as an exam-

ple, interpolation of ERS-1 data for Antarctica using mathematical routines is sufficiently accurate for the very-low angle areas of the Antarctic interior, Bamber 1994). Applications and comparison to field observations have demonstrated that the Atlas maps are correct for the marginal, steeper areas also, and they reveal many interesting features (e.g. Herzfeld and Matassa 1997; Herzfeld et al. 2000b).

For an introduction to geostatistics for resource estimation, the reader is referred to Matheron (1963), Journel and Huijbregts (1989), for geophysical and geodetical data analysis, to Herzfeld (1992), for applications to satellite radar altimeter data analysis, to Herzfeld et al. (1993, 1994), Herzfeld and Matassa (1999) and Herzfeld et al. (2000b). It may be worth mentioning that a book on the utilization of geostatistics in cartography is not available elsewhere. The geostatistical section may also serve as a quick primer for students and researchers in glaciology, cartography, geography and environmental sciences.

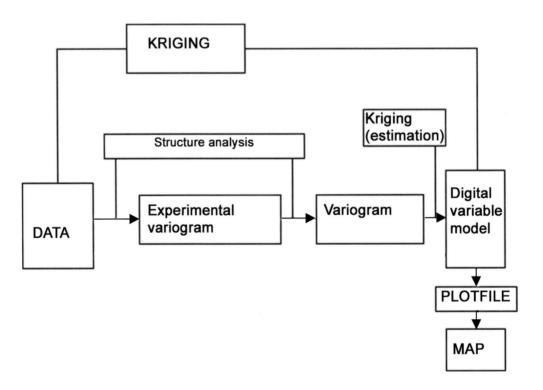

Figure C.3-1. Kriging Flow Diagram. Notice that both the estimation step and entire process of generating a digital terrain model are termed "Kriging".

(C.3.1) Concept of the Regionalized Variable and Principles of Variography

Procedures used by miners to estimate ore resources were formalized by Krige (*cf.* Krige 1951, 1966) and put into a theoretical framework by Matheron (*cf.* Matheron 1963, 1971, Journel and Huijbregts 1989 (1978)): Thus geostatistics, the theory of regionalized variables, sometimes simply referred to as "Kriging" by the generic name for the estimation methods used, is a theory rooted in techniques that worked in practice in an economic branch were success is the (only) criterion. The term "Kriging" is derived from the name of Dani Krige, who introduced what were to become geostatistical concepts to gold mining in South Africa. The name "geostatistics" is sometimes confused with the application of statistics in the geosciences, also termed "geostatistics" in some of the (North American) literature. While most commonly a probabilistic framework is preferred in geostatistical theory, it is equally feasible to use an inverse-theoretical deterministic formulation (Herzfeld 1992). In this introduction we use the probabilistic formulation.

Readers who wish to read up on probability and statistics may find Spiegel et al. (2000) and Stirzaker (1999) useful.

The underlying concept of geostatistics is that of the regionalized variable. A regionalized variable is a spatial variable (a variable defined over $\mathcal{R}^2$ or $\mathcal{R}^3$, the two- or three-dimensional space of real numbers) with a transitional behaviour between deterministic and random states. The classic example in ore resource estimation is the grade of mineralization:

The grade of mineralization $z(x)$ at a location x in a deposit D is to some extent determined by the properties of the deposit, but also shows a random variation from one location to another, which increases with distance. So, mineralization grade shows the transitional behaviour between deterministic and random status that is typical for a regionalized variable ($z(x)$).

Historically, geostatistics had been confined to mineral and petroleum exploration for a few decades, before its usefulness for other branches of geology and geophysics was recognized (e.g., Herzfeld 1989a). Similar methods are Gandin's method of optimum interpolation, applied in meteorology (Gandin 1965), least-squares prediction and collocation methods applied in physical geodesy (Kaula 1967, Heiskanen and Moritz 1967, Grafarend 1975, Lauritzen 1977, Moritz 1980) all of which may be considered a form of geophysical inverse theory (Herzfeld 1992, 1996).

The primary variable describing topography is surface elevation. Elevation also has the characteristics of a regionalized variable: Two points that are located close to each other are likely to have a similar elevation, they *have* similar elevation unless there is a cliff, mathematically a discontinuity, which is rare. As one moves farther away from a point with measured elevation, one is less likely able to predict the elevation that will be measured at another site, until a distance is reached beyond which prediction from the first measurement site is no longer possible. This distance is the *range.*

Similarly, ocean depth (bathymetry), the Earth's gravity field, the magnetic field, and concentration of pollutants in the air may be described by regionalized variables, in fact, every property of the Earth that is defined or observed in space or on a map or in an area (i.e. with spatial support) may be considered a regionalized variable.

The measurements $z_i(x)$, where $i = 1, \ldots, n$, on $z(x)$ are thus considered as a realization of a spatial random function $Z(x)$ for $x \in D$ with the property that

$$(Z(x) - Z(x + h))$$

is a second order stationary random function

$$(C.3-1)$$

for fixed $h \in \mathcal{R}^2$ such that $x, x + h \in D$. Here we assume that D is a subset of $\mathcal{R}^2$, but $D \in \mathcal{R}^3$ is also possible. The property given in (C.3-1) is called the *intrinsic hypothesis.* (Note that stationary random functions are intrinsic, but the converse is not true in general.)

We notice that, while mineralization grade, surface elevation and other geo-variables share the characteristics of a regionalized variable, their spatial properties are different, specific for each variable. For instance, in a mountain range, topography varies over short distances (hundreds of meters to kilometers), whereas the gravity field only changes over much larger distances (tens of kilometers). Topography (elevation) generally has a smaller range than gravity (gravity anomaly).

In addition, topography in a mountain range is rougher (topographic relief is higher) than topography on a plain. The method utilized to determine spatial properties of a regionalized variable is *variography*.

Under the intrinsic hypothesis, the

(semi-) variogram

$$\gamma(h) = \frac{1}{2} E[Z(x) - Z(x+h)]^2 \qquad (C.3-2)$$

exists and is finite, γ measures the spatial continuity. (E denotes the mathematical expectation.) In practice, an experimental variogram is calculated from the data set, according to the formula

$$\gamma(h) = \frac{1}{2n} \sum_{i=1}^{n} [z(x_i) - z(x_i + h)]^2 \qquad (C.3-3),$$

where $z(x_i), z(x_i + h)$ are samples taken at locations x_i, $x_i + h \in D$ respectively, where n is the number of pairs separated by $h \in \mathcal{R}^2$.

Experimental variograms can have different forms, as should be expected from the natures of different experiments. Characteristic examples are given in Figure C.3.1-1. A transitional variogram (first panel in Figure C.3.1-1) reflects the properties of a surface with features which are spatially correlated at a certain scale and up to a certain distance and which are not correlated for large distances. The variogram model in the first panel with the nice characteristics has been produced from a synthetic variogram. A variogram from observational data will look more like the experimental variogram in the second panel in Figure C.3.1-1. The experimental variogram values have some ups and downs. Such oscillations may correspond to measurement errors or to geophysical properties of the surface. An experimental variogram with a clear minimum after a maximum is an indicator of a surface with a quasiregular structure (third panel in Figure C.3.1-1).

The analysis of complex variogram structures is the objective of geostatistical classification (Herzfeld and Higginson 1996). In the context of ice surface elevation mapping, geostatistical interpolation/ estimation is applied, and the variogram takes the role of a modeling function, which captures the spatial variability of the surface and, at the same time, satisfies mathematical conditions that are required in the interpolation.

The mathematical condition that needs to be satisfied is termed the positive-definiteness condition of kriging, that is, the kriging system has a unique solution. (Note that the associated kriging matrix is not generally positive definite because of the unbiasedness conditions.) Commonly, variogram models are selected from a collection of model types for which the positive-definiteness condition has been shown to hold, including the spherical model, the linear model (with sill and range), the exponential model, and the Gaussian model (see, for instance, Journel and Huijbregts 1989). The positive-definiteness problem will be revisited after introduction of the variogram models.

The variogram models (see Figure C.3.1-2) may be characterized by their function or model type and a small number of parameters: sill, range, nugget effect (the variogram parameters). The *sill* corresponds to the total variance of the population, the *nugget effect* is the residual variance of repeated sampling in the same location, or of samples spaced closer than the resolution of the survey (notice that the terminology is still reminiscent of gold exploration). The concept of a regionalized variable involves the notion that samples spaced closer together are related closer than samples spaced farther apart, which results in a *transitional* variogram with low variogram values for short distances, and values increasing to the sill value indicative of the random nature of the variable for samples exceeding a certain distance, this distance is termed *range*. This leads to a more mathematical definition of the *range* (which we already introduced above).

For distances increasing from the support to the range parameter a in Equations (C.3-4 – C.3-7), the variogram increases and the regionalization effect decreases; for instance, beyond the range the regionalized variable theoretically behaves like a random variable (probabilistically speaking), and data with a distance larger or equal to the range all have the same influence on the interpolated value (numerically speaking).

Typical Experimental Variograms

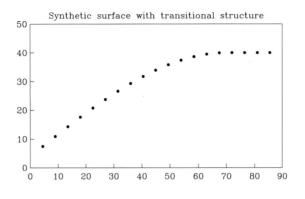

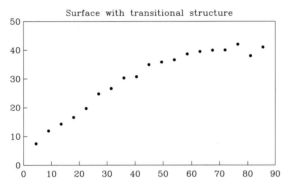

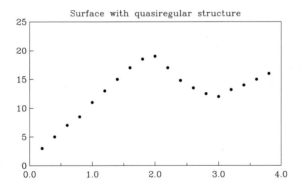

Figure C.3.1-1. Typical experimental variograms.

A *linear variogram model* with total sill $(c_0 + c_1)$ and range a is given by

$$\gamma_{lin-sill}(h) = c_0 + \frac{c_1}{a}|h| \ \text{ if } |h| \le a$$

$$(C.3\text{-}4)$$

$$\gamma_{lin-sill}(h) = c_0 + c_1 \ \text{ if } |h| > a$$

This is the easiest way to model a surface with a transitional structure. The linear variogram model, however, has the disadvantage that it is not differentiable in $|h| = a$, which causes problems in the interpolation if the model is needed for lag values larger and smaller than a.

A *spherical variogram model* is given by

$$\gamma_{sph}(h) = c_0 + c_1(\frac{3|h|}{2a} - \frac{h^3}{2a^3}) \ \text{ if } |h| \le a$$

$$\gamma_{sph}(h) = c_0 + c_1 \ \text{ if } |h| > a \qquad (C.3-5)$$

Variogram Models

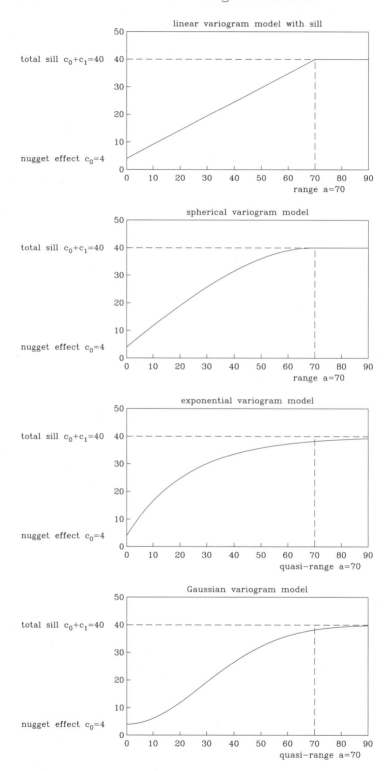

Figure C.3.1-2. Variogram models. Notice that all models have the same sill, range and nugget effect. A variogram (model) models the transitional spatial structure of the underlying regionalized variable.

For the linear and the spherical model, the range is equal to the parameter a. Both models have a linear behaviour in the origin. The spherical model solves the differentiability problem of the linear model with sill (equation (C.3-4)).

An *exponential variogram model*, which is defined by

$$\gamma_{exp}(h) = c_0 + c_1(1 - e^{[-\frac{|3h|}{a}]}) \qquad (C.3-6)$$

has a practical quasi-range of a. An exponential model has a linear behaviour in the origin, but a more gradual change in spatial covariance than the spherical model.

A *Gaussian variogram model*, which is defined by

$$\gamma_{gau}(h) = c_0 + c_1(1 - e^{[-\frac{3h^2}{a^2}]}) \qquad (C.3-7)$$

has a practical or quasi-range of a with $\gamma(a) \approx c_0 + 0.95c_1$.

This is easy to check: For $a = h$ in equation (C.3-7), we get

$$\gamma_{gau}(a) = c_0 + c_1(1 - e^{-3}) \qquad (C.3-8)$$

With $e^{-3} = 0.04947$ we get $1 - e^{-3} = 0.95021$ and hence $\gamma_{gau}(a) = c_0 + 0.95021c_1$ to 5 decimal digits precision (also not the exact value), i.e. at lag $a = h$ the Gaussian model has almost reached the sill value of $c_0 + c_1$. For larger distances, $\gamma_{gau}(h)$ approaches $c_0 + c_1$, but never quite reaches it (the sill is a limit value). For the spherical variogram, the sill is simply reached at $h = a$.

The exponential model shares this property with the Gaussian model: The sill $c_0 + c_1$ is a limit value that is approximated infinitesimally close, but never reached. As for the Gaussian model, $\gamma_{exp}(a) \approx c_0 + 0.95c_1$ (to verify this, set $h = a$ in equation (C.3-6) to get

$$\gamma_{exp}(a) = c_0 + c_1(1 - e^{-3}) \qquad (C.3-9)$$

and calculate as above for the Gaussian model).

Since in practice experimental variogram values are only fitted by a model, but not interpolated exactly, a model that reaches its sill and a model that approximates its sill work equally well. In theory, the distinction between a *range* $a = a_{lin-sill}$ for

the linear model with sill (C.3-4) and $a = a_{sph}$ for the spherical model (C.3-5) on the one hand and a *quasi-range* $a = a_{exp}$ for the exponential model (C.3-6) and $a = a_{gau}$ for the Gaussian model (C.3-7) is made. Some authors also distinguish between *sill* and *quasi-sill* in an analogous sense.

Formulation of the variogram models as in the above equations (C.3-4) to (C.3-7) has the advantage that $a_{lin-sill}$, a_{sph}, a_{exp}, and a_{gau} all have the the same value a, which makes fitting a model to an experimental variogram easy: From a plot of the experimental variogram, sill, range, and nugget effect can be determined visually, and then the function type that matches the data and/ or the variable properties (see note 2) best can be selected.

Note 1 on variogram modeling. In some of the geostatistics literature, the exponential and the Gaussian models are formulated "without the factor 3" as

$$\gamma^*_{exp}(h) = c_0 + c_1(1 - e^{[-\frac{|h|}{a^*_{exp}}]}) \qquad (C.3-10)$$

for the exponential variogram model and

$$\gamma^*_{gau}(h) = c_0 + c_1(1 - e^{[-\frac{h^2}{a^*_{gau}{}^2}]}) \qquad (C.3-11)$$

for the Gaussian variogram model, likely because these are slightly simpler models. For these models, the quasi-ranges are

$$a_{exp} = \frac{a^*_{exp}}{3} \qquad (C.3-12)$$

and

$$a_{gau} = \frac{a^*_{gau}}{\sqrt{3}} \qquad (C.3-13)$$

i.e. the same sills and ranges are determined numerically as in equations (C.3-6) and (C.3-7) but the range parameters a^*_{exp} and a^*_{gau} are not equal to the ranges anymore, and they are different for each model. This has lead to serious misconceptions about the shapes of variogram models. One might be inclined to forget the formulations in (C.3-10) and (C.3-11) altogether, if it were not for the following caveat: If you use a computer program for kriging or variogram modeling, you need to check whether equations (C.3-6) and (C.3-7) "with 3" or equations (C.3-10) and (C.3-11) "without 3" are used in the program code and divide or multiply your range estimates accordingly by 3 for the spherical model and $\sqrt{3}$ for the Gaussian model.

Surface roughness. The roughness of a surface is indicated by the behaviour of the variogram in its origin: A rough surface is one with high topographic relief, and measurements are likely to vary already over short distances. This corresponds to a model with a steep function in the origin (see Figure C.3.1-2).

A smooth surface leads to a variogram model that has a low or zero gradient in the origin. In between these two extremes are surfaces with a variogram that has a linear behaviour in the origin, but of non-zero slope. The slope of the tangent in the origin of the variogram indicates the roughness of the regionalized variable: The steeper the slope, the rougher the variable (in our case, the surface). The lower the slope, the smoother the surface. Of course, roughness depends on scale, but for our application we simply analyze the surface at one scale matching the scale and resolution of the data.

A linear variogram is associated with a regionalized variable that is continuous in the mean-square sense, the same is true for a variable with a spherical variogram, because the spherical variogram has a linear function as a tangent in the origin (with non-trivial slope). An exponential variogram model also has a tangent with a non-zero slope in the origin and consequently is the variogram of a mean-square continuous variable. A regionalized variable with a Gaussian variogram is not only mean-square continuous but also mean-square differentiable. A Gaussian variogram has a tangent with zero slope in the origin (see Fig. C.3.1-2). Use of a smoother variogram in the kriging estimation results in a smoother map.

Role of the linear, spherical, exponential and Gaussian variogram model types in kriging. As will be described in section (C.3.2) on "Kriging", the variogram model plays an important role in the estimation (or interpolation or extrapolation) step. The spherical, exponential and Gaussian variogram models have the property that a kriging equation formulated with use of these models has a unique solution. This is termed "positive-definiteness condition of kriging", and, in an even looser use of terminology, variogram models for which it can be shown that the kriging equation has a solution (although the kriging matrix may not be positive definite) are called "positive-definite variogram models". Proof that a given model is positive-definite is not simple, and it depends on the type of kriging. Therefore it is advisable to use models from the "zoo" of proven models rather than "wild animals" that may also fit the experimental variogram. The "zoo" contains a few more models than the four introduced above (Equations (C.3-4) to (C.3-7)), but those are the most common ones. The linear model with sill (Eqn. (C.3-4)) should be used only for distances smaller than the range because of the differentiability problem at $h = a$. Linear combinations of models in the "zoo" lead to more models in the "zoo", which increases selection of types of acceptable models. A pure nugget-effect model reduces kriging to arithmetic averaging.

Note 2 on variogram modeling. In particular inexperienced practitioners should simply fit the data. It is possible to utilize the variogram as a modeling kernel and this way allow known spatial continuity properties of the variable to enter in the interpolation, but this approach requires an advanced understanding of mathematical optimization and inverse theory as well as experience in geostatistics and knowledge of the geophysical variable that is analyzed. The rewards of the variogram-as-a-modeling-tool approach are more natural-looking maps.

Note 3 on variogram modeling. Another consideration is that a variogram without nugget effect or with a relatively small nugget effect may lead to very noisy maps or even to failure of the inversion step in kriging (see equation (C.3-22) below). It may happen that the kriging matrix is almost singular and thus behaves like a singular matrix numerically. In such a case, adding a larger nugget effect may solve the problem, even if the variable is very smooth and the observations have only very small errors such that a variogram model with zero or almost zero nugget effect is a correct one. Alternatively, a covariance function model rather than a variogram model may be utilized under certain mathematical conditions.

A regionalized variable may have a (global) trend or a (local) drift. The concept of the drift matches the concept of local estimation in kriging (see eqn. (C.3-23) below). To detect a *drift component, m,* calculate

$$m(h) = \frac{1}{n} \sum_{i=1}^{n} [z(x_i) - z(x_i + h)] \qquad (C.3-14)$$

and the *residual variogram*

$$res(h) = \gamma(h) - \frac{1}{2} m(h)^2 \qquad (C.3-15).$$

Note that if the random function Z is stationary, then $m(h) = 0$ and $\gamma(h) = res(h)$ for all h. This is then used as the variogram γ^* of the residuals of a drift, although technically it is not the variogram of the residuals. But since it is not possible to estimate the drift and the variogram of the residuals at the same time, estimating the drift first and using the residual variogram is a practicable work-around.

(C.3.2) Kriging

Estimation of surface elevation at a given location $x_0 \in D$ is performed by kriging. Kriging is a family name for least-squares-based estimators, the general linear estimator being

$$Z_0^* = \alpha_0 + \sum_{i=1}^{n} \alpha_i Z(x_i) \quad \text{with} \quad \alpha_i \in \mathcal{R}$$
$$(C.3 - 16),$$

where weights are determined by a minimum variance criterion

$$E[Z_0^* - Z_0]^2 \quad \text{minimal !} \qquad (C.3 - 17)$$

and with respect to unbiasedness conditions $E[Z_0^* - Z_0] = 0$.

The estimation variance is

$$\begin{aligned} \text{var}[\varepsilon_0] &= \text{var}[Z_0 - Z_0^*] \\ &= \text{var}[Z_0] \qquad (C.3 - 18) \\ &\quad -2\sum_{i=1}^{n} \alpha_i \, \text{cov}[Z_0, Z_i] \\ &\quad + \sum_{i,j=1}^{n} \alpha_i \, \alpha_j \, \text{cov}[Z_i, Z_j] \end{aligned}$$

The covariances in equation (C.3-18), however, are not known. The salient concept in kriging is to replace the unknown covariances by "known" variogram values. The variogram values are taken from the model. Mathematically, this is justified by the intrinsic hypothesis. Using the notation

$$\begin{aligned} \gamma_{0i} &= \gamma(x_0 - x_i) \\ \gamma_{ij} &= \gamma(x_i - x_j) \\ &\qquad \text{for } i,j = 1, \ldots, n \quad (C.3 - 19) \end{aligned}$$

Minimization of the estimation variance yields a matrix equation (or n conditions). The unbiasedness conditions depend on the information available on the expectation of $Z(x)$:

SIMPLE KRIGING: In case the expectations $E[Z(x_i)]$ are known for all $i = 1, \ldots, n$, the minimization is constrained by

$$\alpha_0 = E[Z(x_0)] - \sum_{i=1}^{n} \alpha_i \, E[Z(x_i)] \quad (C.3 - 20).$$

This – practically rare – case is called *simple kriging*.

ORDINARY KRIGING: Unknown but constant expectation $0 \neq E = E[Z(x)]$ for all $x \in D$ leads to the unbiasedness condition

$$\sum_{i=1}^{n} \alpha_i = 1$$

$$(C.3\text{-}21)$$

$$\alpha_0 = 0$$

A solution satisfying the n conditions from equations (C.3-16), (C.3-17), and (C.3-18) and the unbiasedness condition (C.3-21) is obtained from a system of $(n+1)$ linear equations (kriging system), using a Lagrange parameter λ:

$$\begin{pmatrix} \gamma_{11} & \cdots & \gamma_{1n} & 1 \\ \vdots & \ddots & \vdots & \vdots \\ \gamma_{n1} & \cdots & \gamma_{nn} & 1 \\ 1 & \cdots & 1 & 0 \end{pmatrix} \begin{pmatrix} \alpha_1 \\ \vdots \\ \alpha_n \\ \lambda \end{pmatrix} = \begin{pmatrix} \gamma_{01} \\ \vdots \\ \gamma_{0n} \\ 1 \end{pmatrix}$$
$$(C.3 - 22).$$

This is the most commonly applied form of kriging, known as *ordinary kriging*, the matrix in (C.3-22) is the *(ordinary) kriging matrix*.

The kriging equation (C.3-22) has a unique solution, if positive-definite variogram models are used to build up the kriging matrix. Such models include the spherical, exponential and Gaussian variogram models (equations (C.3-5) to (C.3-7)) and the linear model with sill up to its range (eqn (C.3-4)).

UNIVERSAL KRIGING: In case of an underlying trend, the process is split:

$$Z(x) = m(x) + Y(x) \qquad (C.3-23)$$

into a deterministic drift component $m(x)$ and a residual random function $Y(x)$ which is second-order stationary and has zero expectation and variogram γ^*.

Theoretically, it is not possible to estimate the variogram of the residual function $Y(x)$ and the drift from a single realization. In practice, however, usually a way can be found to circumnavigate this problem (*cf.* Matheron (1971, pp. 189-195), Armstrong (1984), Herzfeld (1989a,b), for alternatives see Chiles (1977, pp. 29-48), and Delfiner (1982)).

The drift is supposed to be a linear combination of functions f_l, $l = 1, \ldots, k$:

$$m(x) = \sum_{l=1}^{k} a_l \, f_l(x) \quad \text{for} \quad k \in \mathcal{N}$$

$$\text{and} \quad a_l \in \mathcal{R} \quad (C.3\text{-}24)$$
$$\text{for} \quad l = 1, \ldots, k$$

in practical applications, the functions f_l are most often chosen as low-degree polynomials. Equation (C.3-24) leads to the following k unbiasedness conditions:

$$\sum_{i=1}^{n} \alpha_i \, f_l(x_i) = f_l(x_0) \quad \text{for} \quad l = 1, \ldots, k$$
$$(C.3-25).$$

Kriging weights and drift coefficients are determined simultaneously from a system of $(k + n)$ equations, using Lagrange multipliers. For instance, for a linear drift

$$m(x) = a_1 x_x + a_2 x_y \qquad (C.3-26)$$

(where indices x and y denote components, and a_1 and a_2 are real parameters as in (C.3-26)), the *(universal) kriging system* is

$$
\begin{pmatrix}
\gamma^*_{11} & \cdots & \gamma^*_{1n} & 1 & x_{1x} & x_{1y} \\
\vdots & \ddots & \vdots & \vdots & \vdots & \vdots \\
\gamma^*_{n1} & \cdots & \gamma^*_{nn} & 1 & x_{nx} & x_{ny} \\
1 & \cdots & 1 & 0 & 0 & 0 \\
x_{1x} & \cdots & x_{nx} & 0 & 0 & 0 \\
x_{1y} & \cdots & x_{ny} & 0 & 0 & 0
\end{pmatrix}
\begin{pmatrix}
\alpha_1 \\ \vdots \\ \alpha_n \\ \lambda \\ a_1 \\ a_2
\end{pmatrix}
=
\begin{pmatrix}
\gamma^*_{01} \\ \vdots \\ \gamma^*_{0n} \\ 1 \\ x_{0x} \\ x_{0y}
\end{pmatrix}
$$
$$(C.3-27).$$

The latter method is called *universal kriging* (also *unbiased kriging of order k*).

In universal kriging, variogram models from the "zoo" should be employed, so the kriging system has a solution. With increasing order, numerical instabilites of the solution become more likely. More data points should be used in the kriging step to counteract this effect. In practice, universal kriging with a linear or quadratic drift are the two forms that are commonly used.

The estimation variance is given by

$$s^2 = \sum_{i=1}^{n} \alpha_i \, \gamma^*(x_i, x_0) + \sum_{l=1}^{k} \lambda_l \, f_l(x_0) \quad (C.3-28)$$

where λ_l for $l = 1, \ldots, k$ are Lagrange multipliers. (In the example in (C.3-26) and (C.3-27), $\lambda_1 = \lambda$, $\lambda_2 = a_1$, and $\lambda_3 = a_2$.) It should be noted that the estimation variance depends only on the data distribution (in space) for simple and ordinary kriging, and on both the data distribution and the data values for universal kriging. It is not an error measure in the sense of numerical error analysis. For an error analysis applied to kriging, see section (C.3.5).

These are just the most common and most simple linear kriging methods. In this formulation, the above methods make use of the Gaussian hypothesis (ie., the assumption that the data have a Gaussian distribution in the statistical sense), in the steps where the linearity of the expectation and the covariance operator are used (but see Herzfeld 1992). Other kriging methods are introduced in Journel and Huijbregts (1989) and Wackernagel (1998).

(C.3.3) Variography for Satellite Radar Altimeter Data over Antarctic Ice Surfaces

In the previous sections (C.3.1) and (C.3.2) regionalized variables, variogram analysis and kriging were introduced in general. Next, we derive a form of applying these tools to the analysis of satellite radar altimeter data over Antarctic ice surfaces.

Ice surface elevation is considered a regionalized variable $Z(x)$, and radar altimeter data, corrected as described in section (C.1) and referenced to UTM coordinates as described in section (C.2), are considered a realization of Z. That is, each data point (x_i, z_i) for $i = 1, ..., N$, N the number of points, is given in the form

$$(x_{1_i}, x_{2_i}, z_i) = (x_i, z(x_i)) \qquad (C.3 - 29)$$

where $x_i = (x_{1_i}, x_{2_i})$ is the location with UTM-east coordinate x_{1_i} and UTM-north coordinate x_{2_i} and $z_i = z(x_i)$ is the elevation at location x_i. Both location (center of radar footprint) and elevation are identified by procedures described in section (C.1).

An important step in mapping is the selection of the variogram model that is used in the kriging step to weight distant points relative to close points.

Selection of the variogram depends not only on the data analysis, but also on the mapping purpose. On the one hand, the variogram needs to represent the spatial continuity properties inherent in the data, on the other hand, it needs to represent the surface roughness of the mapped features at the scale of the DTM. For an atlas, it is also important that all maps are constructed using the same variogram to avoid edge effects (see the second atlas property). Hence, a variogram representative of the most important features is selected. For the GEOSAT and the ERS-1 Atlases, two different variograms are used (see section (C.3.6) on the influence of the variogram versus the influence of data types). In the monitoring study of Lambert Glacier / Amery Ice Shelf (Herzfeld et al. 1997; see section (E)), the same variogram is used for all years to facilitate comparison, but this is also justified since SEASAT data are similar to GEOSAT data since the two altimeters were similar, whereas the ERS-1 altimeter yields data of a different characteristic.

The main features of interest in a study of Antarctic (ice) surface elevation are ice streams and glaciers. Consequently, a variogram that is representative of the surface of a large ice stream is selected for mapping. Variograms are computed for 4 subareas (s1, s2, s3, s4) of Lambert Glacier/Amery Ice Shelf, the largest ice-stream/ice-shelf system in East Antarctica (see description of maps m69e61-77n67-721 Lambert Glacier and m69e61-77n71-77 Upper Lambert Glacier, chapter (D), and detail map Lambert Glacier/ Amery Ice Shelf, chapter (F), for glaciology of Lambert Glacier/ Amery Ice Shelf; see Herzfeld et al. (1993) for variogram analysis). These areas are located on the floating tongue (Amery Ice Shelf), in the grounded part (Lambert Glacier), and in the grounding zone where the transition between grounded and floating part occurs (subarea s3). All areas are selected such that they do not contain data from the margin of the ice stream, the overdeepened marginal shear zone, or the mountainous adjacent areas, because (1) the noise level of the altimeter data is largest over those areas with high relief, and (2) the surface roughness is unusually high, and hence the variogram values would be too high, and (3) data from the ice surface, away from the margin, best represent the main features of interest in mapping Antarctic ice elevations. The variogram of the grounding zone is finally selected; data stem from a transitional zone with not too smooth topography. As will be demonstrated later (in section (C.3.5) on error analysis), this selection of a subarea increases the mapping accuracy over ice surfaces. High-relief areas would have a different variogram, but cannot be mapped accurately with altimeter data anyways, as noted in section (C.1). Areas in the interior of the Antarctic ice sheet are generally smoother than outlet glaciers, but mapping those areas is a well-posed problem and the errors are low anyways. Also, the low-relief interior of Antarctica is well-mapped with most interpolation methods. The analytical power of the geostatistical method really makes a difference in extending the limits of altimetry-based mapping along the margins of Antarctica — this is where (1) other methods do not yield useful maps, and (2) flux from the interior to the ocean occurs along ice streams —, and thus selecting the variogram of an outlet ice stream is justified.

Methodologically, we (1) utilize the variogram of a representative subarea, rather than a global variogram which would be computationally prohibitive and glaciologically meaningless, and (2) employ the variogram as a modeling tool in achieving the appropriate continuity of the mapped surface such that it represents the studied geophysical object adequately. Simply, on the resulting maps a glacier looks like a glacier. This approach to variogram modeling has both a data-analysis and a geophysical-knowledge component and can-

not be automated. To the contrary, an automated variogram fitting method would match a function solely to the data.

Next, we look at the derivation of the variogram model:

The variogram of GEOSAT data from subarea s3 (480.000-520.000 E/-7880.000- -7930.000 N) is shown in Figure C.3.3-1 (GEOSAT data here are GEOSAT ERM data from 9 Nov. 1986 – 9 Jan. 1989).

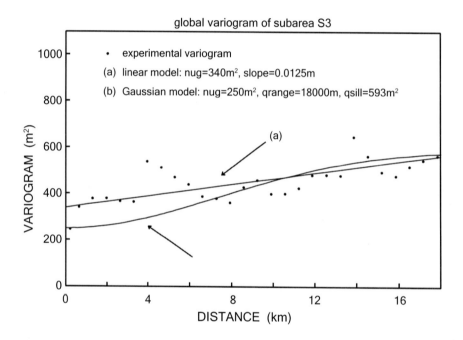

Figure C.3.3-1. Variogram model for GEOSAT Atlas. Experimental variogram calculated for data in the grounding zone of Lambert Glacier/Amery Iceshelf.

Variograms and residual variograms are similar for most lags and can be fitted with the same model. The experimental variogram shows effects that can be attributed to the groundtrack pattern of the ERM data (i.e. jumps or wave-like patterns with approximately constant wavelength, superimposed on increasing variogram values). GEOSAT GM data have closer spaced tracks, so this effect should not be as strong and the noise level lower. These patterns are disregarded in the model fitting. The linear model best describing the spatial structure in subarea s3 is:

$$\gamma(h) = 340 + 0.0125h \quad [m^2] \qquad (C.3-30)$$

The Gaussian model best describing the experimental variogram (with neglect of the track effects) has a nugget effect $c_0 = 250m^2$, a total sill $c_1 + c_0 = 593m^2$, hence a value $c_1 = 343m^2$, and a quasi-range $a = 18000m$.

A Gaussian model is smoother near the origin (it is the model of a mean-square differentiable process), which matches the fact that an ice stream near the gounding zone is smooth at 3 km resolution. Hence the Gaussian model is an appropriate model for mapping ice-stream/ice-shelf systems. The variogram model has a relatively high nugget effect, because altimeter data at this resolution have a low signal-to-noise ratio.

A new variogram analysis is carried out for ERS-1 GM data used in the Atlas mapping. The resultant model is shown in Figure C.3.3-2.

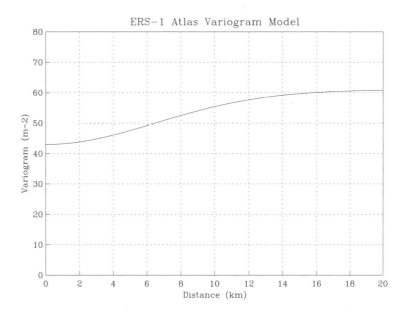

Figure C.3.3-2. Variogram model for ERS-1 Atlas. Gaussian model with nugget effect $c_0=43m^2$, quasisill $c_1=18m^2$ and quasirange $a=16km$.

ATLAS VARIOGRAMS

In summary, the variogram models used in calculation of the ANTARCTIC ATLAS DTMs are Gaussian models with the following parameters:

Table C.3.3-1. Atlas Variogram Parameters.

	c_0	c_1	a
GEOSAT Atlas:	$250m^2$	$343m^2$	$18000m.$
ERS-1 Atlas:	$43m^2$	$18m^2$	$16000m.$

Using equation (C.3-7) of the Gaussian variogram model, we get:

GEOSAT ATLAS VARIOGRAM MODEL:

$$\gamma(h) = 250 + 343(1 - exp[-\frac{3h^2}{18000}]) \quad (C.3-31)$$

and

ERS-1 ATLAS VARIOGRAM MODEL:

$$\gamma(h) = 43 + 18(1 - exp[-\frac{3h^2}{16000}]) \quad (C.3-32)$$

On data noise levels and variogram-model parameters.

Using statistical analysis, the noise level inferred from a zero-lag intercept of 250 m^2 is 22.36 m ($250 \times 2 = 500$; $\sqrt{500} = 22.36$); for an intercept of 43 m^2 it is 9.27 m ($43 \times 2 = 86$; $\sqrt{86} = 9.27$).

However, this noise level is not relevant in geostatistical estimation, because the variogram acts as a weighting function in the inversion, which gives relative weights to values depending solely on their mutual distance and the distance to the grid node. This follows directly from the kriging algorithm and the unbiasedness condition $\sum \alpha_i = 1$ (equation (C.3-21)). The noisiness of the data is not captured in the absolute size of the nugget effect, but in the relative size of the nugget effect compared to the sill: The question is: "What is the error (or subscale variability) of the data relative to the observed maximum structural variability of the data?" Consequently, the noisiness of thedata is captured by the ratio n_0 of the nugget effect c_0

to the total sill $c_0 + c_1$ in the following equation (*nugget-effect ratio*):

$$n_0 = \frac{c_0}{c_0 + c_1} \qquad (C.3 - 33)$$

The nugget-effect ratio n_0 is 0.422 for GEOSAT data and 0.705 for ERS-1 data; so, ERS-1 data are noisier. The effect of using a variogram with a higher nugget-effect ratio is that contour lines are smoother. This is correct, because noisiness in the data does not warrant picking up too many details elsewhere. If the data have low errors, then it is warranted to follow the ups and downs in the data fairly closely with an interpolator (Aside: Kriging is an exact interpolator, but points are rarely on grid nodes, so the fact that kriging is an exact interpolator does not matter in the above context).

A study of the effect of the variogram versus data properties in the final appearance of maps is undertaken in section (C.3.6) for the GEOSAT and ERS-1 maps m3e11wn67-721 Fimbul Ice Shelf. A comparison of a pair of maps ERS-1 – GEOSAT in sections (D.1) and (D.2) demonstrates that noisiness depends on the data sets, but also on specific feautures and on geographic areas.

To improve the representation of ice surface structures in terrain models of individual glaciers, regional variations of the variogram may be taken into account. To this extent, classes of ice with homogeneous surface properties such as floating ice tongues, grounded glacier ice, steep margins, mountainous terrain, and almost flat inland ice are characterized by specific variogram functions (see Stosius and Herzfeld 2004), which may then be used in the interpolation (but that has not been realized yet).

(C.3.4) Application: Search Algorithm and Kriging Parameters for Antarctic Atlas DTMs. Mapping Parameters.

Ordinary kriging (universal kriging of order 0) (equation (C.3-22)) is better suited for interpolation of radar altimeter data than universal kriging of a higher order, because the drift parts modelled by a polynomial component in the higher order universal kriging methods is likely to create artefacts in the gaps.

(C.3.4.1) Search Routine for Geophysical Line Survey Data and Software

Most numerical search routines in interpolation software are written for irregularly spaced data. Irregular, however, often implies that the data are not located on a regular grid, but still distributed throughout the survey area in a somewhat even pattern without too much clustering. Satellite altimeter data, in contrast, are geophysical line survey data (data resultant from surveys that follow track lines). The data distribution in any situation is characterized by dense information along the survey tracks and gaps of information in between. The data may be recorded continuously (radio-echo soundings, magnetic readings) or discretely along a line (single-beam bathymetric surveys, satellite radar altimeter data, gravity anomaly data from ship-towed devices), or on a stripe of varying width (swath data, e.g. multibeam deep sea sonar devices). The necessity of appropriate software to evaluate line survey data was outlined as early as 1974 (Briggs 1974). The specific distribution of survey line data does not allow immediate interpolation by isolines.

A specific search routine that is utilized prior to the kriging operation was designed and implemented in a universal kriging program (Herzfeld 1990). First, the routine checks for the data density in a close neighbourhood of the grid node and uses a criterion to decide whether the grid node is close to a survey track ("on" a survey track or inside a swath). If that is the case, the interpolation is based on a small number of points (sufficiently many points to solve the kriging equations, which depends on the selection of ordinary kriging or universal kriging of order k, k to match first- and second-order polynomials in several dimensions). If the criterion is not met (i.e., if the grid node is located in a data gap), then a larger number of points is searched for and used in the kriging step. All mentioned parameters are user-selected to facilitate adaptation to the survey instrumentation and its resulting data distribution on the surface (see track maps), and in addition can be combined with a number of search types.

The program builds on a basic kriging program developed at the Free University of Berlin, Institute of Geology/Mathematical Geology, by R. Schoele for resource assessment and updated by H. Burger in the 1980's. Geophysical applications required many changes and adaptations by U.C. Herzfeld. A modified version of that program is applied in the Atlas mapping project. The neighbourhood

search can be adjusted to the data distribution, in particular to clustered data or line survey data to avoid resemblance of the survey pattern in the output map. The actual track directions of the line surveys need not be known before program operation.

In the Antarctic Atlas mapping, the four nearest neighbours in each quadrant were used in the interpolation of any given grid node. The coefficients in the equation are determined using the variograms in Equations (C.3-31) and (C.3-32) in section (C.3.3) for GEOSAT GM and ERS-1 data, respectively. This is carried out for every grid node in the map area.

(C.3.4.2) Grid Spacing

Grid spacing is 3 kilometers. Determination of the grid spacing follows geophysical requirements: The DTMs are required to have a resolution that is high enough to permit use of glaciological models that take longitudinal stress gradients into account, which are the more sophisticated glaciological models, but also low enough to avoid unneces-

sary complexity in modeling (as explained in section C.2.3). The resolution is of course limited by the resolution of the observation technique.

Computation has been carried out on a multiprocessor SUN HYPERSPARC or on a multiprocessor SUN SPARC 20 computer.

(C.3.4.3) Mapping Parameters: Contouring and Coloring Scheme

For geophysical modeling, the digital terrain models are utilized. For viewing an area of Antarctica, for geographic investigations, for study of the morphology, and for any kind of a synoptic study, contoured maps of an area are superior to DTMs. Derivation of contour maps from DTMs requires application of a contouring program. Colors or grey shades may be added to facilitate visual interpretation.

The Atlas maps were contoured using the computer programs CONTOUR and COLOR by Robert L. Parker (Institute of Geophysics and Planetary Physics, University of California San Diego). The color and contouring scheme was created such that it resolves the topography of the ice margin and the inland ice sheet at all elevations, and that the same scheme can be applied throughout the Antarctic Atlas. Isohypses are denser in the lower elevations to resolve the topographic relief of outlet glaciers and ice streams, and less dense over the elevations most typical of the in-

land ice. Isohypses are drawn for 40, 50, 60,..., 140 m (in steps of 10 m) and for 200, 400, 600,..., 5600 m (in steps of 200 m). Color up to 70 m is white (this is mostly the area of off-coastal ice shelves), from 70 m to 200 m increasing in darkness of grey shade, at 200 m a new 10-step grey cycle is started, increasing in darkness to 2200 m, this 10-step cycle starts over at 2200 m and at 4200 m.

Elevations are above WGS84 ellipsoid, but sea surface (relative to which ice shelves float) is represented by the geoid. However, the geoid is only poorly supported in many areas of the Antarctic (see section (C.4)). Depending on the actual difference between the geoid and the ellipsoid, which varies locally, contour lines other than the ones used here may be optimal to outline ice tongues or ice shelves. This is discussed in chapter (D) for maps where this effect is most obvious, and contouring is changed accordingly where necessary for detail maps in section (F).

(C.3.5) Error Analysis

The inversion in the geostatistical estimation step yields the estimation variance with little extra computational effort, according to Equation (C.3-28). However, this variance depends on the data distribution only, not on the data values; so, a variance map simply reflects the survey pattern, which is known beforehand.

A numerical error analysis may be approached by propagating noise levels through the kriging equations as follows (Herzfeld et al. 1993), using methods applicable to random error propagation.

Theorem: Propagation of standard error through a function. If U is a quantity derived as a function $U(x_1, ..., x_n)$ of measured quantities $x_1, ..., x_n$ for a natural number n, then the standard deviation σ_U can be expressed in terms of the standard deviations σ_i of x_i for $i = 1, ..., n$ as follows:

$$\sigma_U = \sum_{i=1}^{n} \left(\frac{\partial U}{\partial x_i}\right)^2 \sigma_i^2 \qquad (C.3-34)$$

The theorem is quoted after Moffit and Bouchard (1975, eqn. 4-16, p. 168).

It should be noted that an analytical approach to error approximation, based on absolute errors, would yield a different result. The form of the error equation is the same (based on derivatives) in the analytical case.

The theorem is applied to the kriging equation (eqn. C.3-16), noting unbiasedness conditions (C.3-21) for ordinary kriging. As a first step in creating error maps, noise levels need to be calculated.

(Step 1) Calculation of noise levels

Experimental variograms are calculated, depending on position, on a grid (10 km grid). The nugget effect is then determined by fitting a 4th-order polynomial to the exponential variogram and extrapolating that polynomial to lag $h = 0$ (see also Lingle et al. 1990).

(Step 2) Calculation of error map

From the gridded noise levels, an estimate of the kriging standard deviation is derived by propagating the noise levels through the kriging calculations, using methods applicable to random error propagation, that is, simply by application of the Theorem (with equation (C.3-34)).

Application of the error propagation method (Theorem, equation (C.3-34)) to the kriging estimator equation (C.3-16) (with $\alpha_0 = 0$)

$$Z_0^* = \sum_{i=1}^{n} \alpha_i Z(x_i) \quad \text{with} \quad \alpha_i \in \mathcal{R} \quad (C.3-35),$$

gives

$$s_0^2 = \sigma_0^2 \sum_{i=1}^{n} \alpha_i^2 \qquad (C.3-36)$$

where s_0 is the error estimate associated with elevation $z(x_0)$ in location x_0, Z_0^* denotes the kriging estimator, σ_0 the average noise level in the neighbourhood of x_0 (as determined in step 1), and $\alpha_i, i = 1, ..., n$ are the kriging weights (from ordinary kriging), because

$$\frac{\partial Z_0^*}{\partial Z_i} = \alpha_i \qquad (C.3-37)$$

and the standard error of measurement in each point is the same, denoted as σ_0.

If the weights α_i are not saved with the kriging output, then $s_0 = \frac{\sigma_0}{\sqrt{n}} = \frac{\sigma_0}{4}$, using $\sum \alpha_i = 1$ and assuming all weights are equal (since $n = 16$ here) is a simple estimate. Usually, this is too much of a simplification, though.

A map of the noise levels calculated for GEOSAT ERM data and using residual variograms in step (1) shows that noise levels depend on topographic relief. They are low over the central Amery Ice Shelf and Lambert Glacier (10–50 m) and higher over the adjacent mountains (150 m) (from Herzfeld et al. 1993). This is as expected, as altimeter data collection over terrain with high topographic relief is problematic (the first return of the satellite signal is difficult to identify, so retracking may not always be possible or correct). Also, the altimeter has a large footprint which smoothes out relief signatures. An error map of 1987–1989 GEOSAT ERM data is given in Figure C.3.5-1.

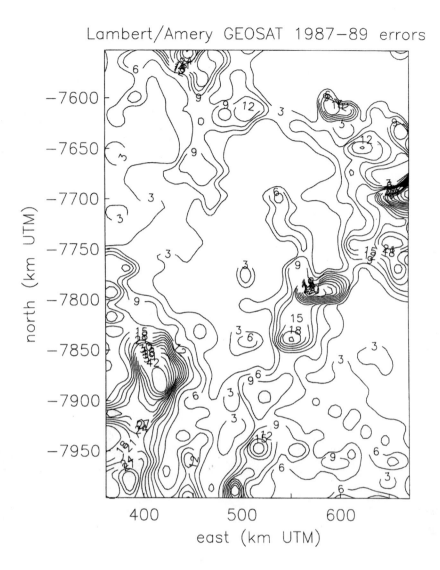

Lambert/Amery GEOSAT 1987−89 errors

Figure C.3.5-1. Error map calculated for 1987-1989 GEOSAT ERM map of Lambert Glacier/Amery Ice Shelf system in Figure E.3-9. Elevation contours are in meters, spacing of contours 3 meters. In the center of the ice-stream/ice shelf system, the error is lowest.

The distribution of the errors mirrors topography. On the flat Amery Ice Shelf, the error is mostly below 3 m (probably lower, but 3 m is the lowest contour used). In the grounding zone, it increases to 3–6 m, with higher values upglacier over the rougher topography of the grounded ice stream.

An important point concerning errors in kriged maps is the following: The error equation gives an approximation of the pointwise error in a grid node. Because of the moving-window averaging technique (kriging with local neighbourhoods), the errors in the ice-surface elevations are related to those in neighbouring elevations. The position of the grounding zone (in elevation), for instance, or the elevation of any region on the map, has an error that is much lower than the approximate standard error derived from equation (C.3.-32) and given in the error map.

The Atlas maps are constructed from GEOSAT GM data, which are much denser than GEOSAT ERM data, and the errors are expected to be lower.

(C.3.6) Influence of the Radar Altimeter Sensor Compared to Influence of the Variogram in Kriging for GEOSAT and ERS-1 Data

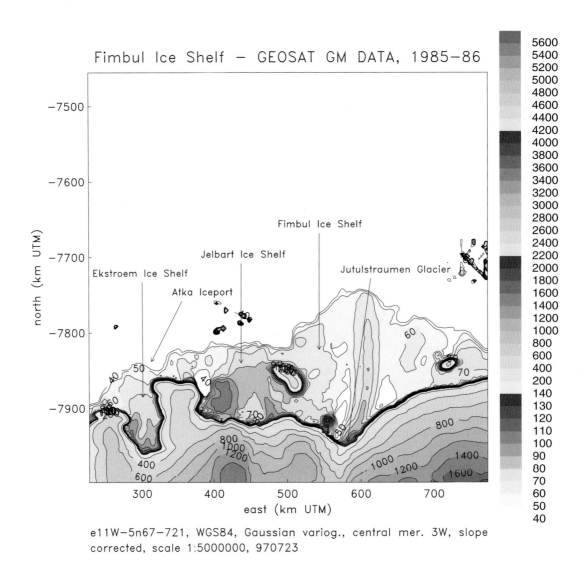

Fimbul Ice Shelf — GEOSAT GM DATA, 1985–86

e11W−5n67−721, WGS84, Gaussian variog., central mer. 3W, slope corrected, scale 1:5000000, 970723

Figure C.3.6-1. GEOSAT Atlas Map m3We11W-5n67-721 Fimbul Ice Shelf. GEOSAT data interpolated using GEOSAT Atlas variogram model (Eqn. (C.3-31)).

The influence of a variogram of the same type (here: the Gaussian type), but a different nugget effect ratio, is investigated for the example of the maps m3we11w-5n67-721 Fimbul Ice Shelf. For the Atlas, the following study also serves to compare the influence of data of different qualities — GEOSAT GM versus ERS-1 GM data — and the influence of different variogram models.

On the one hand, monitoring of ice advance or retreat from evaluation of radar altimeter data at two points in time and from different satellites requires knowledge of the general offset, in this case of GEOSAT and ERS-1. An upper bound for this offset is calculated in section (C.1.1) as 0.72 m or 0.22 m, depending on orbit accuracy values used. On the other hand, accuracy of the mapping using a given altimeter instrument may be assessed from contoured maps as demonstrated in the following

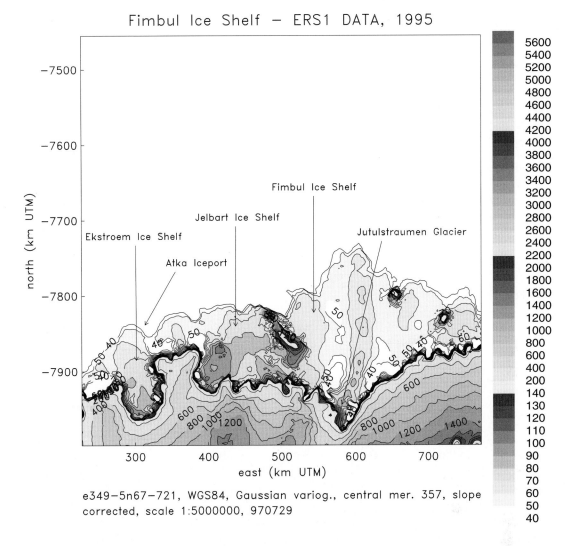

Fimbul Ice Shelf – ERS1 DATA, 1995

e349–5n67–721, WGS84, Gaussian variog., central mer. 357, slope corrected, scale 1:5000000, 970729

Figure C.3.6-2. ERS-1 Atlas Map m3We11W-5n67-721 Fimbul Ice Shelf. ERS-1 data interpolated using ERS-1 Atlas variogram model (Eqn. (C.3-32)).

example. GEOSAT (1985-86) and ERS-1 (1995) Atlas maps m3we11w-5n67-721 "Fimbul Ice Shelf" are given in Figure C.3.6-1 and C.3.6-2, respectively. The large feature is the ice "tongue" of Jutulstraumen extending straight north into Fimbul Ice Shelf. Jutulstraumen Glacier follows a large geologic trough believed to be of structural origin (Decleir and van Autenboer 1982, after Swithinbank 1988).

Contours of the GEOSAT map are much smoother than those of the ERS-1 map, possibly due to more data and higher noise in ERS-1 data than in GEOSAT data. GEOSAT data are prone to noise offshore, caused by signals reflecting off islands and erroneously recorded or corrected. This

problem has been largely eliminated from ERS-1 data. Since the GEOSAT Atlas and the ERS-1 Atlas have been calculated with different variograms (cf. section (C.3.3)), the ice shelves are not immediately comparable from the Atlas maps.

The variogram used for the ERS-1 Atlas has a larger smoothing effect than that used for the GEOSAT Atlas, as quantified by the nugget-effect ratio (eqn. C.3-33). To study the effects of these two variograms on the cartographic output and to eliminate it from a comparison of different years, the GEOSAT data have been rekriged using the ERS-1 variogram, and vice versa (cf. Figures C.3.6-3 and C.3.6-4). Apparently, the effect of the different variograms is fairly small, the GEOSAT

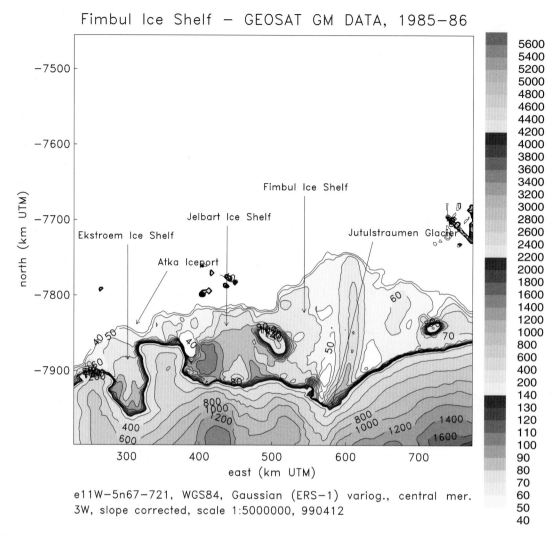

Fimbul Ice Shelf — GEOSAT GM DATA, 1985—86

Figure C.3.6-3. GEOSAT data interpolated using ERS-1 Atlas variogram model (Eqn. (C.3-32)).

map still is smoother than the ERS-1 map, and only very few features have changed after the variogram was altered. Prior to a direct comparison of ice elevation, however, the general offset between GEOSAT and ERS-1 needs to be known. On all maps, the ice-shelf edge is determined by two near-parallel contour lines (10 m difference) close to each other. The "coast" line inland ice/Fimbul Ice Shelf has many indentations, which is probably

due to sampling, because this error is also found on other ERS-1 maps.

Next to Jutulstraumen, the smaller Schytt Glacier may be identified at (490,000;-7,950,000— -7,880,000) (cf. Swithinbank 1988, p. B95). The geographic features pictured on these maps are described in detail in the Atlas map section (D) for the map m3we11w-5n67-721 Fimbul Ice Shelf.

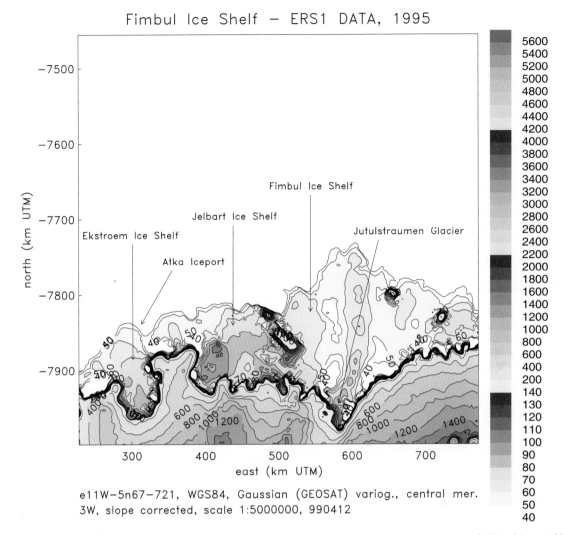

Figure C.3.6-4. ERS-1 data interpolated using GEOSAT Atlas variogram model (Eqn. (C.3-31)).

(C.4) The Role of the Geodetic Reference Surface

(C.4.1) Ellipsoid and Geoid Concepts

In a first approximation, the Earth is ball-shaped. In a next step one considers the fact that the Earth is compressed at the poles relative to the equator due to the centrifugal force resulting from its rotation. This shape is an ellipsoid. WGS84 is an example of an ellipsoid, its exact coefficients were determined in 1984. The geoid is a further refinement of the shape following the gravitational potential on the entire surface of the globe (i.e. extrapolating the ocean surface over the continents). The anomalous gravitational potential field is approximated by spherical harmonics. The resolution of the gravity field model depends on the degree of the spherical harmonical functions, which in turn depend on the density of available data. The OSU91A model, computed at Ohio State University (Rapp 1992) is complete to spherical harmonics of degree 360 (Rapp and Pavlis 1990), which corresponds to a resolution of 50 km. Accuracy varies with data density and quality. The geoid model GEM-T3 from the NASA Goddard Space Flight Center is complete to degree 50, corresponding to a spatial resolution of approximately 400 km. OSU91A is based on satellite and ground observations. The 50-km resolution of OSU91A is a maximum resolution, it is achieved in areas where 30' x 30' mean gravity anomalies were incorporated in the model development. These areas are primarily ocean areas, because satellite radar altimeter data are available there and give geoid height directly (whereas on land, altimeter data are also available, but (a) harder to evaluate, and (b) provide only the terrain correction needed for the geoid determination, but not density). Land areas, for which 30' x 30' anomalies are available, are North America, Europe, Australia, Japan, parts of Africa, South America and India. Areas lacking good gravity data include eastern Europe, Asia and the Arctic and Antarctica. The disadvantage of models with a high degree is that errors in areas that are poorly constrained by data are extra large (polynomial functions).

We use two recently published geoids, "Goddard Earth Model" GEM-T3 (Lerch et al. 1982) and "Ohio State University Model" OSU91A (Rapp 1992, 1994), kindly made available to us by NASA GSFC and by R. H. Rapp (pers. comm., Feb. 1995).

Comparison of the two geoid models in the area of Lambert Glacier/Amery Ice Shelf shows significant differences (see Figures C.4.1-1 and C.4.1-2): The GEM-T3 surface gradually slopes in a northeasterly direction, the OSU91A surface has a more complex topography with a hill. According to R. H. Rapp (pers. comm., Feb. 1995), both models are based on the same set of data of only fair reliability (mean anomalies in 1° by 1° cells), therefore differences between the two models should be attributed to philosophical differences in the interpolation method rather than warrant a geophysical interpretation. For the OSU91A model, global average in estimation of the total undulation error is 0.57 m, but 2 m for land area with no surface gravity data (Rapp 1992).

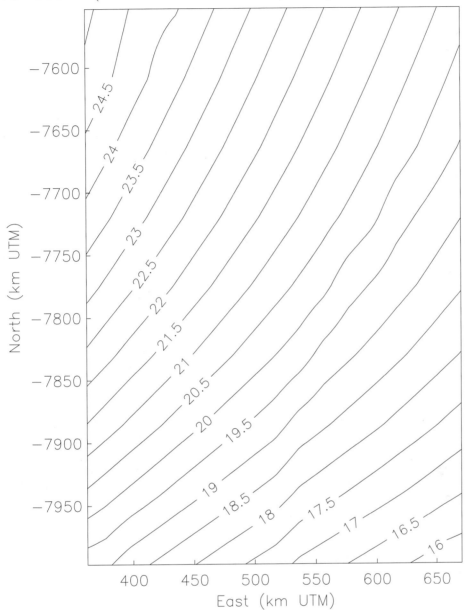

Figure C.4.1-1. Elevation of GEM-T3 geoid relative to WGS84 ellipsoid. Lambert Glacier/Amery Ice Shelf area. Scale 1:3.000.000.

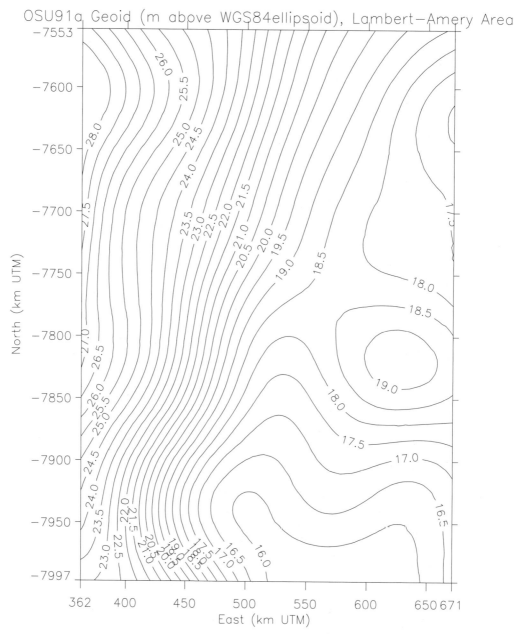

Figure C.4.1-2. Elevation of OSU91a geoid relative to WGS84 ellipsoid. Lambert Glacier/Amery Ice Shelf area. Scale 1:3.000.000.

(C.4.2) Mapping of Ice Surfaces with Reference to Geoid Models

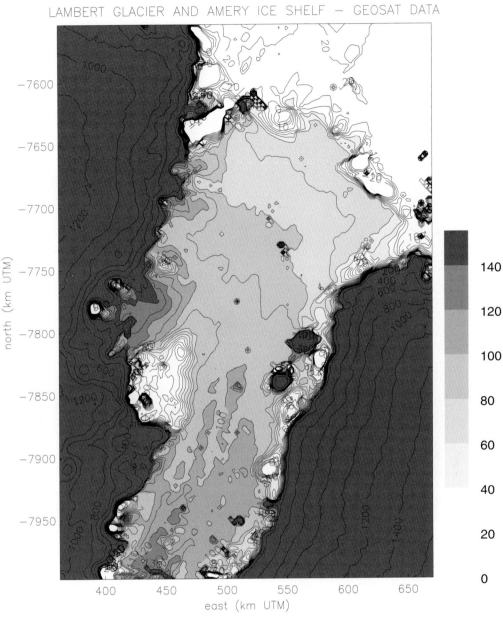

GEM T2—refd., slope corrected, Gaussian variog., ameryallm.dtm, 930729

Figure C.4.2-1. Surface elevation of Lambert Glacier/Amery Ice Shelf area above WGS84 ellipsoid. Scale 1:3.000.000. 1987-1989 GEOSAT ERM altimeter data from time frames corresponding in season to SEASAT data (see chapter (E)).

In the Atlas map section, elevations are referenced to the WGS84 ellipsoid. When studying changes, the reference surface is unimportant. Comparison with published studies and field work may be facilitated by maps referenced to a geoid, although for the Lambert Glacier area elevations

68

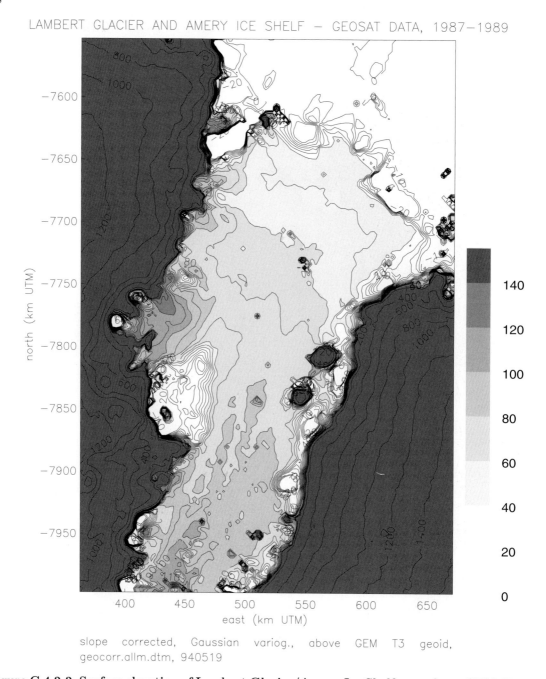

slope corrected, Gaussian variog., above GEM T3 geoid,
geocorr.allm.dtm, 940519

Figure C.4.2-2. Surface elevation of Lambert Glacier/Amery Ice Shelf area above GEM-T3 geoid. Scale 1:3.000.000. 1987-1989 GEOSAT ERM altimeter data from time frames corresponding in season to SEASAT data (see chapter (E)).

are rarely given in the literature. Referencing to a geoid, however, involves geodetic problems of geoid determination, which is difficult in areas of poor availability of gravity data. Mathematical methods for geoid determination are described in Moritz (1984), Tscherning (1984), and Engels and others (1993).

To demonstrate the result of referencing to different geodetic surfaces, we reference the surface elevation in the Lambert Glacier/Amery Ice Shelf area first to the WGS84 ellipsoid (Figure C.4.2-1), next to the GEM-T3 geoid (Figure C.4.2-2) as contoured in Figure C.4.1-1, and last

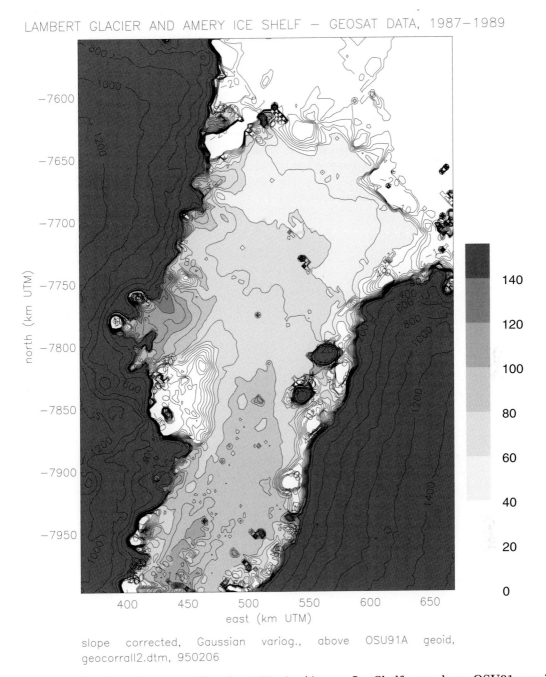

slope corrected, Gaussian variog., above OSU91A geoid, geocorrall2.dtm, 950206

Figure C.4.2-3. Surface elevation of Lambert Glacier/Amery Ice Shelf area above OSU91a geoid. Scale 1:3.000.000. 1987-1989 GEOSAT ERM altimeter data from time frames corresponding in season to SEASAT data (see chapter (E)).

to the OSU91A geoid (Figure C.4.2-3) as contoured in Figure C.4.1-2. Differences in ice-surface morphology are particularly apparent in the ice-stream/ice-shelf system which itself has a low gradient. Whereas the simple surface shape fo the GEM-T3 geoid has a small effect on the appearance of the ice-surface morphology, the more com-

plex surface shape of the OSU91A geoid results in a significantly different ice-surface morphology (see for instance the areas corresponding to the area of the "fingers" in the 100 m contour in the original map (between -7800.000 N and - -7900.000 N)). The elevation relative to any one of the two geoid surfaces is of course lower than the elevation relative to the WGS84 ellipsoid. This relationship does not generally hold for all of Antarctica, in some areas the geoid surface is lower than the ellipsoid surface.

Not only can grid models be used for mapping with reference to the geoid rather than the ellipsoid our ellipsoid-referenced topographic maps can also be used for terrain correction in the inverse gravimetric problem, and thus may facilitate improvements of the geoid in this poorly constrained region. For purposes of field study and comparison to the literature both geoid-referenced maps are sufficiently accurate, as well as the WGS84-referenced maps when a constant of about 20 m is subtracted from the elevation values (the approximate difference between ellipsoid and geoid in the map area).

Apparently, integration of accurate terrain data in geodetic models will yield a much better improvement of the geoid model for Antarctica than application of more sophisticated mathematical models combined with poor data.

Part II

The Atlas

(D) Atlas Maps

(D.0) Map Organization and Description Principles

In section D, maps are ordered in rows from north to south, and within each row, from east to west, starting with the map that is east of the Weddell Sea. In Table D.0-1 maps are listed in the order in that they appear in this atlas. A schematic representation of the relative location of all Atlas maps is given in Figure D.0-1. The proper geographic location of the Atlas maps and their overlaps with neighbouring maps can be inferred from the index maps in Figures C.2.3-1 to C.2.3-3. Hence Figure D.0-1 is best used for finding a map given by name in the scheme, and for identifying the names of neighbouring sheets. If the objective is to determine on which maps a given geographic location is covered, then Figures C.2.3-1 to C.2.3-3 are most useful.

Major geographic features set in italics in the text where they are described in some detail. Places labeled in the maps are indexed (section (I.2)). Areas that occur on more than one map are cross-referenced to the maps on which they are also portrayed, in particular if a more detailed description is given there.

Glaciologic terms are italicized where a definition follows. A glossary of glaciologic terms is given in section I.1. The description focusses on the glaciology of Antarctica. Geologic features are only described briefly; for more detailed information on the geology of Antarctica, the reader is referred to Tingey (1991).

Table D.0-1. List of Antarctic Atlas Maps

Latitude row 63°–68° S (GEOSAT and ERS-1)	
m45e37-53n63-68	Casey Bay
m57e49-65n63-68	Napier Mountains
m69e61-77n63-68	Mawson Coast East
m81e73-89n63-68	Leopold and Astrid Coast
m93e85-101n63-68	Queen Mary Coast
m105e97-113n63-68	Knox Coast
m117e109-125n63-68	Sabrina Coast
m129e121-137n63-68	Clarie Coast
m141e133-149n63-68	Adélie Coast
m153e145-161n63-68	Ninnis Glacier Tongue
m297e289-305n63-68	Antarctic Peninsula (Graham Land)

Latitude row 67°–72.1° S (GEOSAT and ERS-1)	
m15we23W-7Wn67-721	Ekström Ice Shelf
m3we11w-5n67-721	Fimbul Ice Shelf
m9e1-17n67-721	Princess Astrid Coast
m21e13-29n67-721	Erskine Iceport
m33e25-41n67-721	Riiser-Larsen Peninsula
m45e37-53n67-721	Prince Olav Coast
m57e49-65n67-721	Kemp Coast
m69e61-77n67-721	Lambert Glacier
m81e73-89n67-721	Ingrid Christensen Coast
m93e85-101n67-721	Wilkes Land (e85-101n67-721)
m105e97-113n67-721	Wilkes Land (e97-113n67-721)
m117e109-125n67-721	Wilkes Land (e109-125n67-721)
m129e121-137n67-721	Wilkes Land (e121-137n67-721)
m141e133-149n67-721	Wilkes Land (e133-149n67-721)
m153e145-161n67-721	Cook Ice Shelf
m165e157-173n67-721	Pennell Coast
m292e284-300n67-721	Antarctic Peninsula (Palmer Land)

Latitude row 71°–77° S (ERS-1)	
m333e315-351n71-77	Riiser-Larsen Ice Shelf
m357e339-15n71-77	New Schwabenland
m21e3-39n71-77	Sør Rondane Mountains
m45e27-63n71-77	Belgica Mountains
m69e51-87n71-77	Upper Lambert Glacier
m93e75-111n71-77	American Highland
m117e99-135n71-77	Dome Charlie
m141e123-159n71-77	Southern Wilkes Land (e123-159)
m165e147-183n71-77	Victoria Land
m213e195-231n71-77	Ruppert Coast
m237e219-255n71-77	Bakutis Coast
m261e243-279n71-77	Walgreen Coast
m285e267-303n71-77	Ellsworth Land
m309e291-327n71-77	Black Coast

Latitude row 75°–78°S (ERS-1)

m333e315-351n75-80 Coats Land
m357e339-15n75-80 Western Queen Maud Land (North)
m21e3-39n75-80   Central Queen Maud Land (North)
m45e27-63n75-80   Valkyrie Dome
m69e51-87n75-80   South of Lambert Glacier
m93e75-111n75-80   East Antarctica (Sovetskaya)
m117e99-135n75-80 East Antarctica (Vostok)
m141e123-159n75-80 East Antarctica (Mt. Longhurst)
m165e147-183n75-80 Scott Coast
m189e171-207n75-80 Roosevelt Island
m213e195-231n75-80 Saunders Coast
m237e219-255n75-80 Northern Marie Byrd Land
m261e243-279n75-80 Northern Hollick-Kenyon Plateau
m285e267-303n75-80 Zumberge Coast
m309e291-327n75-80 Ronne Ice Shelf

Latitude row 78°–81.5°S (ERS-1)

m333e315-351n78-815 Filchner Ice Shelf
m357e339-15n78-815 Western Queen Maud Land (South)
m21e3-39n78-815   Central Queen Maud Land (South)
m45e27-63n78-815   Eastern Queen Maud Land (South)
m69e51-87n78-815   Dome Argus
m93e75-111n78-815   East Antarctica (e75-111n78-815)
m117e99-135n78-815 East Antarctica (e99-135n78-815)
m141e123-159n78-815 Byrd Glacier
m165e147-183n78-815 Hillary Coast
m189e171-207n78-815 Ross Ice Shelf
m213e195-231n78-815 Shirase Coast
m237e219-255n78-815 Southern Marie Byrd Land
m261e243-279n78-815 Southern Hollick-Kenyon Plateau
m285e267-303n78-815 Ellsworth Mountains
m309e291-327n78-815 Berkner Island

76

						m45 Casey Bay						
					m57 Napier Mtns.	**m69** Mawson Coast and East Coast						
				m45 Kemp Coast		**m81** Leopold and Astrid Coast	**m93** Queen Mary Coast					
Weddell Sea			**m33** Prince Olav Coast		**m69** Lambert Glacier	**m81** Ingrid Christen Coast	(e85-101; n67-72.1)	**m105** Wilkes Land				
	m150 Ekstroem Ice Shelf	**m2A** Erskine Iceport		**m45** Belgica Mountains	**m69** Upper Lambert Glacier	**m93** East Antarctica (Sovetskaya)	**m93** Wilkes Land (e97-113; n67-72.1)	**m105** Knox Coast	**m117** Wilkes Land			
m333 Riiser-Larsen Ice Shelf	**m30** Fimbul Ice Shelf	**m2A** Riiser-Larsen Penin.	**m45** Valkyrie Dome	**m69** South of Lambert Glacier	**m93** American Highland	**m117** East Antarctica (Vostok)	**m105** (e109-121; 125n67 -72.1)	**m117** Sabina Coast	**m129** Wilkes Land	**m141** Wilkes Land		
m333 New Schwabenland	**m9** Princess Astrid Coast	**m357** Sor Rondane Mountains	**m69** Dome Argus	**m93** East Antarctica (o75-111; n78-8;t5)	**m117** Dome Charlie	**m141** East Antarctica (Mt. Longhurst)	**m117** (e121-137n67 149n67 -72.1)	**m129** Clare Coast	**m141** Wilkes Land	**m153** Cook Ice Shelf		
m357 Western Queen Maud Land (North)	**m2A** Central Queen Maud Land (North)	**m357** Central Queen Maud Land (North)		**m117** East Antarctica (o99-135;n78-8;t5)	**m141** Southern Wilkes Land (e123-159)	**m141** Byrd Glacier	**m129** (e133-149; n67-72.1)	**m141** Adelie Coast	**m153** Ninnis Glacier Tongue	**m165** Pennell Coast		
					m165 Victoria Land	**m165** Scott Coast	**m141**	**m165** Hillary Coast				
					(Ross Sea)	**m189** Roosevelt Island	**m189** Ross Ice Shelf					
					m213 Rupert Coast	**m213** Saunders Coast	**m213** Shirase Coast					
					m237 Bakutis Coast	**m237** Northern Marie Byrd Land	**m237** Southern Marie Byrd Land					
					m261 Walgreen Coast	**m261** Northern Hollick-Kenyon Plateau	**m261** Southern Hollick-Kenyon Plateau					
					m285 Ellsworth Land	**m285** Zumberge Coast	**m285** Ellsworth Mountains	**m297** Antarctic Penin. (Graham Land)				
					m309 Black Coast	**m309** Ronne Ice Shelf	**m309** Berkner Island	**m297** Antarctic Penin. (Palmer Land)				

m2A Central Queen Maud Land (South)
m357 Western Queen Maud Land (South)
m2A Central Queen Maud Land (South)
m45 Eastern Queen Maud Land (South)
m333 Filchner Ice Shelf
m357 Western Queen Maud Land (South)

Coats Land

63-68
67-72.1
91-77
75-80
78-81.5

Figure D.0-1 (Facing page). Schematic representation of Antarctic Atlas map locations.

(D.1) Latitude Row 63-68°S: Maps from GEOSAT and ERS-1 Radar Altimeter Data

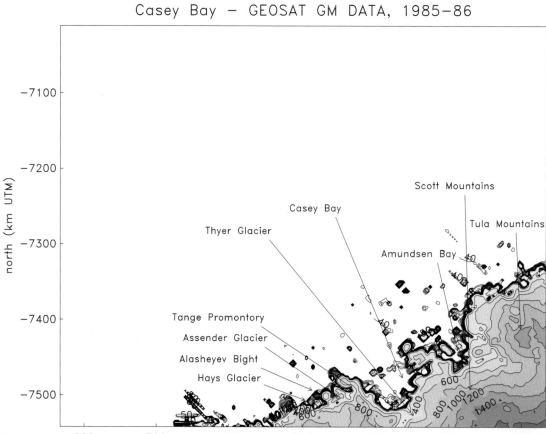

Casey Bay — GEOSAT GM DATA, 1985–86

e37−53n63−68, WGS84, Gaussian variog., central mer. 45, slope corrected, scale 1:5000000, 970723

Map m45e37-53n63-68 Casey Bay

Map m45e37-53n63-68 Casey Bay shows part of the coast of Enderby Land (see also maps m45e37-53n67-721 Prince Olav Coast, m57e49-65n67-721 Kemp Coast, and m57e49-65n63-68 Napier Mountains). As a consequence of the atlas mapping scheme, not much of the hinterland of this stretch of coast is seen on this map.

The contour lines up to the 200 m contour are crowded on this map, and the 400 m contour line is very close to the coast, so the terrain is steep in this area. The coast is an ice wall coast with coastal nunataks and offshore islands. Both the steepness of the coast and the existence of offshore islands complicate mapping with a satellite radar altimeter, as the "snagging effect" of the signal (see sections (B.2), (B.3) on altimetry) leads to artefacts and islands offshore — the latter is most notable on the GEOSAT map and almost entirely corrected for in the ERS-1 altimeter data set). Offshore is so-called fast ice. In summer, rivers

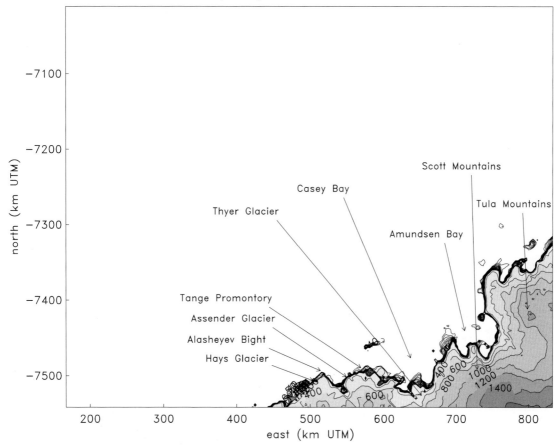

Casey Bay — ERS1 DATA, 1995

e37–53n63–68, WGS84, Gaussian variog., central mer. 45, slope corrected, scale
1:5000000, 970725

of meltwater flow from the Scott Mountains and
Nye Mountains to the sea. The Tula Mountains
are described in the text of map m57e49-65n63-68
Napier Mountains. Bare land is visible along the
coast in LANDSAT imagery recorded in austral
summer (Swithinbank 1988, fig. 58, p. B80).

Assender Glacier and *Hays Glacier* are small
glaciers to the west of *Tange Promontory*. Hays
Glacier is the larger of the two, it flows in from
the south and has a catchment area of 10000 km^2
(Meier 1977, 1983). Assender Glacier flows from
east to west (its front is close to that of Hays
Glacier) in a valley south of Tange Promontory,
the eastern part of the same valley is occupied by
Molle Glacier, which flows west to east into Casey
Bay — since the bottom of the valley is well below sea level, Tange Promontory is an island. This
information is derived from an annotated LAND-
SAT map in Swithinbank (1988, fig. 58, p. B80)

and field observations quoted after Swithinbank
(1988), it cannot be derived from satellite data
alone.

A similar depression that extends to 200–400 m
below sea level connects the valley of Hays Glacier
with the valley of "Campbell Glacier", located
about 25 km farther west, forming another "island". The Russian Research Station Molodezhnaya is situated on this "island", east of "Campbell Glacier" (see also map m45e37-53n67-721
Prince Olav Coast). Snow accumulation is high in
the area of Hays Glacier. A velocity of 1400 m a^{-1}
was measured for Hays Glacier at its grounding
line, located partway up the valley, and a discharge
of 3 km^3 a^{-1} glacier ice (Swithinbank 1988, p.
B77).

The *grounding line of a glacier* is defined as the
transition of grounded ice (ice moving over rock)

and floating ice (ice floating on water) of a glacier that extends from land into the ocean or a fjord. Oftentimes, a glacier has a "grounding zone", a transition zone, rather than a clear grounding line. Observations by Korotkevitch et al. (1977) indicate that the valley of Hays Glacier extends inland at 200–400 m below sea level for more than 100 km and joins the valley of Rayner Glacier; Rayner Glacier drains into Casey Bay 10 km west of Thyer Glacier and over 100 km east of Hays Glacier, on the other side of Tange Promontory — so there is another and larger "island" south of Tange Promontory. Rayner Glacier drains an area of 118000 km^2, it has a velocity of 861 m a^{-1} near the 500 m elevation and an ice thickness of 2300 m (at 30 km upglacier of the 500 m elevation) decreasing to 500 m at the floating ice front, and a mass flux of 10.4 Gt a^{-1} (11.4 km^3 a^{-1}) (after Swithinbank 1988, fig. 58, p. B80).

To the east of Tange Promontory is a small ice shelf, this is Hannan Ice Shelf, which forms in a sheltered bay and is nurtured by several glaciers, including Molle Glacier, and can be seen on the ERS-1 map Casey Bay (grey area extending from the coast at (-7500.000 N/600.000 E) UTM.

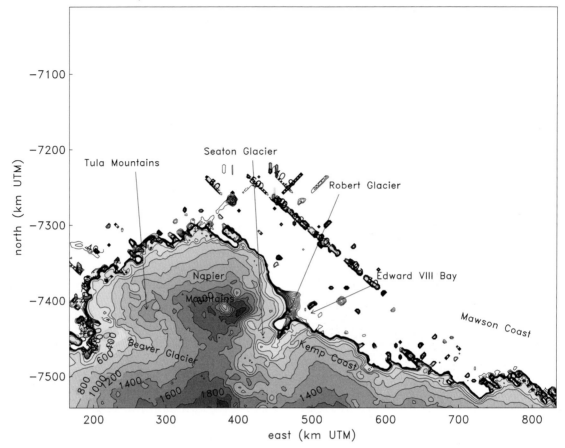

Napier Mountains — GEOSAT GM DATA, 1985–86

e49–65n63–68, WGS84, Gaussian variog., central mer. 57, slope corrected, scale 1:5000000, 970721

Map m57e49-65n63-68 Napier Mountains

This map is the northern extension of map m57e49-65n67-721 Kemp Coast. It shows the northern part of *Enderby Land*, which is a landmass projecting off the Antarctic continent between 44° 38'E and 59° 34'E. The only named coastal part is Kemp Coast. Noteably, the coast and coastal area of Enderby Land between Shinnan Glacier at 44° 38'E and William Scoresby Bay at 59° 34'E does not have a special name (Alberts 1995, p. 221). To the west is Prince Olav Coast (Lützow Holm Bay at 40° E to Shinnan Glacier at 44° 38'E), farther west is *Prince Harald Coast* (Riiser-Larsen Peninsula at 34° E to Lützow Holm Bay at 40° E), to the east of Enderby Land is Mawson Coast (the coast of Mac Robertson Land

between *William Scoresby Bay* (67° 24'S/59° 34'E) and *Murray Monolith* (67° 47'S/66° 54'E)).

Kemp Coast extends from William Scoresby Bay (67° 24'S/59° 34'E) to the head of Edward VIII Bay (66° 50'S/56° 24'E); it is named for a British Whaling Captain, Peter Kemp, who discovered land here in 1833 (Alberts 1995). Offshore of the rocky, 8 km long and 6 km wide bay are *William Scoresby Archipelago* and other islands, the GEOSAT map shows offshore reflections, likely of these islands. the bay was discovered in 1936 by William Scoresby, who also discovered Edward VIII Bay in 1936 and named it after the King

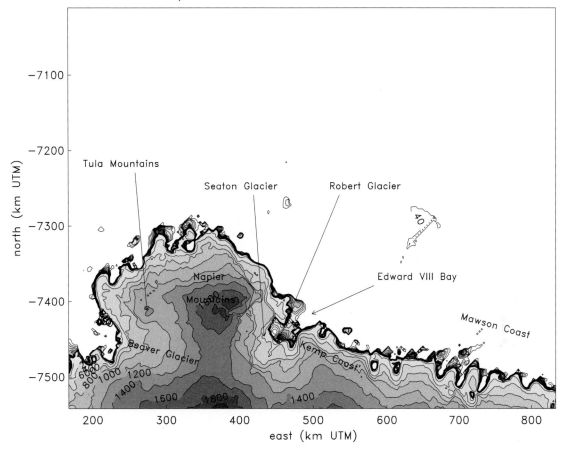

Napier Mountains — ERS1 DATA, 1995

e49−65n63−68, WGS84, Gaussian variog., central mer. 57, slope corrected, scale
1:5000000, 970725

of England, in part of the bay is Edward VIII Ice
Shelf.

In the eastern part of the map is *Mawson Coast*,
the coast of Mac Robertson Land, bounded by
William Scoresby Bay and Murray Monolith (67°
47'S/66° 54'E) (see map m57e49-65n63-68 Mawson Coast).

The *Napier Mountains* (center of the peninsula in
the left half of the map) are much gentler than
the Admiralty Mountains, as is seen by the contours (almost no closed cells). The Tula Mountains are located west of the Napier Mountains at
(-7420.000 N/280.000 E). The valley bordered in
the north by the Tula and Napier Mountains is
occupied by Beaver Glacier, the valley separating
the Napier Mountains peninsula from Kemp Coast
(east of 57°, 500.000) contains *Robert Glacier*
which drains into Edward VIII Bay. At 580.000 E

is Hoseason Glacier, at 720.000 E Stefansson Bay
can be identified. Seaward of Kemp Coast are the
Hobbs Islands which may cause some of the speck-
les on the map. The Australian Mawson Station is
located on the coast at ≈63° East, on the eastern
coast of Holme Bay (≈720.000 E UTM).

The subglacial valley which contains Beaver
Glacier is connected below sea level to the valley of
Wilma Glacier, which flows into Edward VIII Bay
in a location about 150 km east of Beaver Glacier.
Wilma Glacier joins Robert Glacier shortly before
they both drain into Edward VIII Bay (Ferrigno et
al. 1996); the Wilma Glacier valley is the east-west
part, the Robert Glacier valley the SW-NE part
of the valley labeled on our maps — so without
the ice sheet, the peninsula of Enderby Land with
the Tula Mountains and Napier Mountains would
be an island (Korotkevich et al. 1977; Allison et
al. 1982; Swithinbank 1988, p.B77)!

The section 61° E to 63° E of Mawson Coast is also shown in Swithinbank (1988, fig. 56, p. B76). This is an example of a typical *ice-wall coastline* crossing an island archipelago. Here, a narrow band of rock is visible at the foot of an ice cliff (except where ice streams cross the coastline). Further advance of the ice wall is prevented by a relatively high marine melting rate, and consequently the ice-wall coastline is likely to remain in a stable position that is mainly controlled by sea level (Hollin 1962). Ice walls can be in shallow water (*strand ice wall*)(ragged coastline); or submerged (*neritic ice wall*), the latter rest on rock that is up to several hundred meters below sea level. Surface velocities at the ice wall have been measured at 10–20 m a^{-1}, between Casey Range and Mt. Henderson at 20–40 m a^{-1} (SE and SW of Holme Bay, respectively, 700.000–730.000 E UTM). Field measurements in the area were carried out during the International Geophysical Year (July 1957 – December 1958).

LANDSAT satellite images or aerial photographs are necessary to distinguish between mountain ranges, nunataks, and bare-ice ablation areas (*blue-ice areas*), which are extensive in this area. Some glaciers still do not have names. Ablation areas tend to develop downwind of obstacles that break up ice flow. There are even surface meltwater lakes in this area (north of Casey and David Ranges). In the satellite-altimetry-derived maps, the roughness of the land surface and roughness of the coastline are reflected in rough and crowded contours.

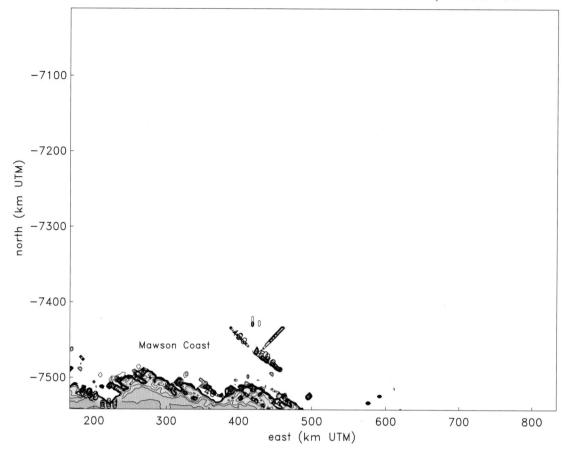

Mawson Coast East – GEOSAT GM DATA, 1985–86

e61–77n63–68, WGS84, Gaussian variog., central mer. 69, slope corrected, scale 1:5000000, 970723

Map m69e61-77n63-68 Mawson Coast East

This map shows a narrow stripe of the eastern part of Mawson Coast and is included in the Atlas mostly for completeness — the area mapped is also shown on map m69e61-77n67-721 Lambert Glacier (eastern part of this map is on the sheet adjoining to the south) or on map m57e49-65n63-68 Napier Mountains (western part of this map is on the sheet adjoining to the west).

Mawson Coast is the coast of Mac Robertson Land between *William Scoresby Bay* (67° 24'S/59° 34'E) and *Murray Monolith* (67° 47'S/66° 54'E). William Scoresby Bay, an 8 km long and 5 km wide bay surrounded by snow-free hills and steep rock headlands (Alberts 1995), is located west of William

Scoresby Archipelago. Murray Monolith, 370 m high, is the detached front of Torlyn Mountain (Alberts 1995). Mawson Coast was first sighted during the BANZARE expedition under Sir Douglas Mawson (1929–1930).

In the area shown on this map Mawson Coast is a rugged stretch of coastline with — from west to east — Utsikkar Bay (with Utsikkar Glacier), Casey Range, "Forbes Glacier", Holme Bay (with the Australian Mawson Station), David Range, the Douglas Islands offshore of the northern extent of Holme Bay, Framnes Mountains, Masson Range, Nilsen Bay, Cape Fletcher, Gustav Bull Mountains (with Mt. Hinks, 630 m), Scullin Mono-

Mawson Coast East — ERS1 DATA, 1995

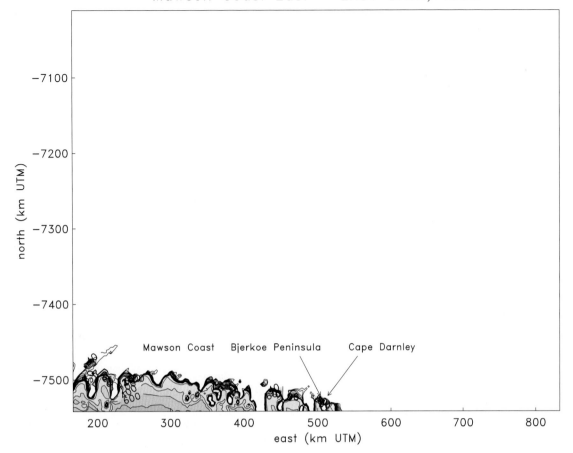

e61—77n63—68, WGS84, Gaussian variog., central mer. 69, slope corrected, scale
1:5000000, 970725

lith, Shallow Bay, Point Williams (an island west of Bjerkø Peninsula), and Bjerkø Peninsula with Cape Darnley.

The coast is mapped with lots of errors on the GEOSAT map (possibly Holme Bay is identified correctly), but better on the ERS-1 map. At *Bjerkø Peninsula*, the coast turns south to Lambert Glacier.

Leopold and Astrid Coast — GEOSAT GM DATA, 1985–86

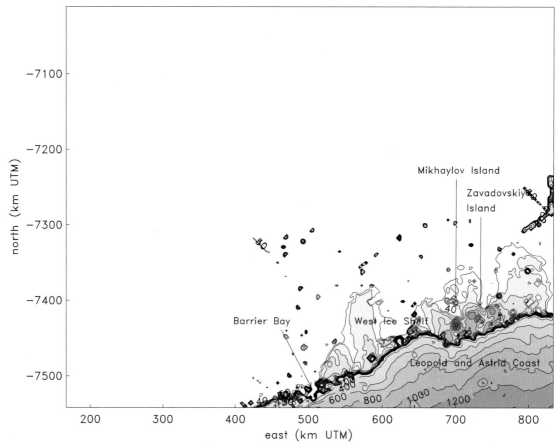

e73–89n63–68, WGS84, Gaussian variog., central mer. 81, slope corrected, scale
1:5000000, 970723

Map m81e73-89n63-68 Leopold and Astrid Coast

Map m81e73-89n63-68 shows *Leopold and Astrid Coast*, a short section of the coast of East Antarctica east of Lambert Glacier and immediately east of *Ingrid Christensen Coast* (see map m81e73-89n67-721 Ingrid Christensen Coast). *Leopold and Astrid Coast* extends from the western extremity of West Ice Shelf (at 81° 24'E) to Cape Penck (66° 43'S/87° 43'E). Cape Penck, an ice-covered point, also fronts West Ice Shelf, but in the east. This part of the coast, named for King Leopold and Queen Astrid of Belgium, was also discovered by the Lars-Christensen Expedition. Leopold and Astrid Coast is bounded by ice shelves on its entire length from Barrier Bay in the west to Philippi Glacier at Cape Penck in the east; Philippi Glacier is on the eastern edge of the ice shelves. To the west is *Ingrid Christensen Coast*, it extends from Jennings Promontory (70° 10'S/72° 33'E), a rock promontory that marks the eastern limit of the *Amery Ice Shelf*, to the western end of the *West Ice Shelf* (at 81° 24'E), visible in the NE part of the map. The coast was discovered and a landing made on Vestfold Hills in 1935 by Captain Mikkelsen of the *Thorshavn*, a vessel owned by the Norwegian whaling magnate Lars Christensen. The coast is named for Ingrid Christensen, the wife of Lars Christensen, who also participated in the whaling expeditions (Alberts 1995, p. 360).

Leopold and Astrid Coast − ERS1 DATA, 1995

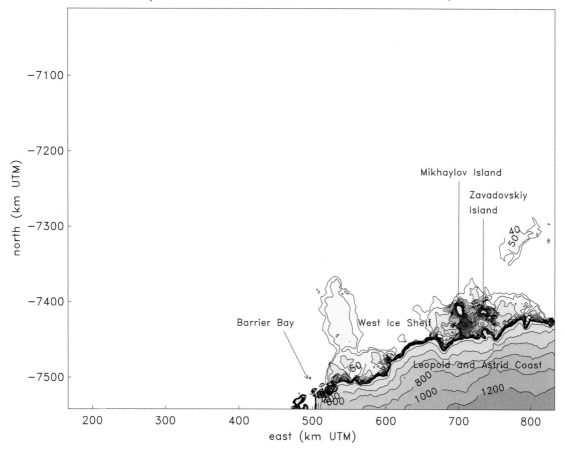

e73−89n63−68, WGS84, Gaussian variog., central mer. 81, slope corrected, scale 1:5000000, 970725

East of Philippi Glacier is *Wilhelm II Coast* (Cape Penck (66°43'S/87°43'E) to Cape Filchner (66°27'S/91°54'E)), named for Kaiser Wilhelm II by Erich von Drygalski.

On the National Geographic Atlas map (National Geographic Society 1992, p. 102), Cape Penck (coordinates (66°43'S/87°43'E) according to Alberts 1995) is just west of Philippi Glacier (center coordinates (66°45'S/88°20'E) according to Alberts 1995), on the USGS satellite image map (Ferrigno et al. 1996), Cape Penck is east of Philippi Glacier (the latter appears to be correct). Cape Penck, an ice-covered point, was charted by the Australian Antarctic Expedition 1911–14 under Mawson and is named for the German geographer Albrecht Penck (Alberts 1995, p. 565).

Another difference between the two quoted maps lies in the extension of *West Ice Shelf*. The most

prominent ice shelf on the USGS map, the name West Ice Shelf is given to the entire ice shelf area off Leopold and Astrid Coast, on the National Geographic map the name applies only to the area west of Mikhaylov Island. Philippi Glacier is a 25 km long glacier, which flows north to the eastern end of West Ice Shelf (25 km west of Gaussberg), named after a geologist of the German Antarctic Expedition 1901–1903 led by Drygalski. Hence, according to Alberts (1995), West Ice Shelf is the entire ice shelf area.

The topography of West Ice Shelf consists of several tongues of ice which extend seaward and decrease in surface elevation (above WGS 84) from 70 m to 40 m, there are two tongues in the western part and three in the eastern part (on the GEOSAT map); the second one from the west is most prominent and deserves the name "tongue". The existence of these tongues indicates that West

Ice Shelf is fed by several glaciers which descend from the rugged and steep coast (for this area, no satellite image is available in Swithinbank 1988). The land climbs rapidly to 400 m, steepest in the eastern part (see detail map 6 West Ice Shelf, section (F.6)).

Comparing the GEOSAT and ERS-1 maps, there are notable differences in the seaward extent of West Ice Shelf: The large ice tongue east of Barrier Bay extends a bit farther on the ERS-1 map (1995) than on the GEOSAT map (1985-86), whereas the eastern section (around Mikhaylov Island and Zavadovski Island and east of the latter) appears to have retreated, most drastically in the eastern-most part: from 100 km (in 1985-86) to less than 50 km (in 1995). Since Philippi Glacier is located here, this could mean a retreat of Philippi Glacier, or less mass discharge by the glacier, or calving of the ice shelf. However, since the ice shelf appears to have lost extent in the entire eastern area, but has several (about three) "tongues" or areas of locally maximal seaward extent, calving in all those areas seems less likely than decreased glacier flow in the eastern part of Leopold and Astrid Coast and in particular in Philippi Glacier. So, we have an area of an advancing glacier next to retreating glaciers (not an unusual, but an interesting phenomenon).

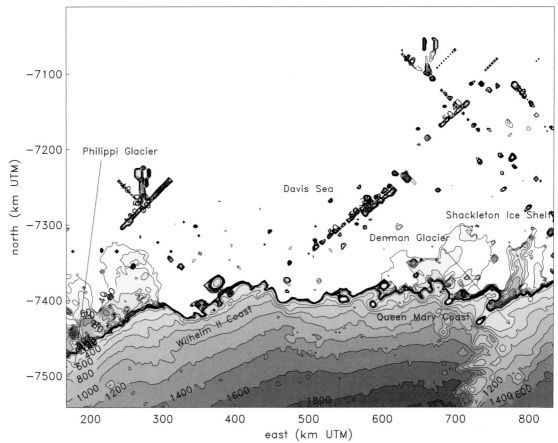

Queen Mary Coast — GEOSAT GM DATA, 1985—86

e85—101n63—68, WGS84, Gaussian variog., central mer. 93, slope corrected, scale 1:5000000, 970723

Map m93e85-101n63-68 Queen Mary Coast

Map m93e85-101n63-68 shows *Wilhelm II Coast* and *Queen Mary Coast*, parts of Leopold and Astrid Coast, and parts of Knox Coast. Leopold and Astrid Coast is west of Wilhelm II Coast. *Wilhelm II Coast* extends from *Cape Penck* (66° 43'S/87° 43'E), Philippi Glacier, to *Cape Filchner* (66° 27'S/91° 54'E), *Queen Mary Coast*, adjacent to the east, reaches from Cape Filchner (named for the leader of the German Antarctic Expedition 1911–1912) to Cape Hordern (66° 15'S/100° 31'E). Queen Mary Coast was discovered by Douglas Mawson, leader of the Australian Antarctic Expedition 1911–14 and named for Queen Mary of England. Easterly adjacent is Knox Coast.

Wilhelm II Coast and *Queen Mary Coast* represent two different types of coast in East Antarctica: Wilhelm II Coast is bounded by Davis Sea without any ice shelves, in contrast, Queen Mary Coast is bounded by the western half of the large Shackleton Ice Shelf (west of Scott Glacier). The eastern half (eastward of Scott Glacier) is part of the same ice shelf (Shackleton Ice Shelf), as determined by the 1956 Soviet Antarctic Expedition (see map m105e97-113n63-68 Knox Coast).

Several glaciers terminate in the Davis Sea, and there are named bays, points and islands: From west to east, there are Philippi Glacier, on Leopold and Astrid Coast, and Cape Penck (see map

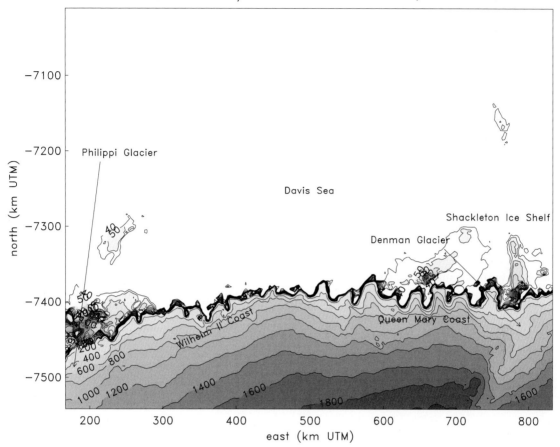

Queen Mary Coast — ERS1 DATA, 1995

e85−101n63−68, WGS84, Gaussian variog., central mer. 93, slope corrected, scale 1:5000000, 970725

m81e73-89n63-68 Leopold and Astrid Coast), Posadowsky Bay, Burton Island Glacier, Krause Point, Cape Filchner (offshore Drygalski Island), on Wilhelm II Coast; McDonald Bay (offshore the Haswell Islands), Helen Glacier, and Farr Bay, on Queen Mary Coast. Since technically, Wilhelm II Coast extends from Cape Penck (87° 43'E) to Cape Filchner (91° 54'E), the western part of Queen Mary Coast is free of ice shelves. The Russian Mirnyy Station is located on the coast west of Helen Glacier. Mirnyy Station was built on two small nunataks (400 m long), the only ice-free areas in this region.

Indentation in the coastline, indicative of glaciers and their valleys, are far more pronounced on the ERS-1 map than on the GEOSAT map. The land rises steeply up to 400 m elevation (200 m in 5 km (4% = 2.25°; 200 m to 400 m), then less steep to 1800 m (on this map).

Shackleton Ice Shelf extends about 360 km east–west from 95° E to 105° E and extends about 140 km into the sea in the western half and 60 km in the eastern half (according to Alberts 1995). Denman Glacier, the largest glacier in this area of Wilkes Land, flows through Shackleton Ice Shelf. On both the GEOSAT map and the ERS-1 map, the extent of Shackleton Ice Shelf is about 100 km in the part west of Denman Glacier. The tongue of Denman Glacier extends about as far (also 100 km, measured from the coastline in the bay) on both maps, but the head of the tongue is shaped differently on the two maps. For more details, see the detail map Denman Glacier (section (F.7)).

Shackleton Ice Shelf was known by the time of the United States Exploring Expedition 1838–1842, led by Charles Wilkes and mapped, in part, from the ship *Vincennes* in 1840, later explored by the Australian Antarctic Expedition under Mawson (1911–1914) and named for Antarctic explorer Sir Ernest Shackleton, and mapped in 1955 by the expedition called "Operation Highjump" (U.S. Navy). *Roscoe Glacier* drains into Shackleton Ice Shelf at its western margin, also several smaller glaciers, as well as *Northcliffe Glacier*, *Denman Glacier* and *Scott Glacier*. There are several islands in the ice shelf, Masson, Henderson, David and Mill Islands.

The section of coast from Krause Point to Shackleton Ice Shelf is also seen on an image in Swithinbank (1988, fig. 48, p. B63). The coastline here is called an ice-wall coastline, which is characterized by small-scale ice-surface topography features visible right up to the coast. The floating termini of *Helen Glacier* and *"Annenkov Glacier"* are exceptions, and both termini are afloat.

Shackleton Ice Shelf is largely flat, with elevations of 40 m above the ellipsoid. On *Helen Glacier*, the line separating an (inland) area of heavy crevassing (sign of rapid movement) from the (seaward) area of heavy rifting (the rifts are water-filled) indicates the grounding line. Similarly, the grounding line of "Annenkov Glacier" can be determined in imaging. In contrast, Shackleton Ice Shelf is an almost featureless, floating ice shelf. Ice velocities of Helen Glacier are 600 ma^{-1} at the floating front (Dolgushin 1966), 300 ma^{-1} at 30–35 km inland (Schmidt and Mellinger, 1966); the adjacent slow ice moves at about 60–80 ma^{-1}.

Drygalski Island is an ice cap on a seafloor shoal, the shoal reaches to higher than 60 m below sea level, but the top of the island is 326 m above sea level. The cap may have started forming from an iceberg and increased by accumulating snow. Maximal snowfall rates are measured 30 km inland from the coast (and on Drygalski Island).

Knox Coast — GEOSAT GM DATA, 1985—86

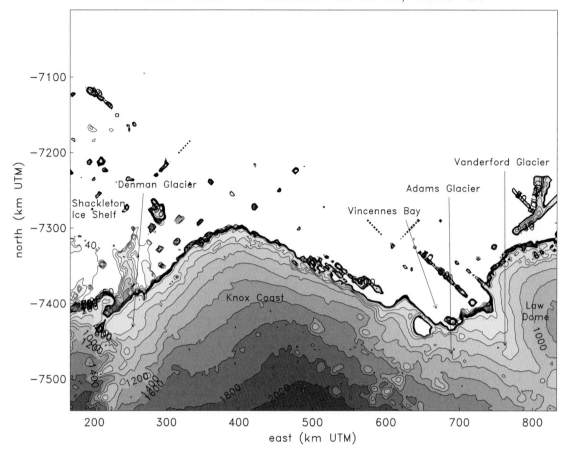

e97—113n63—68, WGS84, Gaussian variog., central mer. 105, slope corrected, scale 1:5000000, 970723

Map m105e97-113n63-68 Knox Coast

Map m105e97-113n63-68 *Knox Coast* shows a section of the coast of Wilkes Land. Knox Coast extends from *Cape Hordern* (66° 15'S/100° 31'E) to *Hatch Islands* (66° 32'S/109° 16'E). Cape Hordern is the northwest end of the Bunger Hills. The Hatch Islands are a group of rocky coastal islands 5 km east of Ivanov Head at Vincennes Bay (near Adams Glacier). The *Bunger Hills* are south of the eastern end of *Shackleton Ice Shelf*. To the east lies *Budd Coast*. Budd Coast extends from Hatch Islands east to Cape Waldron (66° 34'S/115° 33'E), it is named after the master of the ship *Peacock* of the Wilkes Expedition (1838–42). Cape Waldron is an ice-covered cape just west of Totten Glacier (see detail map Totten Glacier, section (F.9)). To

the west of Knox Coast is *Queen Mary Coast*, extending west to Cape Filchner (66° 27'S/91° 54'E).

The largest inland drainage systems are the ones of *Denman Glacier* and *Vanderford Glacier*. The Denman Glacier valley extends at least 180 km inland from the location where Denman Glacier enters Shackleton IceShelf (at (-7400.000 N/220.000 E)) to an elevation of 2200 m (on map m93e85-101n67-721 Wilkes Land). The valley of Vanderford Glacier extends southeast around Law Dome (at the eastern margin of this map). Vanderford Glacier is the largest glacier in Vincennes Bay (650.000 to 750.000 E on the coast). Denman Glacier has a velocity of

Knox Coast — ERS1 DATA, 1995

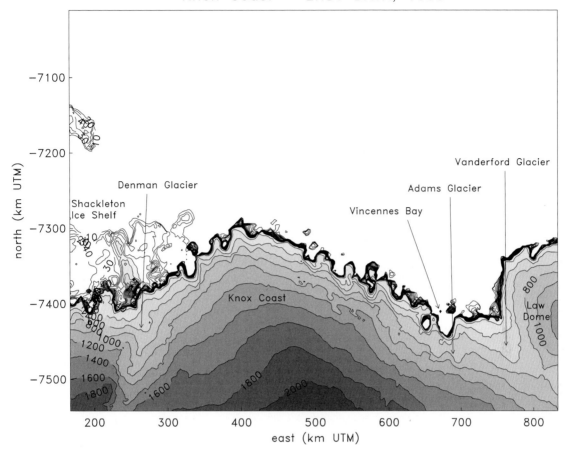

e97–113n63–68, WGS84, Gaussian variog., central mer. 105, slope corrected, scale 1:5000000, 970725

$\approx$1370 ma^{-1} to 1500 ma^{-1} (two sources), Scott Glacier of 1300 ma^{-1}. A detail map of Denman Glacier is given in section (F.7), a detail map of Vanderford Glacier in section (F.8).

On the National Geographic Atlas map (National Geographic Society 1992, p. 102), Shackleton Ice Shelf extends further east around the coast of Knox Coast (to about 106° E) than on our radar altimeter map, but on the USGS satellite image map (Ferrigno et al. 1996), Shackleton Ice Shelf extends to just east of Bunger Hills, to about 101° E (based on a 1983 image in this region), which matches the extent on the ERS-1 map (1995 data). We conclude that (most likely) (1.) the extent of Shackleton Ice Shelf has remained largely unchanged between 1983 and 1995, and (2.) the ERS-1 satellite mapping and processing picks up the ice-shelf edge fairly correctly, whereas the ice data on the National Geographic map is outdated. The

GEOSAT map shows specks off the coast, where the National Geographic map shows many islands, of which *Mill Island* (-7280.000 N/220.000 E) near Shackleton Ice Shelf is the largest. Others include *Merrit Islands* and *Davis Islands* west of Vincennes Bay and *Balaena Islands* east of the bay. The mouth of *Adams Glacier* (at 695.000 E) may also be identified, whereas some smaller named glaciers cannot be resolved.

The named glaciers are, from west to east, *Denman Glacier*, *Scott Glacier*, *Apfel Glacier* (near the Australian Edgeworth David Station and the Russian Bunger Oasis II Station, which was built in 1956 and handed over to Polish Scientists in 1959, then renamed Dobrowolski Station), offshore is *Mill Island*, *Remenchus Glacier* (which flows into Shackleton Ice Shelf), *Du Beau Glacier* at the edge of Shackleton Ice Shelf, *Robinson* and *Underwood Glacier*, *Adams Glacier* (in Vincennes Bay), and *Vanderford Glacier* (offshore are the Windmill Islands).

The *Bunger Hills* are a 1000 km² large snow-and-ice-free area. The area was discovered in 1947 by U.S. Navy Operation Highjump and called an "oasis". Byrd (1947) described the Bunger Hills as "a land of blue and green lakes and brown hills in an otherwise limitless expanse of ice" (after Swithin-bank 1988, p. B59); there were three large lakes of 3 km diameter and 20 smaller lakes, all with open water. The lakes are an exposure of the sea that underlies the ice shelves, but the lake area is separated from the ice edge by the 50 km wide Shackleton Ice Shelf. Bunger Hills and Denman Glacier are also seen in an image in Swithinbank (1988, fig. 47, p. B61).

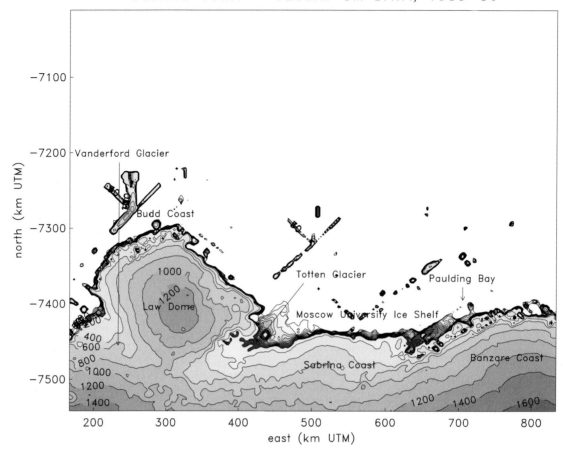

Sabrina Coast – GEOSAT GM DATA, 1985–86

e109–125n63–68, WGS84, Gaussian variog., central mer. 117, slope corrected, scale 1:5000000, 970723

Map m117e109-125n63-68 Sabrina Coast

The map m117e109-125n63-68 Sabrina Coast shows (from west to east) *Budd Coast* with Law Dome, *Sabrina Coast* with Moscow University Ice Shelf and the western half of *Banzare Coast* (to 125° E).

Budd Coast is the coast around Law Dome, between Hatch Islands (66° 32'S/109° 16'E) and Cape Waldron (66° 34'S/115° 33'E). The Hatch Islands are a group of rocky coastal islands 5 km east of Ivanov Head at Vincennes Bay (near Adams Glacier). Budd Coast was discovered in Feb. 1840 by the U.S. Exploring Expedition led by Charles Wilkes and named for Thomas Budd, Acting Master of the *Peacock*, one of the expedition ships.

Law Dome (center coordinates (66° 44'S/112° 50'E)) is a large ice dome which rises to 1395 m. The dome was discovered by the U.S. Expedition "Operation Highjump" (1946–47), it is named for Phillip Law, Director of the Australian Antarctic Division (1949–1966). Intensive glaciological and geophysical surveys were carried out on Law Dome by the ANARE Program in 1962–1965. Notice that the elevation of Law Dome is mapped correctly from GEOSAT and ERS-1 altimeter data (judged by the slope between 1000 m and 1200 m and the diameter of the area above 1200 m). The exact slope of the coastline of Budd Coast varies between GEOSAT and ERS-1 mapping, as has

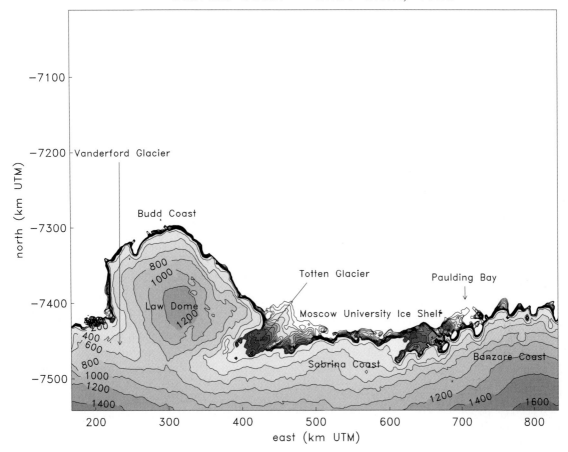

Sabrina Coast — ERS1 DATA, 1995

e109−125n63−68, WGS84, Gaussian variog., central mer. 117, slope corrected, scale 1:5000000, 970725

been noticed for other maps also. The E–W diameter of the Law Dome peninsula is 230 km.

Off Cape Folger are the *Balaena Islands* (possibly the source of reflection offshore on the GEOSAT map), the northernmost point of Budd Coast is *Cape Poinsett*. West of the Balaena Islands are the Windmill Islands (named after the U.S. expedition "Operation Windmill").

The valley of *Vanderford Glacier* that reaches around Law Dome from the northwest (120 km) and the valley of *Totten Glacier* that reaches around Law Dome from the northeast (130 km) meet south of the dome at an elevation of 800 m. Whereas Vanderford Glacier does not extend much into the ocean, Totten Glacier has a large offshore part (95 km long on the ERS-1 map, 85 km long on the GEOSAT map). For more information, see the text and maps in the detail map section for Vanderford Glacier (F.8) and Totten Glacier (F.9).

Williamson Glacier is a small glacier just west of Cape Waldron, at the west edge of Totten Glacier; Fox Glacier is similarly small and 20 km further west of Williamson Glacier (according to Swithinbank 1988, fig. 46, p. B60; and other than indicated on the National Geographic map where Totten Glacier is not named but has the name "Williamson Glacier" printed across it - and ending at the small Williamson Glacier).

Totten Glacier drains an area of 150.000 km^2 and has a balance discharge of 43 km^{3}a^{-1} (McIntyre, after Swithinbank 1988, p. B57). Radio-echo soundings prove the existence of a subglacial valley. The velocity of ice entering Totten Glacier from the south (where the glacier flows in a east-northeasterly direction) is 280 ma^{-1} (near the

1000 m contour on the Australian elevation map; spot soundings of ice thickness in the same area are 1450, 2000, 1900, 1500, and 1600 m. Ice-front velocities are 850–1200 ma^{-1} (Dolgushin 1966).

On the USGS satellite image map (Ferrigno et al. 1996), Totten Glacier extends only about 20 km into the sea; the image in Swithinbank (1988, fig. 46, p. B60) is a composite of two LANDSAT 1MSS images, both from 26 October 1973. Comparing that to the extent in the GEOSAT (85 km) and ERS-1 (95 km) maps suggests that Totten Glacier has been advancing. Taking corrected positions into account, it can be concluded that Totten Glacier has been advancing at a rate of aproximately 1 km per year between 1973 and 1995 (for derivation of this result, see the Totten Glacier detail map, section (F.9)).

The Australian Casey Station is located on Budd Coast near Cape Folger. Wilkes Station was used by the U.S. Antarctic Expedition "Operation Windmill" in the International Geophysical Year (1957/58), then given to Australian scientists in 1960, then rebuilt (Casey).

Sabrina Coast extends from Cape Waldron (at (66° 34'S/115° 33'E), between Williamson Glacier and Totten Glacier) to Cape Southard (66° 32'S/122° 05'E). John Ballemy saw land here in March 1839 at 117°E. The land was mapped as "Totten High Land" by the U.S. Exploring Expedition (1838–1842) led by Charles Wilkes who explored it in 1840. In 1931 the British-Australian-New Zealand Antarctic Research Ex-

pedition (BANZARE) led by Douglas Mawson saw land 1 degree further south than reported by Ballemy, but kept the name "Sabrina Coast", named after a ship of Ballemy that was lost in a storm in 1839.

Moscow University Ice Shelf covers the offshore area of Sabrina Coast, the ice shelf reaches from the Totten Glacier ice to ice in Paulding Bay, Paulding Bay is already on Banzare Coast. The National Geographic Atlas map shows *Dalton Iceberg Tongue* (center coordinates (66° 15' S/121° 30' E) extending north from the eastern end of Moscow University Ice Shelf; the iceberg tongue is not visible on the USGS Satellite Image Map of Antarctica (Ferrigno et al. 1996). Dalton Iceberg Tongue was photographed from the air by U.S. expedition "Operation Highjump" (1946–47) and also observed by the ANARE expeditions in 1958 and in 1960.

Banzare Coast lies between Cape Southard (66° 32'S/122° 05'E) and Cape Morse (66° 15'S/130° 10'E), it was seen from the air by the British-Australian-New Zealand Antarctic Research Expedition in 1930–31, led by Douglas Mawson and named BANZARE like the expedition acronym. West of 125° E (the map edge) are Voyeykov Ice Shelf which protrudes just east of Paulding Bay and is barely identifiable on the GEOSAT and the ERS-1 maps, and there are several small glaciers.

The contour lines indicate rough and steep terrain south of the coast on this map.

104

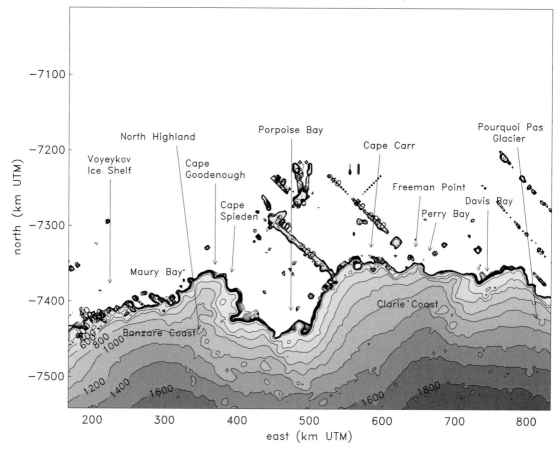

e121−137n63−68, WGS84, Gaussian variog., central mer. 129, slope corrected, scale 1:5000000, 970723

Map m129e121-137n63-68 Clarie Coast

The map m129e121-137n63-68 Clarie Coast shows sections of the coast of *Wilkes Land*, East Antarctica: From west to east, (1.) *Banzare Coast* (*Cape Southard* at (66° 34'S/122° 05'E) to *Cape Morse* at (66° 15'S/130° 10'E) near *Cape Carr*, and (2.) *Clarie Coast* (Cape Morse to *Pourquoi Pas Point* at 136° 11'E). There are small sections of *Sabrina Coast* at the western end of the map and of *Adélie Coast* on the eastern end of the map, but these are better covered on maps m117e109-125n63-68 Sabrina Coast and m141e133-149n63-68 Adélie Coast, respectively.

Cape Morse is a low, ice-covered cape and the eastern side of the entrance of Porpoise Bay. A channel

glacier with the same name (Morse Glacier) enters Porpoise Bay about 3 miles southwest of Cape Morse (Alberts 1995). *Cape Carr* (66° 09'S/130° 42'E) is the first labeled feature on Clarie Coast; it is a prominent, ice-covered cape 15 miles northeast of Cape Morse. Correct identification of the capes was only possible by comparison of the maps from several expeditions — the old maps are of different accuracy, the coastline changes, and so it has not always been clear to which location a certain name was given by the discovering expedition. Clarie Coast was discovered in January 1840 by Captain Jules Dumont d'Urville who noticed land south of the ice cliffs he named "Cote Clarie"

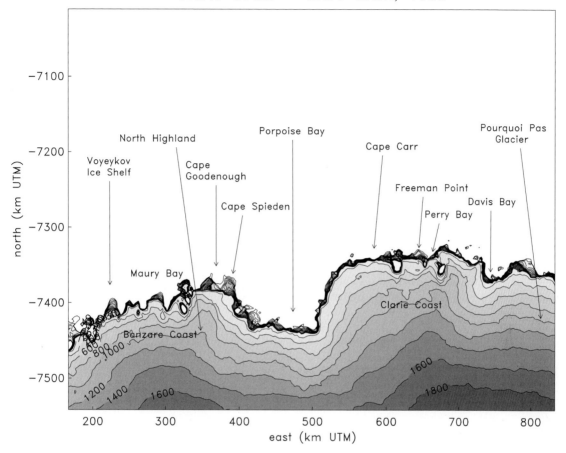

Clarie Coast – ERS1 DATA, 1995

e121–137n63–68, WGS84, Gaussian variog., central mer. 129, slope corrected, scale 1:5000000, 970725

(after the wife of the captain of his second ship "Zélée" (Alberts 1995)).

The land on this map descends steeply towards the coast, with a gradient of 200 m over 30 km, corresponding to $0.66\% = 0.38°$ slope between 1400 m and 1600 m elevation to a gradient of 200 m over 5 km (locally) to 10 km ($2\% = 1.15°$) between 600 m and 800 m. The overall gradient and steepening is similar throughout most of this map and adjacent coastal Wilkes Land maps; the gradient of 200 m in 5 km, equal to a 4% slope or 2.29°, occurs in the northeastern slope of North Highland. Hence it is likely that local slopes vary and are even steeper, because a consequence of the smoothing effects of altimetry mapping and kriging are smoother contours and locally lower apparent elevations.

Several small ice shelves are mapped along *Banzare Coast*: *Voyeykov Ice Shelf* at (-7400.000 N/ 210.000 E), ice shelves east and west of Maury Bay, at Cape Goodenough and Cape Spieden and in Porpoise Bay. The near-coastal topography indicates the existence of several small valley glaciers, which match the signatures on the National Geographic Atlas map (National Geographic Society 1992, p. 102). Named glaciers on the USGS Satellite Image Map (Ferrigno et al. 1996) are *Thompson Glacier* ending in Paulding Bay, *Bell Glacier* in Maury Bay, *Holmes Glacier* in the North Highland region, and *Frost Glacier* south of Porpoise Bay.

Between Cape Morse and Cape Carr is a small glacier with a tongue (*Blodgett Iceberg Tongue*) on the National Geographic Map, not much of it is shown on the GEOSAT and ERS-1 maps, an indication of the changing ice extent along the coast.

On *Clarie Coast*, the only glacier named on the USGS Satellite Image Map is *Dibble Glacier* at *Davis Bay*. A comparison of the USGS Satellite Image Map and the ERS-1 map and the GEOSAT map indicates that Dibble Glacier may occupy the valley south of Davis Bay that extends over 100 km inland. According to Alberts (1995, p. 188), Dibble Glacier is a channel glacier that terminates on the east side of Davis Bay and has a prominent glacier tongue (*Dibble Glacier Tongue* (65° 50'S/ 135° E)) from which an iceberg tongue (*Dibble Iceberg Tongue* (65° 30'S/135° E), mappable in 1946 and 1956) extends even further seaward. The locations of ice extending seaward on the ERS-1 map and on the USGS Satellite Image Map indicate that Dibble Glacier may not occupy the prominent valley south of Davis Bay but flow slightly further east towards the headland east of Davis Bay. A *glacier tongue* consists of floating glacier ice that is still connected (but possibly heavily crevassed) and extends seaward from a glacier; an *iceberg tongue* consists of icebergs, all stemming from the same glacier, but the ice is broken up into larger and smaller blocks (icebergs); icebergs calve off along the crevasses visible in the glacier tongue, as the glacier tongue moves seaward.

Pourquoi Pas Glacier (center coordinates (66° 15'S/135° 55'E)) is the easternmost significant glacier on this map, it is a 6 km wide and 23 km long outlet of the East Antarctic Ice Sheet. Pourquoi Pas Glacier flows in a northnorthwesterly direction and terminates in a prominent glacier tongue (*Pourquoi Pas Glacier Tongue*, which was 6 km wide and 10 km long in 1946–47). Some of the ice east of Davis Bay may stem from Pourquoi Pas Glacier. "Pourquoi Pas?" ("why not?") was the name of the polar vessel of the French Antarctic expedition led by Jules Dumont d'Urville.

108

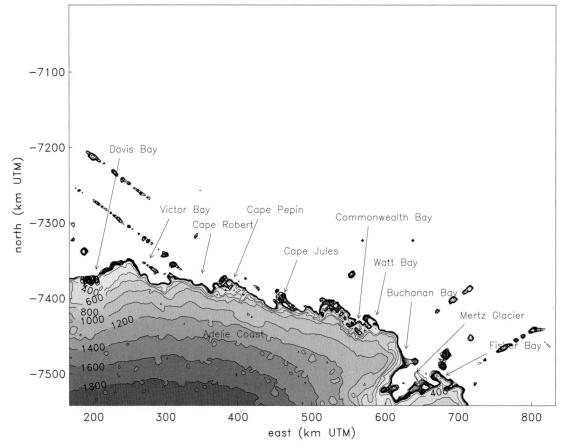

Adelie Coast — GEOSAT GM DATA, 1985—86

e133—149n63—68, WGS84, Gaussian variog., central mer. 141, slope corrected, scale 1:5000000, 970723

Map m141e133-149n63-68 Adélie Coast

This map shows a section of Wilkes Land, East Antarctica, centered around *Adélie Coast* with parts of *Clarie Coast in the west* (cf. map m129e121-137n63-68 Clarie Coast) and *George V Coast in the east* (Mertz Glacier and surroundings, see map m141e133-149n67-721 Wilkes Land e133-149n67-721 and the Mertz and Ninnis Glacier detail maps, section (F.10)).

At the headland northwest of Mertz Glacier, the coast of Wilkes Land turns south, the large part of Wilkes Land that extends to near the polar circle or north of it ends here. With George V Coast, Oates Coast and Pennell Coast, the coast draws southward for 5 degrees of latitude. The name *Adélie Coast* applies to the coast between Pourquoi Pas Point (66° 12'S/136° 11'E) and Point Alden (66° 48'S/142° 02'E), the coast was named "Terre Adélie" by Capt. Jules Dumont d'Urville who discovered the coast in January 1840, for his wife (Alberts 1995). *Point Alden* is an ice-covered point with rock exposure on the seaward side, located at the western side of the entrance to Commonwealth Bay (discovered in January 1840 by Ch. Wilkes). *Pourquoi Pas Point* is the point between Davis Bay and Victor Bay.

Dibble Glacier ends in the eastern side of Davis Bay (see description to map m129e121-137n63-68 Clarie Coast). The next named glacier to the east

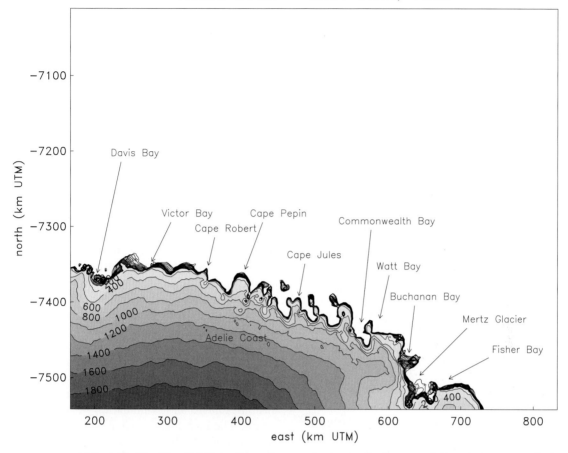

Adelie Coast – ERS1 DATA, 1995

e133–149n63–68, WGS84, Gaussian variog., central mer. 141, slope corrected,
scale 1:5000000, 970725

is *Pourquoi Pas Glacier* (see also map m129e121-137n63-68 Clarie Coast). Several kilometers further east is *Commandant Charcot Glacier* (center coordinate 66° 25'S/136° 35'E) which terminates in the head of Victor Bay.

Commandant Charcot Glacier is 5 km wide and 18 km long, it is an outlet glacier of the East Antarctic Ice Sheet and flows northnorthwest, terminating in a broad and about 3 km long (1950–52) glacier tongue (*Commandant Charcot Glacier Tongue* (66° 22'S/136° 35'E)); offshore ice is seen in the ERS-1 map. "Commandant Charcot" was the name of the ship of the 1950–52 French Antarctic Expedition (Alberts 1995, p. 147). Next to the east are *Cape Robert* (66° 23'S/137° 39'E), *Franąis Glacier*, *Cape Pépin*, and several smaller glaciers, the French overwintering Station Dumont d'Urville near *Astrolabe Glacier, Cape Jules, Point*

Alden, and *Commonwealth Bay* (with the Australian station of the same name).

Cape Robert (66° 23'S/137° 39'E) is an ice-covered point at the west side of Marret Glacier, which is a small 6 km wide, 6 km long channel glacier and an outlet glacier. *Franąis Glacier* (center coordinate (66° 33'S/138° 15'E)) is a 6 km wide and 18 km long outlet glacier with a 5 km long glacier tongue (Franąis Glacier Tongue) in Ravin Bay, it was likely discovered in 1837-1840 (Franąis was the name of a ship of that expedition). *Cape Pépin* (66° 32'S/138° 34'E) is an ice-covered cape between Ravin Bay and *Barré Glacier* (named for Adélie Pépin, wife of J. Dumont d'Urville). Barré Glacier is a 8 km wide and 8 km long channel glacier and an outlet glacier.

Astrolabe Glacier (66° 45'S/139° 55'S) is a 6 km wide and 16 km long outlet glacier, flowing northnortheast and terminating in a prominent (5 km wide, 6 km long) tongue, the *Astrolabe Glacier Tongue*, the latter is located at the eastern end of the Géologie Archipelago.

Cape Jules (66° 44'S/140° 55'E) is a rocky cape with a small cove at its north end, located 5 km west of *Zélée Glacier Tongue* (Jules is the first name of Dumont d'Urville and of his son). *Zélée Glacier* (66° 52'S/141° 10'E) is a small 5 km wide and 10 km long glacier, which flows northnorthwest — like most of its neighbours — along the western side of Lacroix Nunatak and terminates at the western side of Port Martin in a large tongue (3 km wide, 11 km long).

We note that in this part of Wilkes Land even the small and short glaciers are outlet glaciers of the East Antarctic Ice Sheet.

The morphologic feature named *Zélée Subglacial Trench* (center coordinate (68° S/144° E)), however, is not occupied by Zélée Glacier but by *Mertz Glacier* (Alberts 1995, p. 832), the trench or trough runs NNE–SSW.

At Point Alden (66° 48'S/142° 02'E), *George V Coast* begins. Commonwealth Bay is located between Point Alden (W) and Cape Gray (66° 51'S/143° 22'E) (E), a couple of small glaciers end in the 50 km wide bay. The bay was discovered in 1912 by the Australian Antarctic Expedition led by Douglas Mawson, who established the main base of the expedition (Commonwealth Station) at Cape Denison at the head of the bay (Alberts 1995, p. 147). East of Commonwealth Bay is Watt Bay (67° 02'S/144° E), an about 25 km wide bay. Offshore between the two bays is Way Archipelago. On the GEOSAT map, there are offshore reflections, some of which stem from islands.

The area of George V Coast around Commonwealth Bay, and Watt Bay is called "home of the blizzard" in Swithinbank (1988, p. B55–B57 and fig. 44), because it is the windiest place on Earth (or, the windiest place with meteorologic records).

Measured wind speeds are 19 m s^{-1} as an average over 22 months in 1912–13, 25 m s^{-1} as the highest monthly mean (July 1913), and 43 m s^{-1} as the highest daily mean, and peak velocities of wind gusts are, of course, a lot higher (after Madigan 1929). Evidence of wind is seen in satellite images (such as fig. 44, Swithinbank 1988), the directions of snow transported by the wind and visible in the image demonstrates that it is katabatic wind that rages as a blizzard. Katabatic wind is caused by pressure gradients over an ice sheet, higher pressures exist high on the center of the ice sheet, lower pressures over the coast and the ocean, consequently the katabatic wind flows offshore in directions normal to contour lines. Katabatic winds carry large amounts of snow from the ice sheet to the sea, this effect needs to be taken into account in glaciologic mass balance calculations. Loewe (1956), an overwinterer at the French Station Port Martin in 1951, estimated that 240.000 tons of drift snow crossed each kilometer of coast on a blizzard day; this corresponds to removal of half of the estimated accumulation. The katabatic winds do not extend far out to the sea (10–20 km). The snow freezes together with crystals of rapidly freezing seawater, forming *Shuga*, an accumulation of spongy white lumps. The heat loss contributes to the formation of cold Antarctic bottom water.

The largest glacier in the area of this map is *Mertz Glacier*, it is described with the southerly adjacent map m141e133-149n67-721 Cook Ice Shelf, and with the Mertz Glacier detail maps (F.10). *Buchanan Bay* (67° 05'S/144° 40'E) is a sheltered bay formed by the junction of the western side of Mertz Glacier Tongue and the mainland. *Cape de la Motte* marks the western entrance point.

Fisher Bay is already in *George V Coast*, at (67° 31'S/145° 45'E), located between the eastern side of Mertz Glacier Tongue and the mainland, this is 20 km wide. The land slopes with increasing gradients to the coast.

112

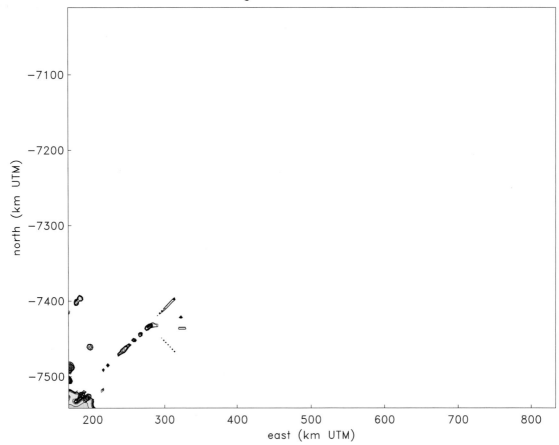

e145–161n63–68, WGS84, Gaussian variog., central mer. 153, slope corrected, scale 1:5000000, 970723

Map m153e145-161n63-68 Ninnis Glacier Tongue

Ninnis Glacier is an outlet glacier of the East Antarctic Ice Sheet, on George V Coast of Wilkes Land. The Ninnis Glacier Tongue was so prominent at the time of the old expeditions that the name of the tongue rather than the name of the glacier is listed in some maps (e.g. National Geographic Atlas map, National Geographic Society 1992, p. 102). Now the glacier tongue has receded. Ninnis Glacier and Ninnis Glacier Tongue were discovered by the Australian Antarctic Expedition (1911–1914) led by D. Mawson, Ninnis was a member of the expedition who lost his life in a sledge journey. For a discussion of glaciers with tongues, their advance and retreat, see the description of the detail maps Mertz and Ninnis Glaciers (section (F.10)).

This map is a good example of the disadvantage of regular map tiling in sheets as used in the Atlas scheme — then and now a map with a poor cut results, one with hardly anything on it (this map is the only one in this Atlas). Ideally, one would always like the feature of interest to be in the center of the map and the neighbourhood around it, but that is not always possible with regular map tiling. For this reason, special detail maps are produced for areas of particular interest — this is chapter (F) of this book.

Ninnis Glacier Tongue — ERS1 DATA, 1995

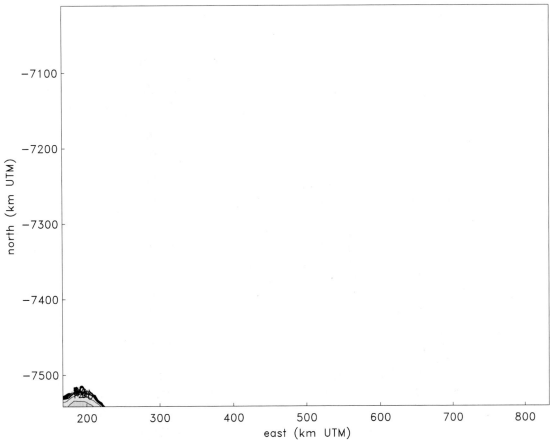

e145-161n63-68, WGS84, Gaussian variog., central mer. 153, slope corrected, scale 1:5000000, 970725

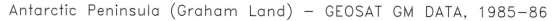

Antarctic Peninsula (Graham Land) — GEOSAT GM DATA, 1985–86

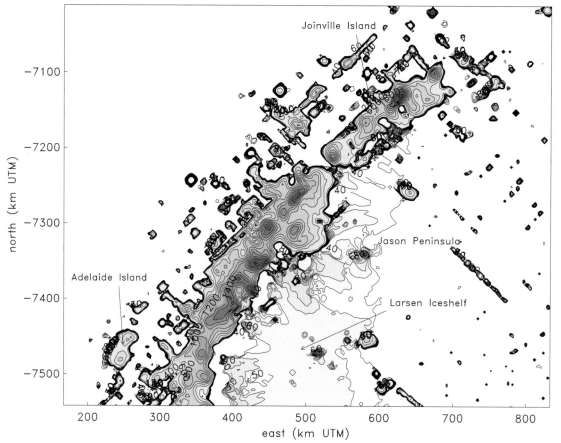

e289–305n63–68, WGS84, Gaussian variog., central mer. 297, slope corrected, scale 1:5000000, 970723

Map m297e289-305n63-68 Antarctic Peninsula (Graham Land)

(Coordinates e289-305 correspond to w71-55)

Because of its high topographic relief, the *Antarctic Peninsula* is not well represented in satellite-radar-altimeter data, but it is revealed better in satellite images (see USGS Satellite Image Map of Antarctica (Ferrigno et al. 1996) and Swithinbank, 1988, p. B105–B118). This is most obvious in the GEOSAT maps of the peninsula (this map and m292e284-300n67-721 Antarctic Peninsula (Palmer Land)). The retracking algorithms do not work properly, and signal returns picked from islands appear as long features in sub-satellite track locations. The ERS-1 maps, however, are

very good and give a largely realistic picture of the Antarctic Peninsula's topography.

Larsen Ice Shelf, the largest ice shelf of the peninsula is mapped correctly, and several of the glaciers that discharge ice from the mountain ranges that form the backbone of the peninsula into Larsen Ice Shelf. *Jason Peninsula* marks the northern limit of Larsen Ice Shelf, but the bay on the opposite (western) side of the Peninsula is smaller. *Joinville Island*, the largest island of the Joinville Island group, lying off the northeastern tip of the peninsula at 63° 15'S/55° 45'W and 60km long, 19 km

Antarctic Peninsula (Graham Land) — ERS1 DATA, 1995

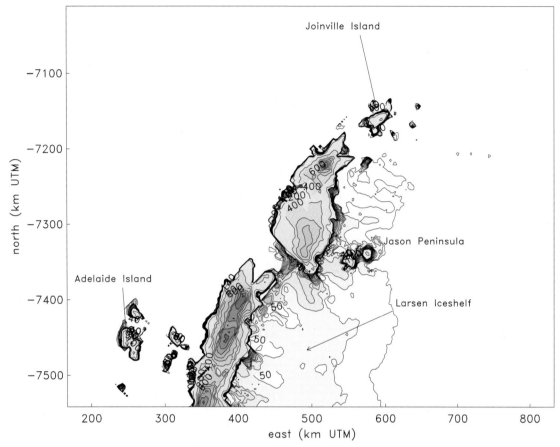

e289−305n63−68, WGS84, Gaussian variog., central mer. 297, slope corrected, scale 1:5000000, 970725

wide, should be mapped farther north. It was discovered in 1838 by the French Antarctic Expedition led by Jules Dumont d'Urville. The backbone of the peninsula extends farther northeastward than on the ERS-1 map (see GEOSAT map), and Adelaide Island is larger, but in the correct location. So it appears that the ERS-1 algorithm cuts out too many data, whereas the GEOSAT algorithm retains some erroneously retracked data.

The ice shelf north of Jason Peninsula does not have a separate name, *Drygalski Glacier* and *Evans Glacier* drain into it. From north to south, the coasts of the peninsula are, with some of their major features: Trinity Peninsula (63° 37'S/58° 20'W) is the northernmost part of the peninsula, separated from it by the Antarctic Sound is Joinville Island.

There are many research stations on the Peninsula, Bernardo O'Higgins (Chile), Hope Bay (UK) and Esperanza (Argentina) on Trinity Peninsula. Southeast of Trinity Peninsula is James Ross Island (near 64° S/57.5° West), separated by Prince Gustav Channel. *Davis Coast* (from Cape Kjellman (63° 44'S/59° 24'W), E side of Charcot Bay, to Cape Sterneck (64° 04'S/61° 02'W), N side of Hughes Bay) is on the western coast, *Nordenskjöld Coast* is the eastern coast. Between the northern part of the peninsula (Davis Coast, Trinity Peninsula) and the South Shetland Islands is *Bransfield Strait*, the islands are north of 63° North, hence not on our maps.

South of Davis Coast is *Danco Coast* (Cape Sterneck (N) to Cape Renard (65° 01'S/63° 47'W), south side of Flandres Bay) and south of that *Graham Coast* (Cape Renard (N) to Cape Bellue (66° 18'S/65° 53'W), north side of Darbel Bay), *Loubet*

Coast (Cape Bellue (N) to Bourgeois Fjord (67° 40'S/67° 05'W), between Pourquoi Pas and Blaiklock Islands), *Fallieres Coast* (Bourgeois Fjord (N) to Cape Jeremy (69° 24'S/68° 51'W)).

Cape Jeremy marks the boundary between Graham Land and Palmer Land, and also the north entrance of George VI Sound. Most names were given by the Belgian Antarctic Expedition led by Gerlache 1898. The west coast is rugged, with mountain ranges and glaciers falling steeply into the sea, many islands are located offshore, the largest is *Adelaide Island* (with Rothera Station, UK) with a 2565 m high point (Mt. Gandry) and an ice piedmont (Fuchs Ice Piedmont) on the western side.

On the east coast of Graham Land are *Nordenskjöld Coast* (Cape Longing (64° 33'S/58° 50'W), S entrance of Prince Gustav Channel, to Cape Fairweather (700 m high) at (65° 00'S/61° 01'W)).

Oscar II Coast extends from Cape Fairweather to Cape Alexander (66° 44'S/62° 37'W) at the head of Churchill Peninsula, just south of Jason Peninsula in Larsen Ice Shelf and north of Cabinet Inlet. *Foyn Coast* lies between Cape Alexander and Cape Northrop (67° 24'S/65° 16'W, 1160 m), north of Whirlwind Inlet and opposite Adelaide Island. *Bowman Coast* extends from Cape Northrop (N) to Cape Agassiz (68° 29'S/62° 56'W), the tip of Hollick-Kenyon Peninsula at the south end of Larsen Peninsula. The Antarctic Peninsula was discovered in the 19th century and at the turn of the 20th century.

The Weddell Sea, between the Antarctic Peninsula and Cape Norvegia, Queen Maud Land, was discovered in 1823 by James Weddell, who named it George IV Sea. The name was changed in 1900.

The east coast of the Antarctic Peninsula, the west coast of the Weddell Sea, is bounded by large ice shelves. The west coast of *Graham Land* is just outside the climatic limit for ice shelves but the east coast is just within this limit, which coincides with the -5° C mean annual isotherm at sea level (Reynolds 1981, after Swithinbank 1988, p. B105). The ice in *Prince Gustav Channel* is the northernmost ice shelf of Antarctica.

Steep glaciers tumble into the ice shelves from *Detroit Plateau* (1500–2000 m) north of Jason Peninsula and to the south into Larsen Inlet. *Larsen Ice Shelf* is around 400 m high at the grounding line and 200 m or less at the ice front; the grounding line is visible in satellite imagery (Swithinbank 1988, p. B105). In warm summers, the ice shelf becomes covered with a well-developed pattern of melt streams (Swithinbank 1988, p. B105). Larsen Ice Shelf suddenly disintegrated in spring 2002, between January and March most of the ice shelf broke up into icebergs.

In the *Marguerite Bay* area precipitous mountain glaciers descend from alpine peaks, there are also crevassed ice piedmonts, calving ice cliffs and hanging glaciers, and an ice-filled sea. The ice on Graham Land Plateau (1500–1700 m a.s.l.) is 500–700 m thick. Swithinbank Glacier (named after British glaciologist Charles Swithinbank) flows into Marguerite Bay at the latitude of Pourquoi Pas Island. A map of the glacier region near San Martin, Marguerite Bay, was generated by Wrobel et al. (2000) using photogrammetry.

Müller Ice Shelf (Lallemand Fjord) in Marguerite Bay could be the northernmost ice shelf on the west side, Jones is a bit further south (Bourgeois Fjord), both are thin and exist only because they are sheltered by fjords (1978 satellite image, Swithinbank 1988, p. B110, fig. 81). Meanwhile Müller Ice Shelf has broken up.

Rapid breakup of *Wordie Ice Shelf* occurred a few years earlier (Vaughan 1993). It appears that the present time is a time of rapid disintegration of ice shelves (Vaughan et al. 2001; Rott et al. 2002; MacAyeal et al. 2002).

The breakup of the northerly ice shelves is a consequence of climatic warming. The triggers of breakup and the physical mechanisms of ice shelf disintegration are not known yet and are an objective of current research. Crevasse formation and propagation, as well as iceberg calving mechanisms play a role, while surficial meltwater ponds and streams, as observed on Larsen Ice Shelf, are indicators of a warm climate.

(D.2) Latitude Row 67-72.1°S: Maps from GEOSAT and ERS-1 Radar Altimeter Data

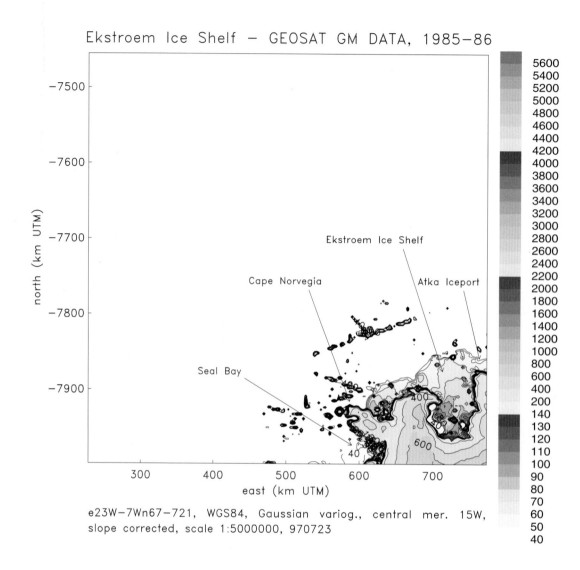

Ekstroem Ice Shelf — GEOSAT GM DATA, 1985—86

e23W—7Wn67—721, WGS84, Gaussian variog., central mer. 15W, slope corrected, scale 1:5000000, 970723

Map m15we23w-7wn67-721 Ekström Ice Shelf

This map, showing the northwestern section of the coast of western Queen Maud Land, is best viewed together with the adjacent map, m3we11w-5n67-721 Fimbul Ice Shelf, to get a complete picture of the ice shelves from *Cape Norvegia* to *Fimbul Ice Shelf*, all north of *Princess Martha Coast*. Princess Martha Coast extends from the terminus of Stancomb-Wills Glacier at 20° W to 5° E. *Cape Norvegia* separates the northern ice shelves from the southerly-extending *Riiser-Larsen Ice Shelf* (with Seal Bay at its northern end), and, in a larger geographic sense, the Atlantic sector of the circum-Antarctic ocean to the east and north and the Weddell Sea to the west and southwest. Between *Ekström Ice Shelf* and Cape Norvegia, offshore of the tip of the large peninsula, is the small *Quar Ice Shelf*. Ekström Ice Shelf is very well studied, mostly by German Antarctic Expeditions, be-

Ekstroem Ice Shelf — ERS1 DATA, 1995

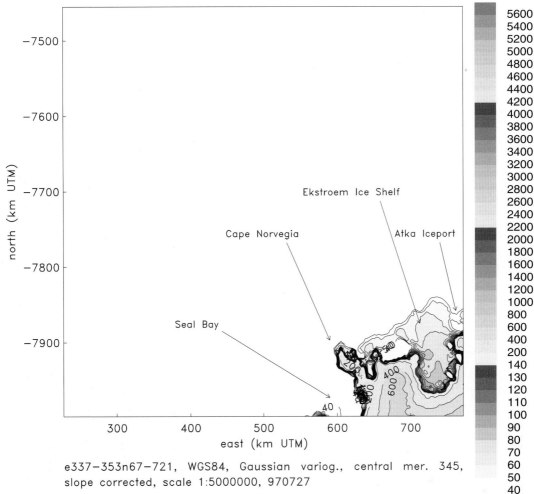

e337-353n67-721, WGS84, Gaussian variog., central mer. 345,
slope corrected, scale 1:5000000, 970727

cause the German overwintering station Georg von Neumayer is situated on this ice shelf in the vicinity of Atka Bay (approximately 24 km from the edge of the ice shelf). Already the second station has been built, since the first station was lost in the ice (intentionally, because of its construction in the ice, the station was sinking each year, and given up after about 15 years of use).

The southerly continuation of Riiser-Larsen Ice Shelf is seen on map m357e339-15n71-77 New Schwabenland, the following section on m333e315-351n71-77 Riiser-Larsen Ice Shelf.

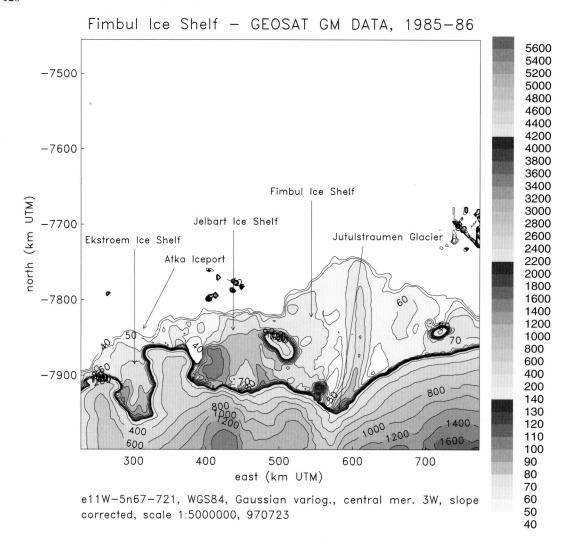

Fimbul Ice Shelf – GEOSAT GM DATA, 1985–86

e11W–5n67–721, WGS84, Gaussian variog., central mer. 3W, slope corrected, scale 1:5000000, 970723

Map m3we11w-5n67-721 Fimbul Ice Shelf

This map shows the coast of western Queen Maud Land, called *Princess Martha Coast*, which has a mountainous hinterland and is bounded by large ice shelves - *Ekström Ice Shelf, Jelbart Ice Shelf* and *Fimbul Ice Shelf*. Princess Martha Coast extends from the terminus of Stancomb-Wills Glacier at 20° W to 5° E. The name Princess Martha Coast was originally given only to the section of coast near Cape Norvegia, which was discovered by Capt. Hjalmar Riiser-Larsen and charted from the air during that expedition in 1930.

The peninsula west of *Ekström Ice Shelf* (center coordinate (71° S/8° W) is Sørasen Ridge; the peninsula separating Ekström Ice Shelf and Jelbart Ice Shelf is Halvfarryggen Ridge ("Ridge" is superfluous in the name, as "ryggen" means "ridge"; center coordinate (71° 10'S/6° 40'W), a broad snow-covered ridge; a large island and several smaller islands and snow ridges are situated between Jelbart Ice Shelf and Fimbul Ice Shelf.

West of Ekström Ice Shelf is the small *Quar Ice Shelf*, located at center coordinate (71° 20'S/ 11° W) and visible in the map; Quar Ice Shelf is separated from the large Riiser-Larsen Ice Shelf by

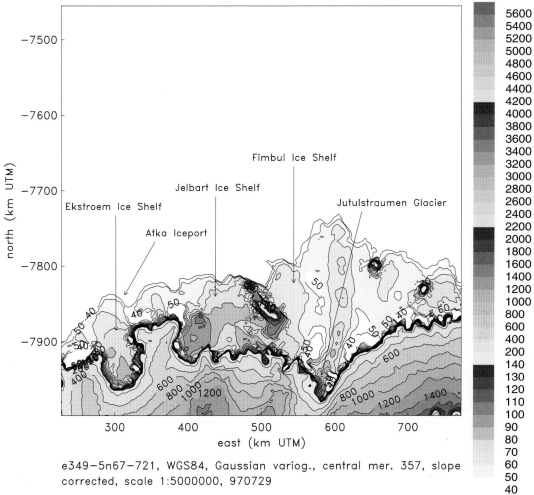

Fimbul Ice Shelf — ERS1 DATA, 1995

e349-5n67-721, WGS84, Gaussian variog., central mer. 357, slope corrected, scale 1:5000000, 970729

the peninsula with *Cape Norvegia* at its head. The westernmost section of the ice shelves is mapped on sheet m345e337-353n67-721 Ekström Ice Shelf.

Ekström Ice Shelf is named for Bertil Ekström, a Swedish engineer who drowned when the weasel he was driving plunged off the edge of Quar Ice Shelf on 24 February 1951, *Quar Ice Shelf* is named for Leslie Quar, a British radio mechanic and electrician, and *Jelbart Ice Shelf* is named for Australian observer John E. Jelbart, both drowned on 24 February 1951 in the same accident as Ekström. This occurred during the Norwegian-British-Swedish Antarctic Expedition 1949–52, led by John Giaever, after whom the ridge south of Jelbart Ice Shelf is named.

The South African Sana Station is located at the western edge of Fimbul Ice Shelf; the German Overwintering Station Georg von Neumayer is located near Atka Iceport on the Ekström Ice Shelf. The research stations Drushnaya 3 (USSR) and Maudheim (Norwegian) were built on the Quar Ice Shelf, the small ice shelf west of Ekström Ice Shelf.

The most impressive feature in the map area is *Jutulstraumen Glacier* (see also detail map in section (F.3)), whose tongue extends 240 km north into the Fimbul Ice Shelf. Jutulstraumen Glacier is the largest glacier between longitude 15° E and 20° W, it drains an area of 124.000 km^2, has a discharge of 12.5 km^3 at 72° 15'S and an ice velocity at the front of 1 km a^{-1}. It follows a geologic trough, the northern part of which is clearly seen on this map; the southerly continuation, the westerly branch in Penck Trough and the easterly branch that joins from east of the Neumayer Cliffs, and the upper parts of the drainage basin are mapped on m357e339-15n71-77 New Schwabenland and described there in more detail. At the western edge of the terminus of Jutulstraumen Glacier are Passat Nunatak and Boreas Nunatak (220 m) on Giaever Ridge, the eastern edge is Roberts Knoll on Ahlmann Ridge.

To the south, the land of this section of Princess Martha Coast ascends to (W to E) *Ritscher Upland*, *Borg Massif* (south of Giaever Ridge, which is south of Jelbart Ice Shelf), and the *Mühlig-Hoffman Mountains* south of eastern Fimbul Ice Shelf.

Features in this map sheet appear more clearly on the GEOSAT map. Ice elevations in the area appear to have decreased from GEOSAT 1985/86 to ERS-1 1995 by possibly 5 m, as deduced from size reductions of the higher-elevation contours on the ice shelves. Most noteworthy, the tongue of *Jutulstraumen Glacier* has a connected center that is 20 m above the level of the surrounding Fimbul Ice Shelf, in 1985/86 the center elevation is 70 m (above WGS 84) with spots above 80 m, in 1995 the 70 m contour is exceeded only in a few areas. The areas below 40 m and 50 m on the ice shelf may have changed less, so it appears that the offshore part of Jutulstraumen Glacier lost elevation, and hence mass (since the ice shelves are afloat). These observations need to be adjusted for a general offset between elevations from GEOSAT and ERS-1 observations, such an offset has not been calculated (to our knowledge), however, it is on the order of decimeters (less than 72 cm or 22 cm, depending on accuracy values, see (C.1.1)), whereas the observations here are on the order of meters.

Princess Astrid Coast – GEOSAT GM DATA, 1985–86

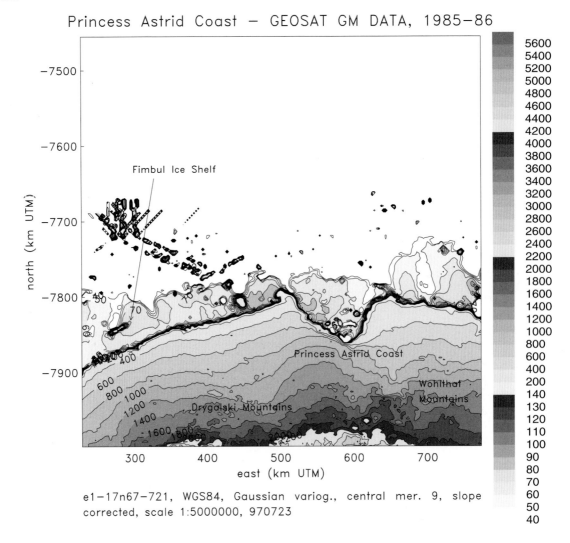

e1–17n67–721, WGS84, Gaussian variog., central mer. 9, slope corrected, scale 1:5000000, 970723

Map m9e1-17n67-721 Princess Astrid Coast

Princess Astrid Coast is a part of the coast of Queen Maud Land, between Fimbul Ice Shelf and Princess Ragnhild Coast (adjacent to the east). The coastline is bounded by 50–70 km wide ice shelves, most of which are unnamed. *Princess Astrid Coast* extends from 5° E to 20° E and is named for Princess Astrid of Norway, this part of the coastline was discovered by Capt. H. Halvorsen of the ship *Sevilla* in March 1931. The name *Princess Ragnhild Coast* applies to the coast of Queen Maud Land between 20° E and Riiser-Larsen Peninsula at 34° E. West of Princess Astrid Coast is *Princess Martha Coast*, which extends from 5° E westward to the terminus of Stancomb-Wills Glacier at 20° W. This coast is also bouned

by ice shelves, and ice cliffs in the area are 20–35 m high. The name Princess Martha Coast was originally given only to the section of coast near Cape Norvegia, which was discovered by Capt. Hjalmar Riiser-Larsen and charted from the air during that expedition in 1930.

About 100–150 km south of the coast, several mountain ranges are located (*Wohlthat Mountains, Drygalski Mountains, Gjelsvik Mountains, Mühlig-Hofmann Mountains, Orvin Mountains,* and *Hoel Mountains*), which create a fairly steep slope toward the coast (1000 m in 100 km, corresponding to a slope of 0.57°) on the large scale and deflect the ice flow of the inland ice locally.

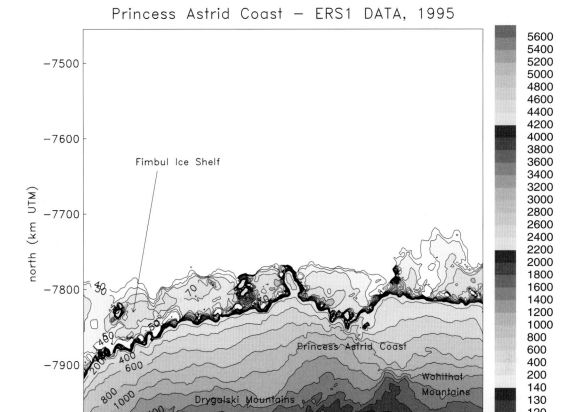

Princess Astrid Coast — ERS1 DATA, 1995

e1—17n67—721, WGS84, Gaussian variog., central mer. 9, slope corrected, scale 1:5000000, 970729

As is typical for GEOSAT data, there are lots of artefacts and errors mapped off the coast, fortunately, in this sheet not obliterating the mapped ice shelves. Several small glaciers drain from the coastal area to the ice shelves. The area is shown, in part, on an annotated NOAA-7 AVHRR image mosaic (IfAG 1982; see Swithinbank 1988, p. B93).

In the northeast of the map are the Schirmacher Hills. The *Schirmacher Hills* (center coordinate (70° 45'S/11° 40'E)) are an 18 km long, narrow range of coastal mountains, located 60 km north of the Humboldt Mountains along the coast of Queen Maud Land, that have received special attention in research since their discovery by the German Antarctic Expedition led by Ritscher in 1938–1939. The Schirmacher Hills are named after Richard Heinrich Schirmacher, pilot of Borea, one of the expedition's seaplanes. Numerous meltwater

ponds named Schirmacher Ponds are found in the Schirmacher Hills thus the earlier names "Schirmacheroase" (oasis, because the meltwater ponds indicate a local climatic optimum) or Schirmacher Seenplatte.

The Schirmacher "Oasis" is a climatic outlier, as indicated by the abundance of ablation features, which include hundreds of melt ponds in summer, raging torrents of water flow in shallow ravines over the ice shelf (Swithinbank 1988, p. B86), and tidal freshwater sea lakes. The Russian Station Novolazarevskaya is located in the eastern part of the Schirmacher Hills. An ice stream passes around the Hills at a velocity of 324 m a^{-1} (Swithinbank 1966; Kruchinin et al. 1967). The Schirmacher Hills sit right on the (estimated) grounding line of the adjacent ice shelf. There are ice dolines on the ice shelf.

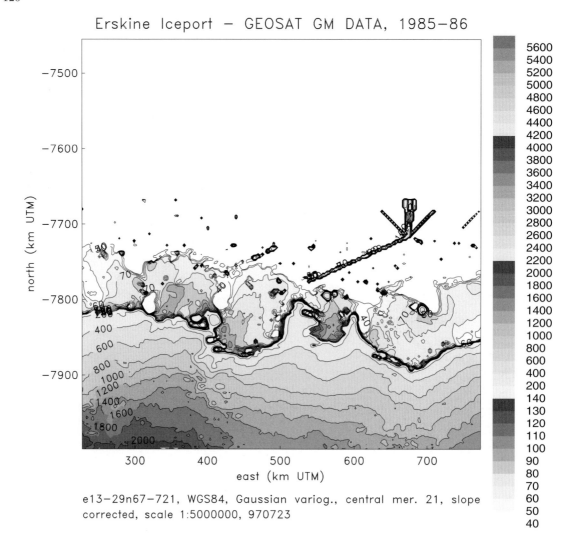

Erskine Iceport — GEOSAT GM DATA, 1985–86

e13–29n67–721, WGS84, Gaussian variog., central mer. 21, slope
corrected, scale 1:5000000, 970723

Map m21e13-29n67-721 Erskine Iceport

This map sheet shows a stretch of Princess Ragn-hild Coast (in the eastern part of our map) and Princess Astrid Coast (in the western part of our map), Queen Maud Land, with several extensive ice shelves. The name *Princess Ragnhild Coast* applies to the coast of Queen Maud Land between 20° E and Riiser-Larsen Peninsula at 34° E, the ocast was discovered by Riiser Larsen and Captain Nils Larsen in aerial flights from the ship *Norvegia* in February 1931 and is named for Princess Ragn-hild of Norway (Alberts 1995, p. 592). *Princess Astrid Coast* extends from 5° E to 20° E and is named for Princess Astrid of Norway, this part of the coastline was discovered by Capt. H. Halvorsen

of the ship *Sevilla* in March 1931. It is interesting to note that in Queen Maud Land the boundaries of the coasts have been set simply by degrees of longitude, whereas in most other areas of Antarctica distinct boundary points such as capes have been determined, the former is an indication that the boundaries were defined much later than the name was applied to the coastal stretches.

The coastal area and the ice shelves are mapped clearly on the ERS-1 map, and with errors on the GEOSAT map. The errors along the coastline and offshore are typical for GEOSAT altimeter data. The best procedure is to work geographically with

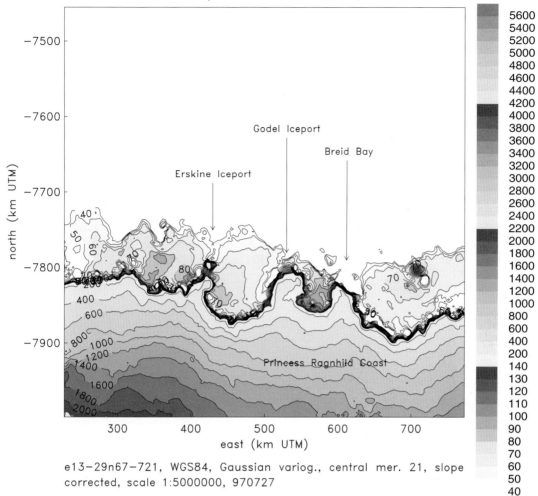

Erskine Iceport — ERS1 DATA, 1995

e13–29n67–721, WGS84, Gaussian variog., central mer. 21, slope corrected, scale 1:5000000, 970727

the ERS-1 map, then compare coastal features and extent of ice shelves on the GEOSAT map. Since the errors are clearly identifiable as such (little black dots, concentric, dense contour lines around small spots, and errors in line in track directions and offshore), the maps are still useful, albeit unpleasing.

The fact that the coast and its ice shelves are so well-mapped on the ERS-1 map, but features in the area inculding five promontories are even unnamed on the USGS satellite image map (Ferrigno et al. 1996), motivates closer investigation of the mapped geographic features.

There are three prominent peninsulas, with bays or iceports offshore of their heads: Erskine Iceport, Godel Iceport, and Breid Bay. The Belgian Station Roi Baudoin is located in Breid Bay at ≈25° E. Be-

tween the iceports lie sectors of ice shelf. A smaller promontory extends north at about 310.000 E separating the ice shelf into two sectors.

For lack of names, let us imagine labeling the sectors of the ice shelf A (area west of small headland), B (area between small headland and Erskine Iceport headland), C (area between Erskine Iceport headland and Godel headland), D (area between Godel headland and Breid Bay), and E (east of Breid Bay).

Comparison between GEOSAT and ERS-1 maps indicate that ice in A stayed about the same from 1985/86 to 1995, the glacier tongue in B decreased somewhat (5–10m) in elevation and the shelf receded or calved off into a bay ≈ (-7760.000 N/350.000 E).

There is a small outlet glacier flowing northwest from the headland (Erskine) at its western side, and a larger one (or two) on its eastern side (decreased in elevation), the latter may correspond to a trough in the coastal topography. There is also a glacier off the north tip of the headland with Godel Iceport, and higher ice elevations in area D, possibly ice rises in area E near (-7800.000 N/700.000 E) with lower elevations close to the coast.

The inland territory increases in relief towards the south, because the *Sør Rondane Mountains* lie south of Princess Ragnhild Coast (21° – 29° E), and the *Wohlthat Mountains*, *Petermann Ranges*, *Weyprecht Mountains* and *Payer Mountains* lie south of Princess Astrid Coast. The mountain ranges extend beyond 72.1° S and are described for the map adjacent to the south (m21e3-39n71-77 Sør Rondane Mountains).

The Wohlthat Mountains have been studied during the GEISHA expedition 1987/88 (German expedition into the Shackleton Range, 1987/88, see, for example Kleinschmidt et al. 1988, Kleinschmidt and Buggisch 1993, Buggisch and Kleinschmidt 1999; see also report on the EU-ROSHACK expedition, Tessensohn et al. 1999), for their role in the Gondwanaland breakup. In this context, the Sør Rondane Mountains have also been investigated. The highest peak in the Sør Rondane Mtns. is Isachsen Mountain (3425 m) at (72° 11'S/26° 15'E), near it is the named glacier Byrd Breen. The Japanese Asuka Station is in the northern Sør Rondane Mountains near Mt. Romnaes (1500 m).

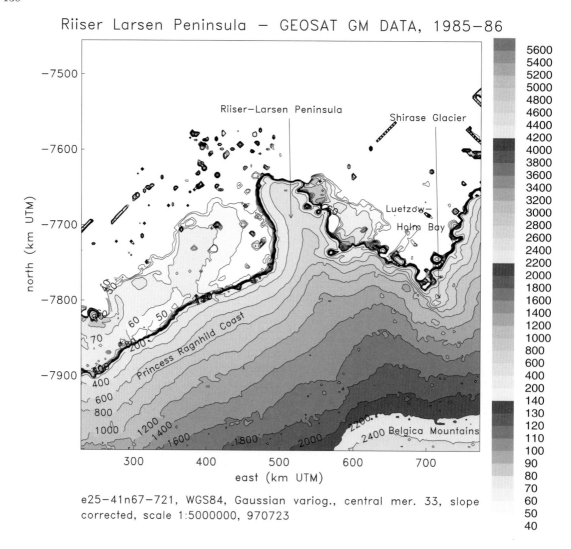

Riiser Larsen Peninsula − GEOSAT GM DATA, 1985−86

e25−41n67−721, WGS84, Gaussian variog., central mer. 33, slope
corrected, scale 1:5000000, 970723

Map m33e25-41n67-721 Riiser-Larsen Peninsula

The *Riiser-Larsen Peninsula* is the large peninsula in the center of the map. To the west of it lies Princess Ragnhild Coast. *Princess Ragnhild Coast* is the coast of Queen Maud Land between 20° E and 34° E (Riiser Larsen Peninsula). It was discovered by Riiser-Larsen and Captain Nils Larsen on flights from the ship *Norvegia* in February 1931 and named for Princess Ragnhild of Norway. The westernmost part of the map shows parts of *Princess Astrid Coast*, which extends from 20° E to 5° E, is named for Princess Astrid of Norway and was discovered by Captain Halvorsen in March 1931. To the east of Riiser-Larsen Peninsula (34° E) is *Prince Harald Coast*, which extends

to *Luetzow-Holm Bay* (40° E), adjacent is *Prince Olav Coast* (40° E to 44° 38' E).

The coastline on our map is farther inland than on the National Geographic map (National Geographic Society 1992, p. 102) (the coastline is identified by closely spaced contour lines between 200 and 60 m). Within the range of the width of this margin the coastline is probably correct (otherwise it would be fuzzy). The contours offshore of the peninsula may correspond to shelf ice, on this as on many other maps their location matches the location of ice shelves. Dots, stripes and dots in stripes are likely caused by retracking errors, or

Riiser Larsen Peninsula — ERS1 DATA, 1995

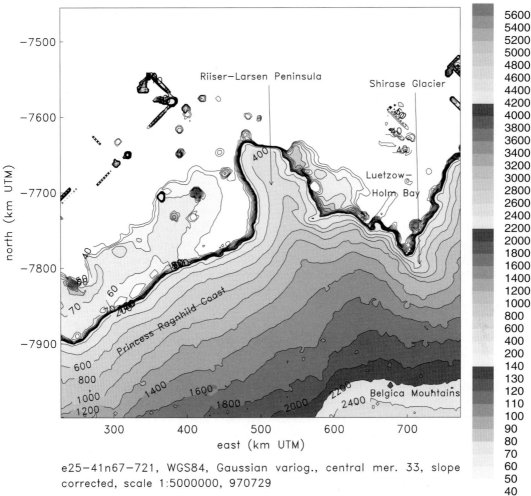

e25-41n67-721, WGS84, Gaussian variog., central mer. 33, slope
corrected, scale 1:5000000, 970729

by errors in the altimetry ("snagging" of the signal on existing islands that leave a trace along the
groundtrack; cf. Herzfeld et al. 1993). The contours offshore on the eastern side of the peninsula
in Luetzow-Holm Bay also match shelf-ice locations.

Extending inland southsouthwestward from the
eastern Luetzow-Holm Bay (Havsbotn) is a large
trough visible. This is the valley of *Shirase Glacier*.
Shirase Glacier is the largest ice stream west
of Rayner Glacier (see maps m45e37-53n67-721
Prince Olav Coast and m45e37-53n63-68 Casey
Bay on Rayner Glacier), it drains an area of
165.000 km² and extends 500 km inland (quoted
after Swithinbank 1988, p. B79), the glacier follows a 9 km wide valley down from the plateau
of the inland ice, this subglacial trough extends
200 km inland from the coast (Korotkevich et al.

1977) and continues in a submarine canyon (the
bed of the ice sheet is close to sea level). Shirase
Glacier is larger than Rayner Glacier and has a
larger drainage basin. In the LANDSAT image,
Shirase Glacier has a floating ice tongue that extends 65 km from the grounding line into Luetzow-
Holm Bay, the tongue flows between Padda Island
and a small island to the east.

Shirase Glacier is the fastest (surveyed as of 1988)
glacier in Antarctica, measurements of velocity
vary: 2500 m a^{-1} at the ice front according to
Nakawo et al. (1978), 2900 m a^{-1} in some locations according to Fujii (1981). The glacier was
already mapped by Hansen (1946) from 1936–37
aerial photography. Estimates of the flux of ice
vary almost by a factor of two, 7.4 Gt a^{-1} (8.1
km³ a^{-1}) according to Nakawo et al. (1978) vs. 14
Gt a^{-1} (15.4 km³ a^{-1}) according to Fujii (1981) —

differences in the estimates are caused by different assumptions on ice thickness and the correctness of the maps in Hansen (1946). A detail map of Shirase Glacier is given in section (F.4).

Mountainous terrain is characterized by large changes in elevation over short distances — this is termed "high relief energy" in geography and is a better characterization than absolute elevation. On the altimetry-derived maps, individual mountains are not outlined because of the relatively low resolution of the altimeter compared to the size (area) of a mountain. The relief energy, however, still translates into rough terrain at this scale and type of mapping. Consequently, the rougher contours on the maps are indicative of mountainous terrain. For example, the rough contour lines extending north from the label 2400 correspond to the *Queen Fabiola Mountains* (also called *Yamato Mountains*, located at 35° E, 73°–71° S according to the USGS Satellite Image Map (Ferrigno et al. 1996)). The highest elevations in the southern part of the map correspond to the *Belgica Mountains*. The highest peak in the Belgica Mountains is Mt. Victor (2588 m), the highest peak in the Queen Fabiola Mountains is Mt. Derom (2390 m). The absolute elevation of a mountain range is not tracked correctly by the altimeter, and, in addition, the kriging operation used for interpolation is an averaging operation; consequently, the elevations on the maps are usually lower than the highest peaks in the mountain range. On this map, however, the difference appears to be not too bad.

The Queen Fabiola Mountains contain extensive blue-ice areas (areas of bare glacier ice surfaces) around nunataks and associated glacial debris. Since 1969, Japanese scientists have collected 4813 meteorites within these blue-ice areas, which is about 25% of the world-wide collection of meteorites (Swithinbank 1988, p. B82, fig. 60) and the largest concentration of meteorites in Antarctica. This concentration may have been caused by ablation and consequent exposure, or by a concentration of falling meteorites.

Comparison with the ERS-1 map shows that the main geographic features are mapped as on the GEOSAT map. Contours are generally smoother, and the contours in the areas identified as mountainous are rougher than contours elsewhere on this map also. The coastline appears a lot smoother on the ERS-1 map than on the GEOSAT map. The outlines of the Ice Shelf have not changed significantly between 1985 and 1995. Notably, a rectangular contour line is extending northeastward into Luetzow-Holm Bay on both maps, this is an ice tongue extending from a glacier, which is also indicated on the USGS map. Higher shelf ice elevations offshore of two smaller bays on the eastern shore of Riiser-Larsen Peninsula are probably associated with glaciers flowing out of the valleys of the Peninsula (note the corresponding valleys along the eastern margin of the Peninsula). The Japanese Station Syowa is located on the coast (on Ongul Island, see Swithinbank 1988, fig. 59, p. B81) near the eastern edge of the map (and, of course, not visible).

A satellite image map of the eastern part of Luetzow-Holm Bay is pictured in Swithinbank (1988, fig. 59, p. B81). The ice tongue on our maps is not identical with the ice tongue of Shirase Glacier (which extends into the fast ice of Luetzow-Holm Bay at least 50 km northeast).

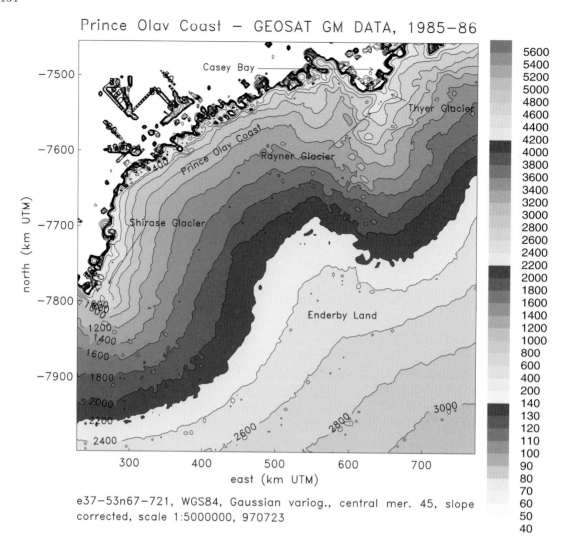

Prince Olav Coast — GEOSAT GM DATA, 1985–86

e37–53n67–721, WGS84, Gaussian variog., central mer. 45, slope corrected, scale 1:5000000, 970723

Map m45e37-53n67-721 Prince Olav Coast

This map shows Queen Maud Land (Prince Harald Coast, Prince Olav Coast) and Enderby Land up to an elevation of 3000 m. The western part of this map shows Luetzow-Holm Bay, which gives a good idea of the amount of overlap between maps. The central and eastern parts of the Shirase Glacier drainage are seen on this map, the western and central parts on map m33e25-41n67-721 Riiser-Larsen Peninsula.

Prince Olav Coast is the coast of Queen Maud Land between 44° E and 44° 38'E, Shinnan Glacier (67° 55'S/44° 38'E) marks the eastern limit of this coast as well as the boundary between *Queen*

Maud Land and *Enderby Land*. Prince Olav Coast was discovered by Captain Hjalmar Riiser-Larsen in January 1930 on a flight from the ship *Norvegia* and is named for Prince Olav of Norway. The features offshore Prince Olav Coast are at least in part artefacts.

Further east than Shinnan Glacier is *Tange Promontory* at (67° 27'S/46° 45'E) or (-7520.000 N/600 on Enderby Land, directly east of the promotory lies Casey Bay. For description of the coastal features, see also map m45e37-53n63-68 Casey Bay. The Russian Station Molodezhnaya is located west of Tange Promontory. The small glacier at the foot

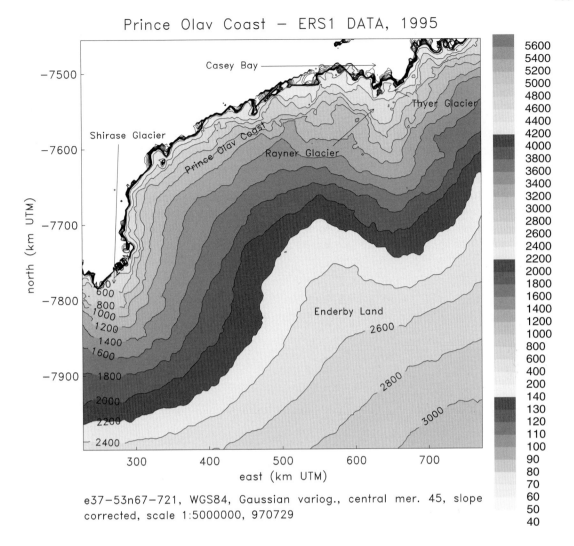

Prince Olav Coast — ERS1 DATA, 1995

e37−53n67−721, WGS84, Gaussian variog., central mer. 45, slope corrected, scale 1:5000000, 970729

of Tange Promontory has no noticeable continuation inland, whereas Rayner Glacier drains a trough that extends at least 200–300 km inland in a southerly direction. Thyer Glacier joins Rayner Glacier shortly before reaching Casey Bay; a small trough of Thyer Glacier is indicated on the map at (-7520.000 N/630.000 E). The rough terrain south of Casey Bay corresponds to the Nye Mountains.

The coast of Queen Maud Land west of Prince Olav Coast is *Prince Harald Coast*, which includes Luetzow-Holm Bay and extends west to *Riiser-Larsen Peninsula*. Prince Harald Coast was discovered by V. Widerøe, N. Romnaes and Ingrid Christensen of the Lars Christensen Expedition 1936–1937 from the air and is named after the infant son of the crown prince of Norway at the time (Alberts 1995).

Notably, the coast of *Enderby Land* does not have specific coastal names (except for the section of Kemp Coast, see map m57e49-65n67-721 Kemp Coast). Enderby Land extends from Shinnan Glacier (67° 55'S/44° 38'E) to William Scoresby Bay (67° 24'S/59° 34'E).

The southern reaches of the large drainage basin of *Rayner Glacier* can be seen in the two southerly adjacent maps m45e27-63n71-77 Belgica Mountains and m45e27-63n75-80 Valkyrie Dome. There is an ice divide that separates the basin of Rayner Glacier (E) from that of Shirase Glacier (W) and extends to the 2400-m contour (at least). According to McIntyre (after Swithinbank 1988), the drainage basin of Rayner Glacier is 118.000 km^2, corresponding to a 200 km inland extension (smaller than observed from the Atlas maps). Thyer Glacier is a smaller glacier that enters

Rayner Glacier from the east. Rayner Glacier is 11 km wide at its confluence with Thyer Glacier. The last 20 km of Rayner Glacier are afloat. The velocity of Rayner Glacier is 861 m a^{-1} near the 500 m surface elevation, mass flux is 10.4 Gt a^{-1} (11.4 km^3 a^{-1}).

An interesting feature is the following: Hays Glacier, located near Tange Promontory, is joined with the valley of Rayner Glacier through a valley that extends inland at elevations of 200–400 m below sea level for more than 100 km. Another valley connection exists between Assender and Molle Glaciers that makes Tange Promontory an "island". A third valley connects "Campbell Glacier" and Hays Glacier at 200–400 m below sea level.

The same features may be identified on the ERS-1 map, which has generally smoother contours and no offshore artefacts. As is best seen in the LANDSAT satellite image (fig. 57, p. B78) in Swithinbank (1988), the coast along Casey Bay and Amundsen Bay (see map m57e49-65n63-68 Napier Mountains) is very mountainous and rugged (which causes a rough coastline in our altimetry-based maps). Some of the area west around Tange Promontory and west of it is seen on the LANDSAT image in fig. 58, p. B80 (Swithinbank 1988), this coast is characterized by a steady descent from the inland ice.

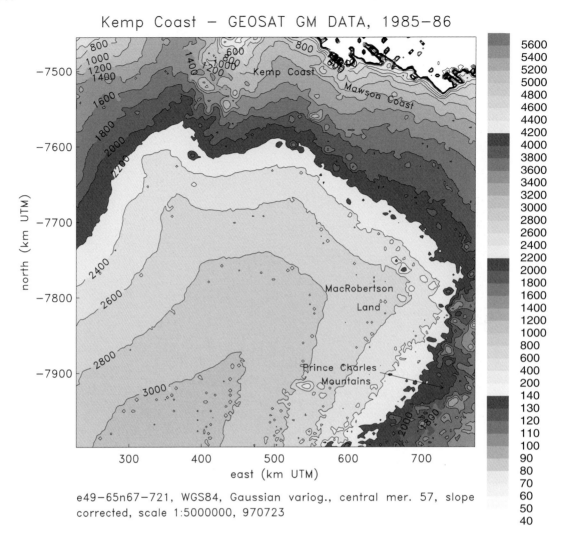

Kemp Coast – GEOSAT GM DATA, 1985–86

e49–65n67–721, WGS84, Gaussian variog., central mer. 57, slope corrected, scale 1:5000000, 970723

Map m57e49-65n67-721 Kemp Coast

Map m57e49-65n67-721 Kemp Coast shows the land south of *Kemp Coast* (the northwestern part of Kemp Coast and Mawson Coast (West) is shown on map m57e49-65n63-68 Napier Mountains, where the coastal features are described), a section of Mawson Coast and MacRobertson Land, dipping in the east to Prince Charles Mountains, the range that borders Lambert Glacier on the west. The area in the southeastern corner of the map has a much "noisier" appearance. This is caused by the rugged terrain of the mountains forming the edge of the Lambert Glacier trough. Although it may be difficult to identify individual mountains, it is possible to distinguish rugged terrain from smooth topography in the maps. Robert Glacier has a 150 km long drainage extending southsouthwest from (-7450,000 N/480,000 E). All these geographic features are shown on both the GEOSAT and the ERS-1 maps.

Kemp Coast extends from William Scoresby Bay (67° 24'S/59° 34'E) to the head of Edward VIII Bay (66° 50'S/56° 24'E); it is named for a British Whaling Captain, Peter Kemp, who discovered land here in 1833 (Alberts 1995). Offshore of the rocky, 8 km long and 6 km wide bay are William Scoresby Archipelago and other islands, the GEOSAT map shows offshore reflections, likely

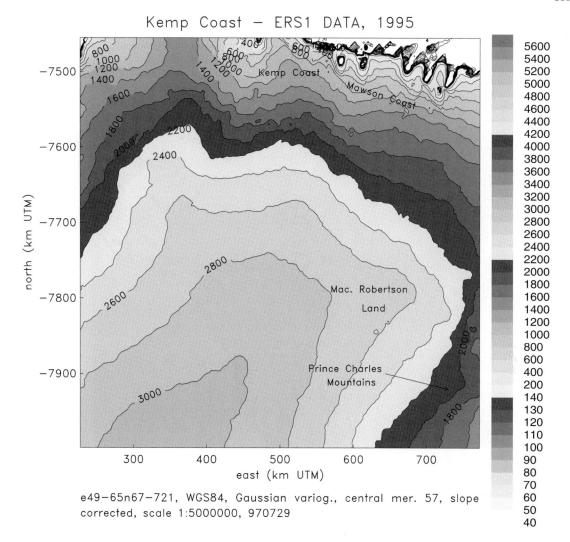

Kemp Coast – ERS1 DATA, 1995

e49–65n67–721, WGS84, Gaussian variog., central mer. 57, slope corrected, scale 1:5000000, 970729

of these islands. The bay was discovered in 1936 by William Scoresby, who also discovered Edward VIII Bay in 1936 and named it after the King of England, in part of the bay is Edward VIII Ice Shelf.

Mawson Coast is the coast of Mac Robertson Land between William Scoresby Bay and Murray Monolith (67° 47'S/66° 54'E) (see map m57e49-65n63-68 Mawson Coast).

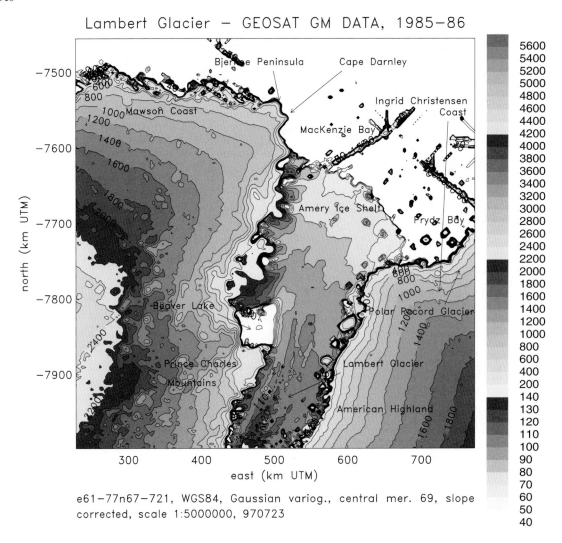

Lambert Glacier – GEOSAT GM DATA, 1985–86

e61−77n67−721, WGS84, Gaussian variog., central mer. 69, slope corrected, scale 1:5000000, 970723

Map m69e61-77n67-721 Lambert Glacier

Lambert Glacier/Amery Ice Shelf is the largest ice-stream/ice-shelf system in East Antarctica, it drains about 10% of the East Antarctic Ice Shield (Drewry 1983). Lambert Glacier is the first area that we investigated with the method presented here (and used to create the entire Antarctic Atlas) — it was this map that was used to demonstrate that mapping ice surfaces accurately from altimeter data is possible at all, using geostatistical methods adapted to the task. The most imminent scientific question was: "Can the grounding line be determined from radar altimeter data?"

Why is the determination of the grounding line such an important question that it is worth de-

signing a new method for data analysis for its solution?

Changes in climate lead to changes in ice elevation and in ice mass — albeit with a large time lag, which depends on the size of the glacier. Good candidates to investigate changes in the Antarctic Ice Sheet are the large outlet glaciers and ice-stream/ice-shelf systems. Not only is the Lambert Glacier/Amery Ice Shelf system large, it is also located relatively far north, and hence more sensitive to warming than more southerly ice-stream/ice-shelf systems such as the Ross Ice Shelf and the Filchner and Ronne Ice Shelves and ice streams leading into those. In a warming climate, northerly

Lambert Glacier – ERS1 DATA, 1995

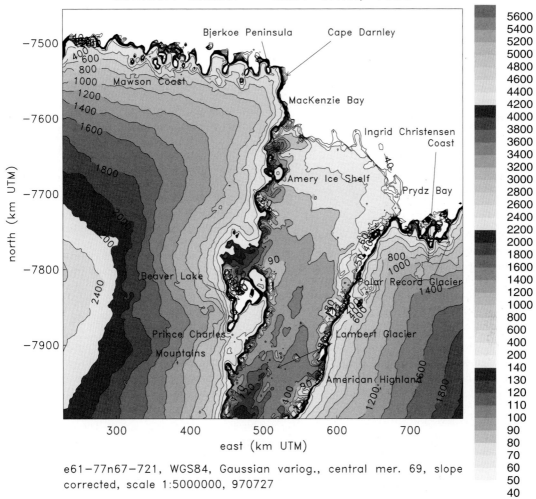

e61–77n67–721, WGS84, Gaussian variog., central mer. 69, slope corrected, scale 1:5000000, 970727

Antarctic ice shelves will disintegrate first, and some along the Antarctic Peninsula have already disintegrated in recent years (see chapter (A)).

Having established that as a large and northerly ice-stream/ice-shelf system Lambert Glacier/Amery Ice Shelf is an important objective of study, the next question is: "How do we investigate possible advance or retreat best?"

The first place that comes to mind as a candidate for monitoring advance or retreat is the front of the glacier. But while the front is a good place for an alpine valley glacier, the front of an ice shelf changes more erratically, large icebergs may calve off suddenly, or there may be long times of slow advance, or small icebergs may break off frequently. For example, in a single calving event (in 1963) 11000 km^2 [that is 110 km by 100 km

in area!] of ice broke off the front of Amery Ice Shelf (Ledenev and Yevdokimov 1965; Swithinbank 1969). Changes in the ice front of Amery Ice Shelf from 1936 to 1965 are described in Budd (1966). At present, a large rift is developing and extending in the front of Amery Ice Shelf, and almost certainly a huge iceberg will calve off in the next one or two years.

The *grounding line*, defined as the place where the glacier becomes afloat, is a better place to monitor advance and retreat. Small changes in ice thickness translate into large changes in grounding line position (for geometrical reasons because of the small surface slope typical of Antarctic ice streams). The principles are the following: Based on a model that assumes idealized bedrock and perfect plasticity of ice, Weertman (1974) showed that the surface slope of an ice sheet decreases with the transition from grounded to floating ice. Using this result, a break in slope may be taken as a grounding-line indicator.

A second indicator of the grounding-line position is the surface roughness of the glacier and the ice stream: Sliding over the rough glacier bed induces a rougher surface topography than floating on the smooth water surface; this means that grounded glacier ice is rough and floating ice-shelf ice is smooth.

The grounding line is a feature suitable to monitor changes – if it can be observed.

A first important step in mapping is determination of an appropriate variogram that represents surface roughness in the grounding zone. The approximate location of the grounding zone is known from field work (Budd et al. 1982). As described in section (C.3) on variography, a detailed structural analysis of Lambert Glacier grounding zone was undertaken. Because ice-stream/ice-shelf systems are areas of major interest, and grounding zones have an intermediate, but low, surface roughness, the Lambert Glacier variogram was found suitable to be selected for mapping the Atlas.

The contours below 140 m were chosen at a 10 m interval to map the ice surface of the ice-stream/ice-shelf system such that it is possible to distinguish between rough topography typical of grounded ice and smooth topography typical of floating ice, which corresponds roughly to the 100 m contour. (Above 140 m, 200, 400 m etc. steps of 200 m are used to map any other elevation occurring in Antarctica.) In addition, the 100 m contour defines a break in slope suggestive of the grounding line. (The fact that the 100 m contour is identified is a coincidence, after producing maps with 95, 96, 97, 98, 99, and 100, 101, 102, 103 m contour lines, we found that the 100 m contour best satisfies the conditions of identifying (a) the break in slope and (b) the boundary between rough and smooth that are associated with the position of the grounding line.)

Indeed, the grounding zone can be mapped with our method (Herzfeld et al. 1993), which constitutes a significant advance in the use of satellite radar altimeter data for investigation of Antarctic outlet glaciers and ice-stream/ice-shelf systems. Monitoring changes in such systems has also become possible with the geostatistical method

for mapping from satellite radar altimeter data (Herzfeld et al. 1997). A comparison of 1978 and 1987/89 satellite data maps demonstrated that Lambert Glacier had advanced ≈10 km in 10 years — while some other glaciers had retreated. A special section in this Atlas is dedicated to monitoring changes in the Lambert Glacier / Amery Ice Shelf System (chapter (E)).

Most geographic features of the Lambert Glacier / Amery Ice Shelf System are also represented on the (larger) special map "Lambert Glacier/Amery Ice Shelf System" (section (F.5)), which has been constructed because the regular Atlas sheet does not contain upper Lambert Glacier south of 72.1° S and important parts of the drainage basin.

There are a few complicating steps associated with the identification of grounding lines. One is that the apparent grounding line is located a little bit upglacier from the grounding line at the bottom of the glacier simply for geometric reasons.

The next issue is the discrimination between "grounding line" and *"grounding zone"*. Usage of the term "grounding line" assumes that a clear line runs more or less straight across the glacier, identifying the boundary between grounded ice on one side and floating ice on the other. On Lambert Glacier (and many other glaciers) the apparent grounding line meanders irregularly across the glacier, as is obvious in the maps. The patchy areas above the 100-m contour farther north suggest the presence of seafloor shoals intersecting the bottom of the ice shelf immediately downstream. The entire area, from north coordinate -7870000 to -7920000 (on the western side of the glacier), or to -7950000 (on the eastern side of the glacier) can probably be considered the grounding zone.

During an Australian expedition to Lambert Glacier, Budd et al. (1982) identified a position at N -7900000/E 480000 as on the grounding line, using optical leveling. This position is in the area identified as the grounding zone by us, which gives assurance that we have identified the grounding zone correctly.

Partington et al. (1987) point out that the altimeter "sees" the grounding line before crossing it, when the direction of satellite motion is up-glacier, and conversely, after crossing it, when the direction of motion is down-glacier, i.e. in both cases the position observed by the satellite is downglacier of the true grounding line position; the difference can be 2–4 km depending on angle between satellite groundtracks and grounding line. The data slope-correction procedure should counteract this effect (but there may still be some error).

The thickness of Lambert Glacier in the grounding zone is about 1000 m (or 770 m or 900 m depending on the reference, see Hambrey 1991, Hambrey and Dowdeswell 1994, Swithinbank 1988). The 3-km grid size of the Atlas maps corresponds to three times the ice thickness in the vicinity of the grounding line. As a consequence, the grid resolution of the DTM is sufficient for numerical models of the Lambert Glacier / Amery Ice Shelf System that take into account longitudinal stress gradients, but at the same time the DTM is coarse enough to warrant neglecting the so-called "T-term" (a double integral over a coordinate perpendicular to the glacier surface of the second derivative of the shear stress, in the longitudinal direction) in the longitudinal stress equilibrium equation (Kamb and Echelmeyer 1986). More simply, this means that the Atlas DTMs provide a reliable and accurate data basis which permits solid geophysical modeling for glaciers of the thickness of Lambert Glacier and more.

Lambert Glacier follows a deep graben. Ice from a large catchment area drains into the Lambert Glacier trough, accelerating and forming an ice stream.

The grounded part is Lambert Glacier, the floating part is Amery Ice Shelf. Along the margin, both Lambert Glacier and Amery Ice Shelf have lowered surface elevation — lower than both the glacier surface and the surface of the adjacent slow-moving ice that is not part of the ice stream of American Highland to the East and the rugged Prince Charles Mountains on the western side of Lambert Glacier. A similar lowered area runs along the side of every glacier (e.g. in the Alps) and is a consequence of increased vertical strain rates in the marginal shear zones of the glacier (for ice physics explaining this effect, see Herzfeld et al. 1993).

However, the overdeepenings visible in the Atlas maps are larger than the physically explainable surface lowering. A second effect that results in an apparent trench alongside the margin of a glacier in a valley is caused by the technology of satellite altimetry — the effect is called "snagging" and describes the tendency of the altimeter to "lock on" to a reflector while crossing a break in slope before regaining track (Partington et al. 1987) — this is the same effect that caused the mislocation of the break in slope associated with the position of the grounding line mentioned earlier. While the apparent marginal surface lowering is about 50 m in our maps, the lowering in the shear zones is 10–30 m according to Budd et al. (1982); so both explanations may account for about the same amount. Along the right glacier margin on the GEOSAT map, we also see several concentric lower spots (rather than one continuous trench) — not all of these are results of locally varying snagging effects — the largest one (at -7820.000 N/600.000 E) is Gillock Island — a high, not an overdeepening — and an island in the glacier, and Single Island (-7750.000 N/480.000 E) near the western margin.

On the ERS-1 maps, the marginal surface lowering is observable, but not as pronounced as on the GEOSAT maps, which gives another indication of the size of the physical lowering versus the apparent lowering (physical plus snagging effect) — now the marginal lowering is closer to 30 meters (on the right side of the glacier and ice shelf).

The mountain ranges to the west of Lambert Glacier are called (from north to south): *Mt. Meredith* (just south of 71° S), *Fisher Massif, Mt. Collins, Mt. Willing, Shaw Massif* (at ≈72° S). In the middle of Lambert Glacier at ≈72° S is *Clemence Massif* (a large nunatak massif) (note southern edge of our map is 72.1° S). The mountains are seen on the map of the Australia Division of National Mapping (1969).

East of Mt. Meredith (≈35 km east) is a so-called *ice doline* (Mellor and McKinnon 1960; Mellor 1960), a 5 km x 1.3 km meltwater lake. It is hypothesized that this meltwater lake may occasionally (over years or tens of years) drain through fractures into the seawater below, as in 1960 its surface was 80 m below the ice surface (field report), in 1973 (satellite image only) its surface was as high as the surrounding ice (after Swithinbank 1988, p. B71). The assumption that there is seawater below the ice doline coincides with our mapping of the (100 m proxy of the) grounding line, which is indented to the south here.

Significant differences between maps from different satellites — see also the time series including the SEASAT map in chapter (E) — exist in the area of Beaver Lake on the western side of Lambert Glacier. There is a peninsula between Beaver Lake and Lambert Glacier, as seen in the SEASAT map and in the ERS-1 map, but in the GEOSAT ice record data sets, the peninsula is missing (due to data processing errors in the initial stages of correction at NASA GSFC).

The *Beaver Lake* is an area of great geologic interest and has been studied intensively. Beaver Lake is an ice-covered freshwater sea lake, into which icebergs calve (Mellor and McKinnon 1960; cited after Swithinbank 1988). The Beaver Lake area is a valley rimmed by mountains (*Else Platform* on the tip of the "peninsula" between Amery Ice Shelf and Beaver Lake; *Loewe Massif* on the opposite side). An arm of the ice shelf that is fed by *Charybdis Glacier* reaches into the Beaver Lake area. (Charybdis Glacier has been suspected to surge.) The term "freshwater sea lake" may require some explanation. Beaver Lake is a 10 km x 10 km stably stratified tidal lake, its halocline is depressed by the addition of meltwater to the surface. The halocline probably lies at the depth of the draft of the ice tongue that blocks the outlet to the sea.

Sedimentary layers of the 800 m thick, uplifted Miocene and Pliocene strata of the Pagodroma Group in the northern Prince Charles Mountains (near Beaver Lake) provide evidence of a dynamic East Antarctic ice sheet during the Neogene. The Neogene Lambert Glacier appears to have had thermal and dynamic characteristics similar to those of some East Greenland ice streams nowadays (Hambrey and McKelvey 2000a,b; McKelvey et al. 2001). Lambert Glacier advanced and retreated several times in the geologic past.

The ice masses of Charybdis Glacier are seen as an increase in the surface elevation of Amery Ice Shelf just north of Beaver Lake (Swithinbank 1988, fig. 53, p. B70). Other inflowing glaciers on the northern side of Amery Ice Shelf are identifiable in the Atlas maps and match large inflowing glaciers in position, for which no names are given (in Swithinbank 1988, fig. 54, p. B73). Ice velocity at the ice front is 1200 m a^{-1}.

To the east of the Lambert Glacier graben is *Mawson Escarpment*, a straight escarpment with elevations rising to 1000 m. The escarpment step and the eastern side of the trough together form a 3000 m high wall. Northeast of Mawson Escarpment are the *Grove Mountains* (Hambrey 1991). A large ice stream and several smaller glaciers transcend Mawson Escarpment and join Lambert Glacier and Amery Ice Shelf.

The even gradient of the floating ice shelf is well visible, as are the irregular nature of the ice front, the overdeepening along the western and eastern margins of the ice stream due to shear stress (actual part) and to satellite snagging effect (apparent overdeepening, larger than in reality). The inflow of several glaciers from the west (near the front of Amery Ice Shelf) can be traced by the higher surface elevations on Amery Ice Shelf:

(1) 50 m higher for 70-km-E-W Glacier (northernmost) at -7640.000 N

(2) 40 m higher for 35-km-E-W Glacier at -7660.000 N

(3) 40 m higher for 30-km-E-W Glacier at -7700.000 N

(4) 30 m higher for 15-km-E-W Glacier

(5) 40 m higher for 10-km-E-W Glacier

(6) 40 m higher for (80-km-E-W) Charybdis Glacier at -7780.000 N, with large valley.

In between these are rock promontories (mapped as eastward extensions of the 200 m contour). Rock promontories are distinguished from glacier inflows as follows. Inflowing glaciers have contours in steps of 10's increasing from the elevation of the ice shelf, whereas rock promontories are seen in the 200 m contour and have a crowding of the next lower contours. Some inflowing glaciers are also seen on a LANDSAT image (Swithinbank 1988, fig. 54, p. B73), where they are unnamed, as well as several promontories, including Foley Promontory and Ladan Promontory. The advantage of elevation mapping is obvious in this area. The extent of the inflowing glaciers cannot be seen on the LANDSAT images.

A discussion of the flow units of the Lambert Glacier/Amery Ice Shelf system is given in the description of the detail map Lambert Glacier (section (F.5)).

Other main features that are visible in the Lambert Glacier map are sections of Mawson Coast to the west of Amery Ice Shelf, with Cape Darnley and MacKenzie Bay, and Ingrid Christensen Coast to the east, with *Prydz Bay* and *Polar Record Glacier*.

Mawson Coast is the coast of Mac Robertson Land between William Scoresby Bay (67° 24'S/59° 34'E) and Murray Monolith (67° 47'S/66° 54'E).

Ingrid Christensen Coast lies north of American Highland and extends from Jennings Promontory (70° 10'S/72° 33'E), a rock promontory that marks the eastern limit of the Amery Ice Shelf, to the western end of the *West Ice Shelf* (at 81° 24'E) (see map m81e73-89n67-721 Ingrid Christensen Coast).

Offshore of Amery Ice Shelf there are errors on the GEOSAT map and not on the ERS-1 map, as is often the case. Mawson Coast appears to be more rugged than Ingrid Christensen Coast. At 700.000 East it may be possible to identify Polar Record Glacier entering Prydz Bay. On the Mawson Coast, *Bjerkoe Peninsula* and a small island offshore (NW of it) and *MacKenzie Bay* can be located.

The differences between the mountainous areas with nunataks, steep ranges, and several side glaciers not individually portrayed here in the west and the gently rising inland ice in the east can clearly be distinguished. Indentation in three consecutive contours (elevation difference 600 m) may be *Scylla* (at -7790.000 N) and *Charybdis Glaciers* (at -7820.000 N) on the western edge of Lambert Glacier. A good map of this area has been produced from satellite images by the Australia Division of National Mapping (1969).

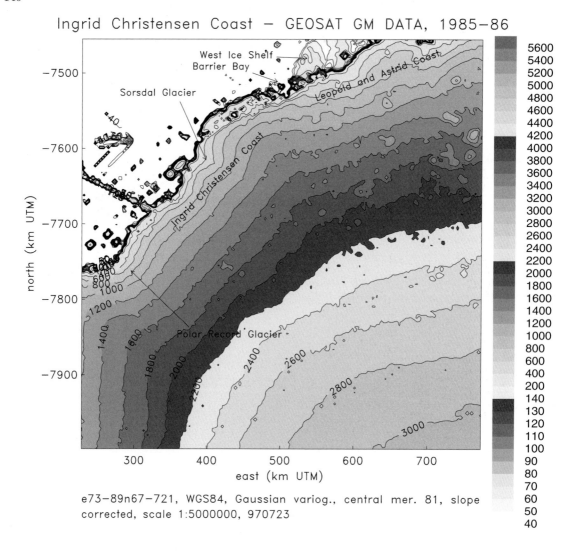

Ingrid Christensen Coast — GEOSAT GM DATA, 1985—86

e73—89n67—721, WGS84, Gaussian variog., central mer. 81, slope corrected, scale 1:5000000, 970723

Map m81e73-89n67-721 Ingrid Christensen Coast

North of the *American Highland* lies Ingrid Christensen Coast. The map extends from Prydz Bay at the western map edge along Ingrid Christensen Coast to Leopold and Astrid Coast (east of 81° E, 500.000–700.000 E).

Ingrid Christensen Coast extends from Jennings Promontory (70° 10'S/72° 33'E), a rock promontory that marks the eastern limit of the *Amery Ice Shelf*, to the western end of the *West Ice Shelf* (at 81° 24'E), visible in the NE part of the map. The coast was discovered and a landing made on Vestfold Hills in 1935 by Captain Mikkelsen of the *Thorshavn*, a vessel owned by the Norwegian whaling magnate Lars Christensen. The coast is named for Ingrid Christensen, the wife of Lars

Christensen, who also participated in the whaling expeditions (Alberts 1995, p. 360).

To the east of Ingrid Christensen Coast lies *Leopold and Astrid Coast* , which extends from the western extremity of West Ice Shelf (at 81° 24'E) to Cape Penck (66° 43'S/87° 43'E). Cape Penck, an ice-covered point, also fronts West Ice Shelf, but in the east. This part of the coast, named for King Leopold and Queen Astrid of Belgium, was also discovered by the Lars Christensen Expedition.

The inland ice rises almost uniformly from the coast. North of the coastline are several features: (1.) errors (west of 300.000 E in track form), (2.)

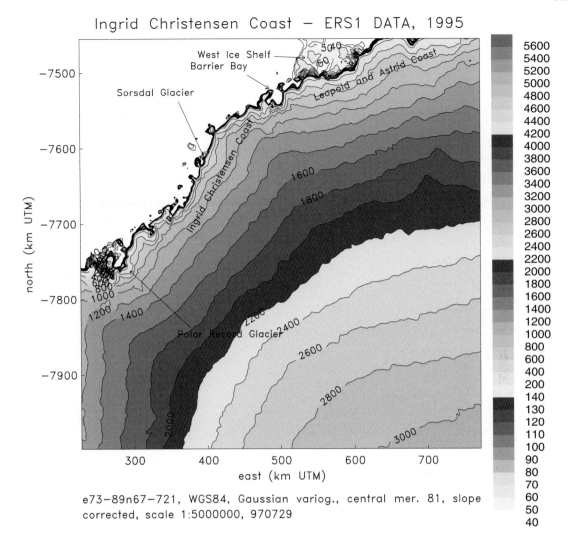

Ingrid Christensen Coast — ERS1 DATA, 1995

e73—89n67—721, WGS84, Gaussian variog., central mer. 81, slope corrected, scale 1:5000000, 970729

islands (Rauer Islands at 360.000 E), (3.) West Ice Shelf (40–70 m contours) located east of *Barrier Bay* (500.000 E) off Leopold and Astrid Coast. The most noticeable trough is that of *Sørsdal Glacier* extending about 80 km inland east of Rauer Island (380.000 E). On the ERS-1 map, there are no artefacts offshore, the trough of Polar Record Glacier is well distinguished, the location of Sørsdal Glacier is identifiable in comparison with the GEOSAT map. Barrier Bay and West Ice Shelf are well mapped. The Rauer Islands are located southwest of Sørsdal Glacier, the land to the northeast of Sørsdal Glacier are the *Vestfold Hills* (see also fig. 50, p. B66 in Swithinbank 1988), which is an ice-free area of low, hummocky hills up to 160 m high, with moraine- and lake-filled valleys, offshore islands, and fjord-like inlets (after Tierney 1975). The lakes are freshwater or

saline lakes. Two long fjords reach all the way to the inland ice (Krok and Langnes Fjords). The land (ice) rises to over 3000 meters in the southeast of the map. The ice in the southern part of the map drains to the Lambert Glacier/Amery Ice Shelf system. In the southwestern part of the map is American Highland, located east of the Lambert Glacier/Amery Ice Shelf system (see also the description of map m69e61-77n67-721 Lambert Glacier, and the special Lambert Glacier map in detail map section (F.5)).

There are several research stations located in the area: Davis Station (Australia), north of Sørsdal Glacier on the edge of the Vestfold Hills, and — close to each other — Law (Australia), Zhang Shan (China), Progress (USSR), all east of Larsemann Hills ≈(76° E, 69.5° S).

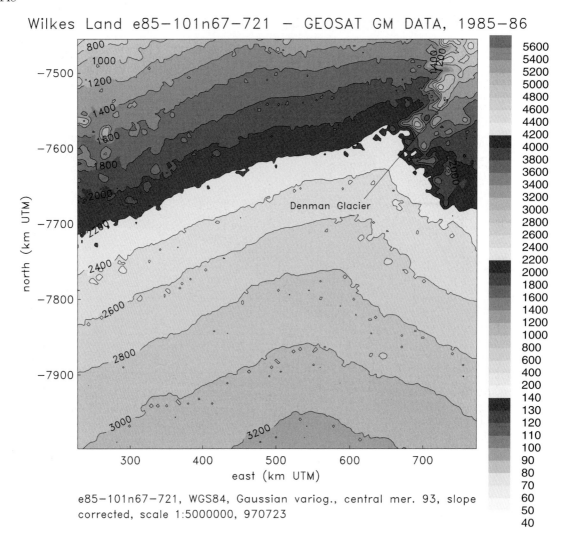

Wilkes Land e85–101n67–721 – GEOSAT GM DATA, 1985–86

e85–101n67–721, WGS84, Gaussian variog., central mer. 93, slope corrected, scale 1:5000000, 970723

Map m93e85-101n67-721 Wilkes Land (e85-101n67-721)

This map shows an area of Wilkes Land, East Antarctica, close to the coast (*Queen Mary Coast* is shown on the northerly adjacent map, m93e85-101n63-68), with elevations of 800 m to above 3200 m. The large scale feature is a ridge-line that runs northnortheast across the whole map area, from the 3200 m contour down to the 800 m contour on the western side of the *Denman Glacier* valley (see description to Denman Glacier detail map (F.7)); the ridge line is a northerly extension of the ice divide separating the Lambert Glacier Basin and the interior of Wilkes Land (in the upper elevations) and the *Denman Glacier* drainage (in the lower elevations). The slope increases from 200 m over 90 km (0.22% = 0.12°; 3200 to 3000 m) gradually to 200 m over 20 km (1% = 0.57°; 1000 m to 800 m in the northwest corner of the map). Steeper slopes are observed in the eastern margin of the Denman Glacier Valley, where mountain ranges exist (e.g. Mt. Strathcona, 1380 m) as is seen in National Geographic Society (1992, p. 102) and also indicated by the rougher contour lines on our map.

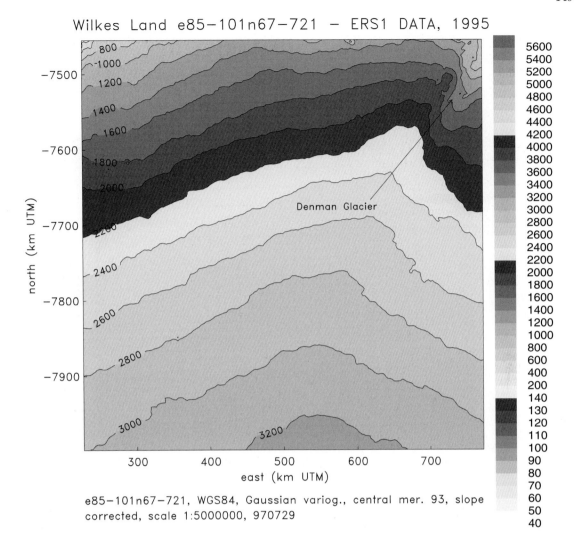

Wilkes Land e85-101n67-721 — ERS1 DATA, 1995

e85-101n67-721, WGS84, Gaussian variog., central mer. 93, slope corrected, scale 1:5000000, 970729

Wilkes Land e97−113n67−721 − GEOSAT GM DATA, 1985−86

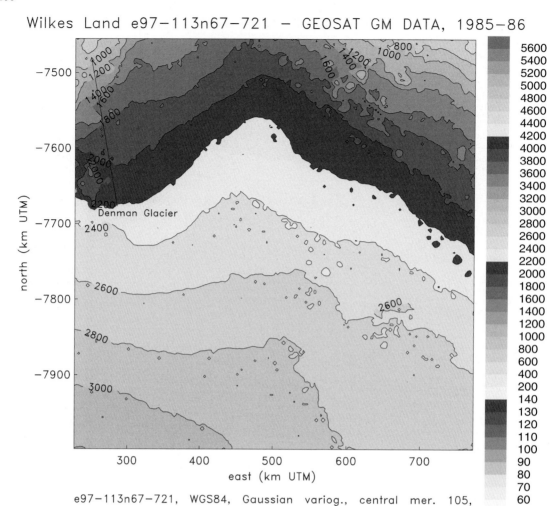

e97−113n67−721, WGS84, Gaussian variog., central mer. 105, slope corrected, scale 1:5000000, 970723

Map m105e97-113n67-721 Wilkes Land (e97-113n67-721)

The map of the area e97-113n67-721 in *Wilkes Land* shows the eastern divide limiting the drainage basin of *Denman Glacier*, it curves around to the north center of the map. At a smaller scale, of course, some of the ice in this drainage flows down smaller glaciers, such as *Scott Glacier* and *Apfel Glacier* (see description of map m105e97-113n63-68 Knox Coast). The divide extends north, ending at the northernmost point of *Knox Coast* which in this location is rather steep. To the northeast of the divide, ice flows to the *Vanderford Glacier* drainage basin, and more lo-cally, to smaller glaciers entering Vincennes Bay, such as Adams Glacier, or Underwood Glacier which enter the sea west of Cape Nutt on Knox Coast.

The surface gradient in the map area is about 200 m in 20 km (1% = 0.57°; near 800 m) on the Vincennes Bay side, 200 m in 40 km (0.5% = 0.29°; 1800 m–2000 m) on that side, and much lower, but variable, above 2200 m elevation. The contours indicate a rough area above 2200 m on the ERS-1 map as well as on the GEOSAT map.

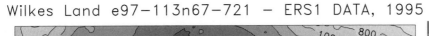

Wilkes Land e97−113n67−721 − ERS1 DATA, 1995

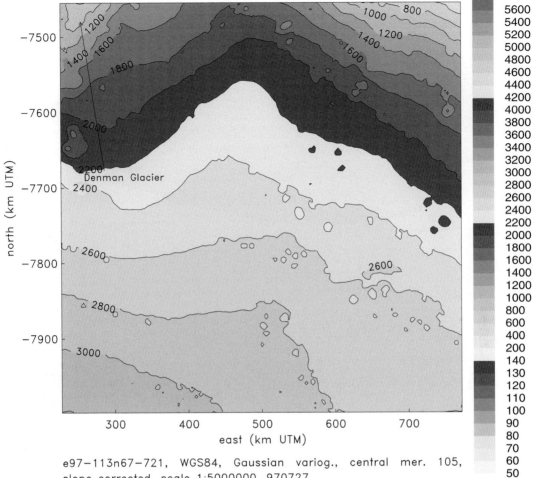

e97−113n67−721, WGS84, Gaussian variog., central mer. 105,
slope corrected, scale 1:5000000, 970727

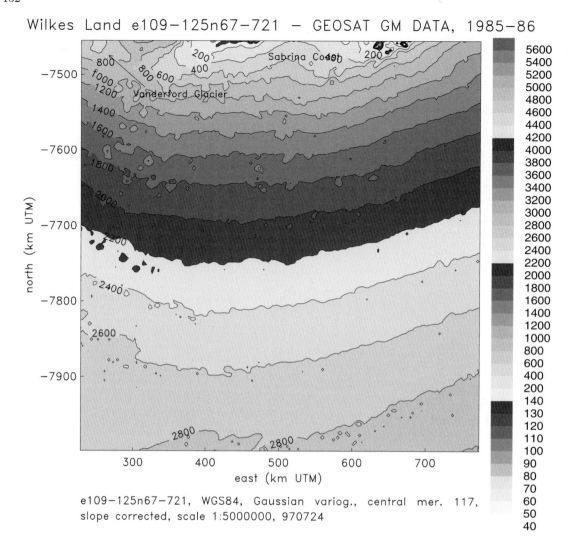

Wilkes Land e109−125n67−721 − GEOSAT GM DATA, 1985−86

e109−125n67−721, WGS84, Gaussian variog., central mer. 117,
slope corrected, scale 1:5000000, 970724

Map m117e109-125n67-721 Wilkes Land (e109-125n67-721)

This map shows Wilkes Land just south of Law Dome, Sabrina Coast and Banzare Coast. All of the near-coastal features are described for map m117e109-125n63-68 Sabrina Coast. Wilkes Land rises steeply to about 800 m elevation (200 m in less than 20 km; steeper than $1\% = 0.57°$) and with a steadily lessening gradient to 2800 m (200 m in 100 km, $0.2\% = 0.1°$, 2600–2800 m). All contours are convex if "seen" from the coast, i.e. from the north.

In the northwestern part of the map, the upper reaches of Vanderford Glacier are seen, which is the largest glacier in Vincennes Bay, located west of Law Dome and east of Knox Coast. The valley of Vanderford Glacier extends for 120 km southeast around Law Dome. (A detail map of Vanderford Glacier is given in section (F.8).)

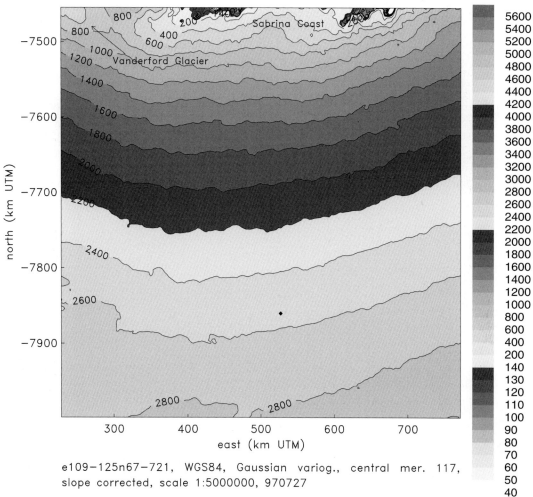

Wilkes Land e109-125n67-721 — ERS1 DATA, 1995

e109-125n67-721, WGS84, Gaussian variog., central mer. 117, slope corrected, scale 1:5000000, 970727

Wilkes Land e121-137n67-721 — GEOSAT GM DATA, 1985-86

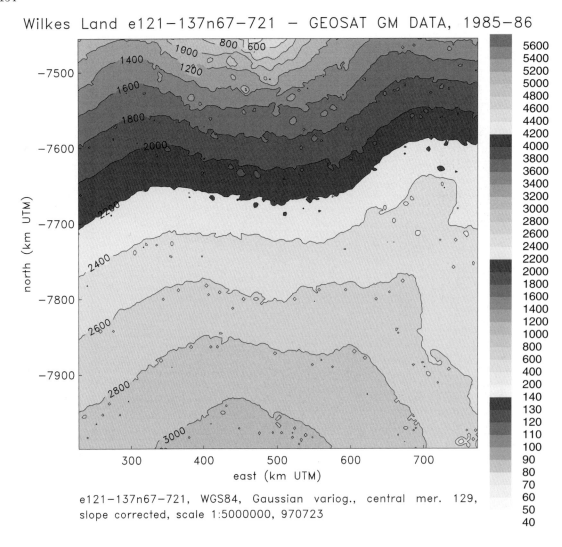

e121-137n67-721, WGS84, Gaussian variog., central mer. 129,
slope corrected, scale 1:5000000, 970723

Map m129e121-137n67-721 Wilkes Land (e121-137n67-721)

This map contains several ice divides in northern Wilkes Land, East Antarctica, that mark the borders of drainage basins at regional and continental scale. In the southern part of the map, a convex 3000 m elevation contour is seen, and a shallow ridge extends from this, in the 2800, 2600, and 2400 m contours. The ridge descends from *Dome Charlie* (map m117e99-135n71-77 Dome Charlie), and above the 3000 m and 2800 m contours, it is part of an ice divide on the continental scale as it separates drainage to the Indian Ocean (N) sector from drainage to the Ross Sea (E) (see map m141e123-159n71-77 Southern Wilkes Land (e123 to 159)). The noses in the 2400 m to 2800 m contours indicate the left branch of a regional ice divide (the right branch extends into Victoria Land and is seen on map m141e123-159n71-77 and on map m165e147-183n71-77 Victoria Land), ice to its west flows towards Banzare Coast and Clarie Coast, ice in the extension of the noses flows toward Adélie Coast, and ice to its east flows towards George V Coast (Fisher Bay, Mertz and Ninnis Glaciers) (see map m141e133-149n67-721 Wilkes Land e133-149n67-721). In the contours from 2200 m to 1200 m the topography of the upland immediately under the coast is captured; in the northwestern part of the map, the surface slopes down to the Banzare Coast. The extension

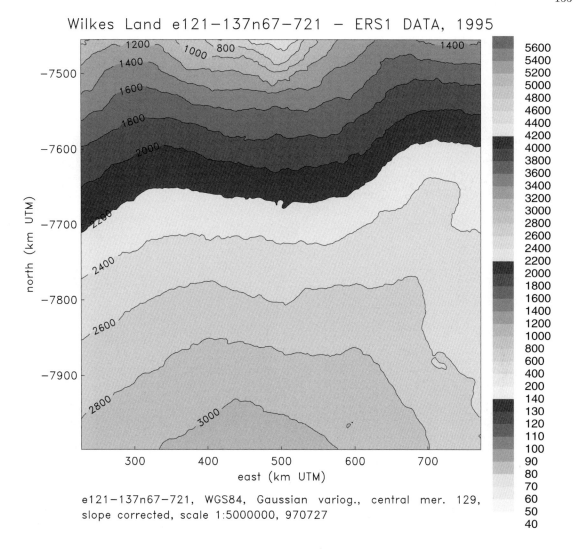

Wilkes Land e121-137n67-721 - ERS1 DATA, 1995

e121-137n67-721, WGS84, Gaussian variog., central mer. 129, slope corrected, scale 1:5000000, 970727

of the convex part of the contours 1200 m and higher is North Highland (south of Cape Goodenough). The concave area leads to Porpoise Bay, and in the 800 m and 1000 m contours we see subdivisions into smaller basins, likely forming valleys of outlet glaciers (see map m129e121-137n67-721 Clarie Coast). The concave area in the 1400-m contour (in the NE corner of the map) is upland of Davis Bay.

In this region of East Antarctica, the topography bifurcates more as the coast is approached, and the ice reaches the coast in many small glaciers, rather than collecting in one large ice stream, as for instance, in the Lambert Glacier drainage basin. The map description has, so far, followed the ERS-1 map. The general topographic features are the same on the GEOSAT map, but the contours are much noisier on the GEOSAT map. The interior of the inland ice is smooth at this elevation and at this scale, hence the little circles on the GEOSAT map are attributed to noise.

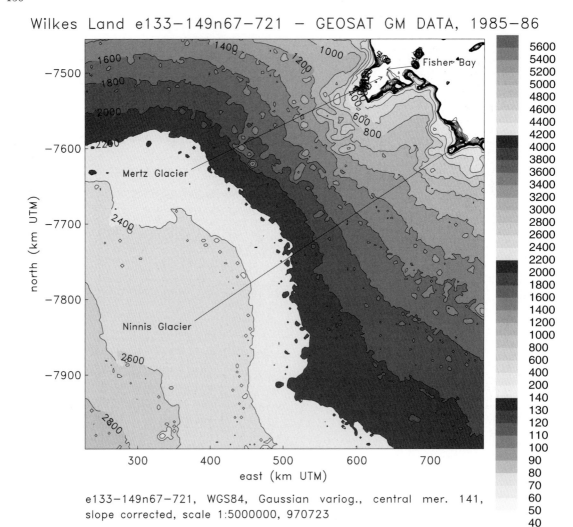

Wilkes Land e133–149n67–721 – GEOSAT GM DATA, 1985–86

e133–149n67–721, WGS84, Gaussian variog., central mer. 141, slope corrected, scale 1:5000000, 970723

Map m141e133-149n67-721 Wilkes Land (e133-149n67-721)

This map shows that part of Wilkes Land, that lies south of Adélie Coast (which itself is seen on the northerly adjacent map m141e133-149n63-68 Adélie Coast), of George V Coast (in part seen on the northerly adjacent map and in part on this map, and in part on the easterly adjacent map m153e145-161n67-721 Cook Ice Shelf). The relationship of the ice topography to the general pattern of ice divides is described in the text for m129e121-137n67-721 Wilkes Land e121-137n67-721 (westerly adjacent map). In brief, on a continental scale, the ice descends from the northwest of a broad ridge feature coming down from Dome Charlie. The contours at the elevations be-

low 2000 m are rougher than on other maps at this elevation, which may be indicative of surface topography.

(On the other hand, a more differentiable variogram model may be used to smooth out the contours in this area.) Existence of subglacial features may be concluded or rather hypothesized from surface-topographic maps, but for a proof radio-echo soundings, which penetrate the ice and reflect off subsurface features, are needed. Such surveys were carried out here.

Around (70° S/136° E) is *Astrolabe Subglacial Basin* (south of Adélie Coast and east of Por-

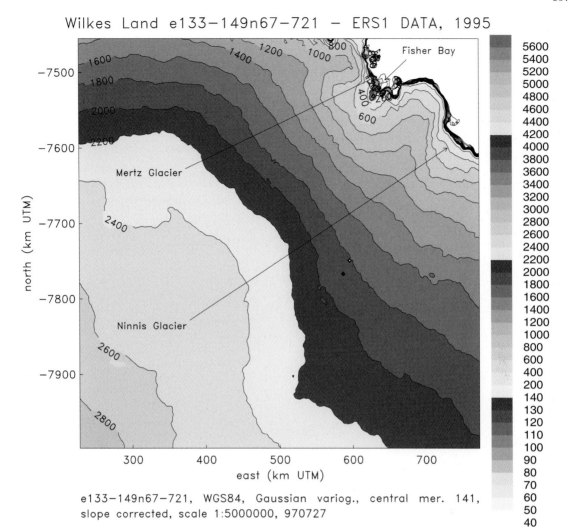

Wilkes Land e133–149n67–721 – ERS1 DATA, 1995

e133–149n67–721, WGS84, Gaussian variog., central mer. 141, slope corrected, scale 1:5000000, 970727

poise Subglacial Highlands (69° 30'S/134° E)), the basin trends N–S and bears the distinction of containing the thickest ice measured in Antarctica, 4776 m thick. Bedrock here is 2341 m below sea level (National Geographic Society 1992, p. 102). *Astrolabe* was the flagship of the 1837–40 French Antarctic Expedition led by J. Dumont d'Urville (Alberts 1995). Both subglacial features were mapped using airborne radio-echosoundings (by SPRI-NSI-TUD) (the *Porpoise* was a ship of the Wilkes Expedition 1838–42).

On the coast, we see the drainage basins of *Mertz Glacier* and of *Ninnis Glacier*.

Mertz Glacier is a heavily crevassed glacier that descends steeply from the highlands near the coast. Calculations of the slope of the Mertz Glacier drainage are given in section (F.10) from

the detail map of Mertz Glacier. Mertz Glacier is about 70 km long and averages 32 km wide. The bays formed by Mertz Glacier Tongue and the western and eastern shores have different names: *Buchanan Bay* (67° 05'S/144° 40'E) is the bay on the western side of Mertz Glacier Tongue, the headland is *Cape de la Motte*. *Fisher Bay* is the bay on the eastern side of the glacier tongue, with the headland *Cape Hurley*.

Ninnis Glacier (center coordinate (68° 22'S/147° E), for area coordinates see detail map 11 Ninnis Glacier, section (F.10)) is a large heavily hummocked and crevassed glacier in a bay about 80 km east of Mertz Glacier, it also descends steeply from the high interior in a broad valley. Ninnis Glacier is approximately half as long and wide as Mertz Glacier (from Ferrigno et al. (1996) and detail maps in section (F.10)).

Both Mertz and Ninnis Glaciers have glacier tongues, but they are in different states of advance and retreat (see discussions of the dynamics of glacier tongues in the detail section (F.10)). Both glaciers were discovered by the Australian Antarctic Expedition 1911–1914, led by Douglas Mawson, and are named after the expedition members Xavier Mertz and B. Ninnis who lost their lives on a sledge journey eastward on 7 January 1913 (Alberts 1995, p.487, p.527).

The appearance of the coastline varies quite a bit between the GEOSAT and the ERS-1 map. The best maps are the GEOSAT detail maps (section (F.10)), which demonstrate that there is a lot more information in the data and in the DTMs of near-coastline areas than is seen in the Atlas maps.

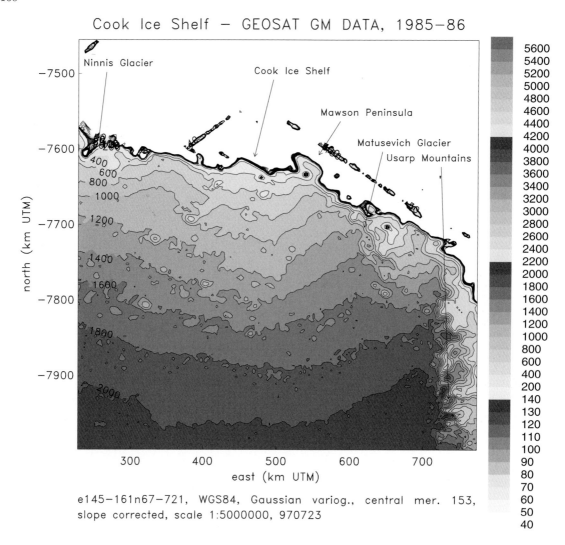

Cook Ice Shelf — GEOSAT GM DATA, 1985—86

e145—161n67—721, WGS84, Gaussian variog., central mer. 153, slope corrected, scale 1:5000000, 970723

Map m153e145-161n67-721 Cook Ice Shelf

This map covers the north coast of East Antarctica from Ninnis Glacier on George V Coast to Wilson Hills on Oates Coast, where the coastline takes a large-scale turn southwards towards Rennick Bay (on eastern Oates Coast) and Pennell Coast (the most easterly part of the north coast). Near the eastern margin of the map are the *Usarp Mountains*, a long and narrow mountain range that trends north-south at 160° eastern latitude, from the coast to Welcome Mountain (2494 m) and Roberts Butte (2828 m). The inland ice south of the coast ascends to above 2000 m.

George V Coast lies between Point Alden (66° 48'S/142° 02'E) (on map m141e133-109n63-68 Adélie Coast) and *Cape Hudson* (68° 20'S/153° 45'E). Cape Hudson is the head of Mawson Peninsula and the eastern boundary point of Cook Ice Shelf; it was explored by members of the Australian Antarctic Expedition (1911–1914), led by D. Mawson, and named for King George V of England (Alberts 1995, p. 275).

Ninnis Glacier, a large glacier in the northwest of the map, is described in the text of maps m141e133-149n67-721 Wilkes Land, m153e145-161n63-68 Ninnis Glacier Tongue (Ninnis Glacier

Cook Ice Shelf — ERS1 DATA, 1995

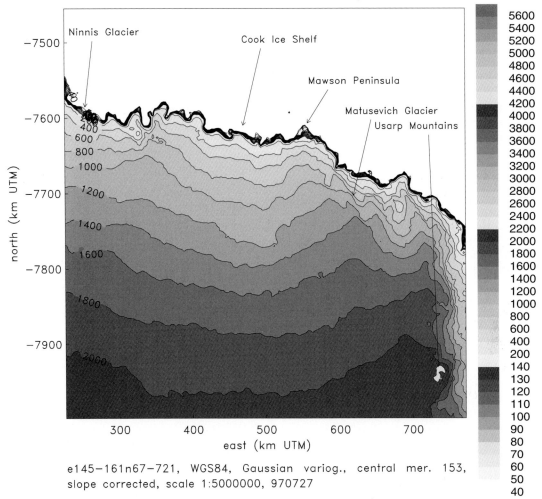

e145-161n67-721, WGS84, Gaussian variog., central mer. 153, slope corrected, scale 1:5000000, 970727

has an interesting retreating glacier tongue), and of the detail maps Mertz and Ninnis Glaciers, section (F.10). Between Ninnis Glacier and Cook Ice Shelf are Buckley Bay, Cape Wild, Deakin Bay, and Cape Freshfield (National Geographic Society 1992, p. 102). *Cook Ice Shelf* is a large ice shelf, about 100 km east to west, bordered by Cape Freshfield on the west and Cape Hudson on the east. Several small glaciers drain into Cook Ice Shelf. (The ice shelves do not show up as well in this part of Antarctica as they do in the Weddell Sea region, because the elevation of the geoid relative to the ellipsoid is different and the same elevation contour scheme is used — see, for instance, the detail maps of this area, section (F.10).) From Cape Hudson, Oates Coast extends east to *Cape Williams* (70°30'S/164°E). Cape Williams is the eastern headland of Lillie Glacier

and an extension of the ANARE Mountains. West of Mawson Peninsula is the small Slava Ice Shelf.

The eastern part of Oates Coast was discovered in February of 1911 by Harry Pennell, commander of the expedition ship *Terra Nova* of the British Antarctic Expedition 1910–1913, led by Robert Scott. Pennell named the coast after Lawrence Oates, who reached the South Pole with Robert Scott and three other companions, who all lost their lives during the return from the Pole in 1912. The story of Scott's expedition to the South Pole is likely known to every reader from his 5th grade English book.

The area around Mawson Peninsula was only delineated from air photos of the U.S. Navy Expedition Operation Highjump 1946–1947. The coast is heavily glaciated, and several smaller and larger glaciers reach the sea in this stretch. *Matusevich Glacier* (center coordinates (69° 20'S/157° 27'E)) is a broad, 80 km long glacier with a prominent 28 km long glacier tongue, it reaches the coast between Lazareo Mountains and Wilson Hills. The glacier was named by the Soviet Antarctic Expedition 1957–58 after N.N. Matusevich, hydrographer and geodesist (it was also named Pennell Glacier). It was explored by this expedition, U.S. Operation Highjump 1946–47, and the ANARE expedition in 1959 and 1962.

Other named features on Oates Coast in the area of this map are (west to east, after National Geographic Society 1992, p. 102) Lauritzen Bay and Mt. Martyn at Matusevich Glacier, Archer Point, Davies Bay, Cape Kinsey, Tomilin Glacier coming out of Wilson Hills, the Russian Station Leningradskaya (no longer used), and Suvorov Glacier (for the latter, see text to next map m165e157-173n67-721 Pennell Coast). The National Geographic map features are sometimes a bit inexact, and no annotated satellite image is avaliable for this area. The area that drains to the east is described on the easterly adjacent map, m165e157-173n67-721 Pennell Coast.

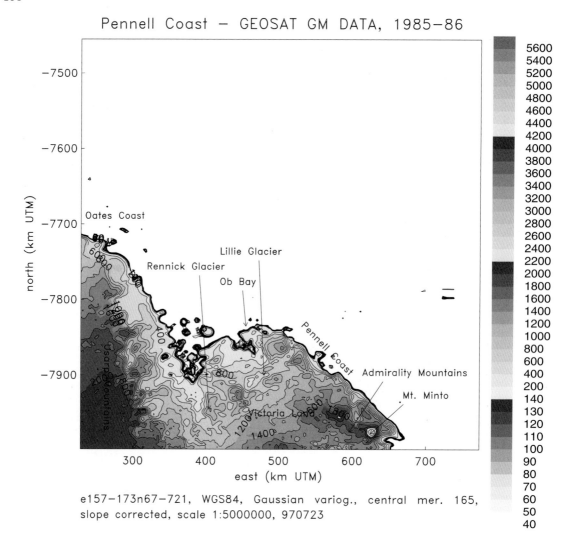

Pennell Coast — GEOSAT GM DATA, 1985—86

e157—173n67—721, WGS84, Gaussian variog., central mer. 165, slope corrected, scale 1:5000000, 970723

Map m165e157-173n67-721 Pennell Coast

This map shows the easternmost part of the northern coast of East Antarctica (*Oates Coast, Pennell Coast*) and the northernmost part of the Ross Sea coast (*Borchgrevink Coast*) and part of the headland of *Victoria Land* formed by the Admiralty Mountains. All the land shown on this map is part of Victoria Land. The mountain ranges are the eastern end of the *Transantarctic Mountains*, which extend from here across the Antarctic Continent to the Shackleton Range south of the Weddell Sea.

Oates Coast extends to *Cape Williams* (on the eastern side of the Lillie Glacier terminus). *Pennell Coast* extends from there eastward to *Cape Adare* (71° 17'S/170° 14'E), the headland that separates the north coast of East Antarctica from the west coast of the Ross Sea. Mt. Minto (4165 m) is located close to Cape Adare in the *Admiralty Mountains*.

We notice that the contour lines on the GEOSAT map show much more topographic relief than those on the ERS-1 map. Since the region has high relief, the GEOSAT contours are probably more correct. However, a correct representation of mountainous terrain from satellite radar altimeter data is not possible, the elevations are

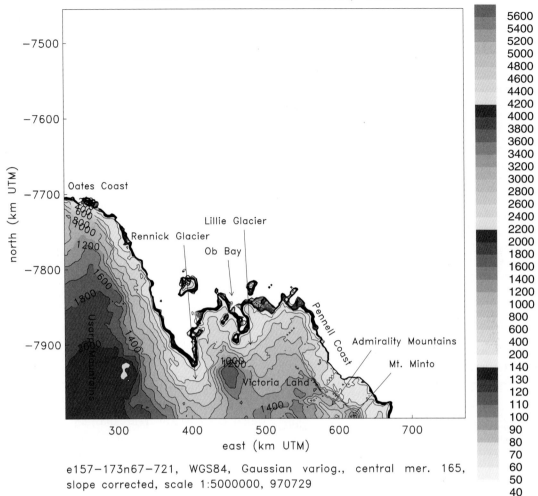

Pennell Coast — ERS1 DATA, 1995

e157–173n67–721, WGS84, Gaussian variog., central mer. 165,
slope corrected, scale 1:5000000, 970729

generally too low and the topographic features smoothed out. Large glaciers (Rennick Glacier, Lillie Glacier) can still be studied to some extent and are represented in the detail maps, chapter (F). The mountain ranges in the western part of the map are the *Usarp Mountains*, named after the acronym of the United States Antarctic Research Program.

North of the Usarp Mountains, and west of Wilson Hills and east of Rennick Glacier are several small glaciers, Noll Glacier near Leningradskaya, Ferguson Glacier (no outlet glacier), Manna Glacier (no outlet glacier) and Suvorov Glacier; Noll and Suvorov Glaciers have small glacier tongues (as seen on satellite image fig. 43, p. B54 in Swithinbank 1988) but the glaciers have no regional significance.

Rennick Glacier is an important glacier with a large drainage valley that extends south between the Usarp Mountains in the west and the Bowers Mountains (2850 m) in the east. On the other side of the Bowers Mountains is *Lillie Glacier*. Both Rennick Glacier and Lillie Glacier have several subsidiary glaciers. Rennick Glacier is joined by Gressit Glacier, Canham Glacier, and Sledgers Glacier and a few smaller ones. (For Rennick Glacier, see figs. 42, 41, 40, for Lillie Glacier, see figs. 39, 40, Swithinbank 1988, p. B49–B53).

Rennick Glacier is one of the largest glaciers of the world, it occupies an almost straight fault-controlled trench that can be followed for 370 km and is 20–30 km wide (Dow and Neall 1974). The trench is the basin of the large north-south trending valley that extends from the coast at (-7900.000 N/400.000 S) south (see also the

166

southerly adjacent map, m165e147-183n71-77 Victoria Land. Rennick Glacier is the only major glacier that flows northward in the Transantarctic Mountains, north of Beadmore Glacier.

Rennick Glacier is likely one of the slowest glaciers in the region, moving only 190 ma^{-1}, as concluded from satellite imagery (Swithinbank 1988, p. B52). According to ground observations (Mayewski et al. 1979), the glacier once covered a larger area in the Transantarctic Mountains and is now in a state of retreat. More details on Rennick Glacier, including a discussion of the grounding line position, are given in the detail map (section (F.11)). Glaciological surveys of the area inland of Rennick Glacier indicate that the glacier receives very little, if any, ice from the interior of the East Antarctic Ice Sheet, so it is not an outlet glacier, but may have been one in the past.

Offshore and east of the floating tongue of Rennick Glacier is Znamenski Island (70° 14'S/161° 51'E), a high, nearly round (4 km long), ice-covered island. The island was charted by the Soviet Antarctic Expedition in 1958 and named after hydrographer K.J. Znamenski, 1903–1941 (Alberts 1995, p. 833).

The next important glacier in the area is *Lillie Glacier*, which terminates in *Ob Bay*. Lillie Glacier is not an outlet glacier of the Antarctic Ice Sheet, all its ice is local, from the main Lillie Glacier and tributaries in the Transantarctic Mountains.

Lillie Glacier is likely the longest local glacier on Earth, it is 220 km long and 19 km wide below the confluence with *Ebbe Glacier*, a major tributary 50 km from the coast (Swithinbank 1988, p. B57 and fig. 39). Bering Glacier in Alaska is more than 200 km long, with Bagley Icefield, and depending on where one places the head of Bering Glacier, it is a contender for the world's largest glacier as well.

Between Lillie Glacier and Ebbe Glacier is the *Everett Range*, east of the northern part of Lillie Glacier are the *ANARE Mountains*.

Graveson Glacier enters Lillie Glacier from the western, Bowers Mountains side. All the mountain ranges appear rugged and cut by deep, glaciated valleys on a LANDSAT 3 RBV image (Swithinbank 1988, fig. 39, p. B49); the rugged mountain topography of Victoria Land also shows on the altimetry maps.

Lillie Glacier Tongue extends less than 10 km beyond Cape Williams. The grounding line is not determined in Swithinbank (1988).

Cape Adare marks the boundary to the Ross Sea. At this point, the first landing in Antarctica was made in 1895 by Borchgrevink, and the first overwintering took place here in 1899. Only 75 years later another expedition came here and built the U.S./New Zealand Hallet Station (at Cape Hallet), approximately 100 km east in the International Geophysical Year 1957–58. The area between the capes and around Mt. Minto is also of high topographic relief and home of many local glaciers, flowing north to south *and* east to west! A larger glacier is *Tucker Glacier* (barely seen on this map), approximately 15 km wide and 100 km long (Swithinbank 1988, fig. 38). Tucker Glacier is also a local glacier. It appears to have a 60 km long floating tongue. Fast ice and pack ice fill enlargements of the coastline.

168

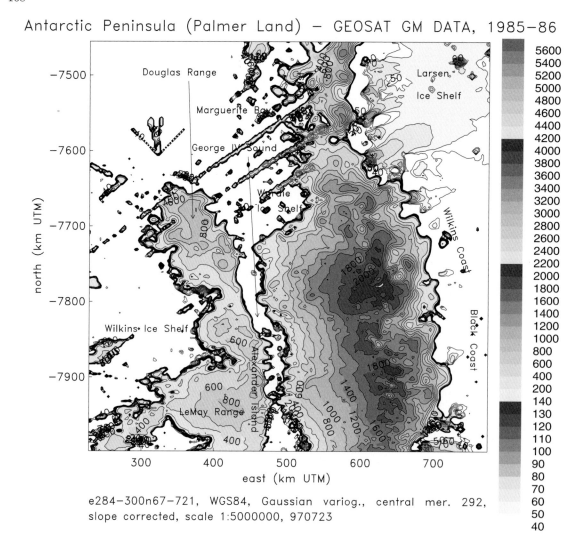

e284—300n67—721, WGS84, Gaussian variog., central mer. 292, slope corrected, scale 1:5000000, 970723

Map m292e284-300n67-721 Antarctic Peninsula (Palmer Land)

(Coordinates e284-300 correspond to w76-60)

This map shows *Palmer Land*, the southern half of the Antarctic Peninsula, and a bit of Graham Land, which is the northern half of the Peninsula, and (most of) Alexander Island. To the west of the Antarctic is the *Bellingshausen Sea*, to the east the *Weddell Sea*.

This area of the Antarctic Peninsula is also — as Graham Land further north — mapped better by ERS-1 radar altimetry than by GEOSAT radar altimetry (see text to map m297e289-305n63-68 Antarctic Peninsula (Graham Land)). A notable exception is that Alexander Island appears as two islands on the ERS-1 map, where the Douglas Range part is separated from the southern Le May Range part, which are connected on the GEOSAT map (and in reality). *Alexander Island* is separated from the mainland of the Antarctic Peninsula by *George VI Sound*, and, in the south, by *George VI Ice Shelf*. On its west side, Alexander Island is bounded by *Wilkins Ice Shelf*, which also shows signs of disintegration, Charcot Island and Latady Island protect Wilkins Ice Shelf somewhat.

Antarctic Peninsula (Palmer Land) — ERS1 DATA, 1995

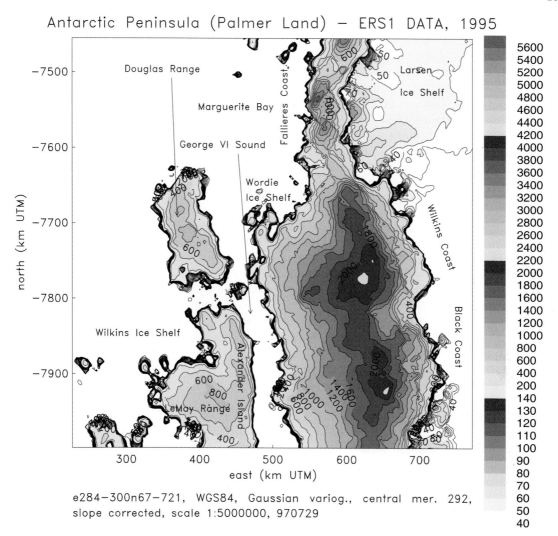

e284–300n67–721, WGS84, Gaussian variog., central mer. 292,
slope corrected, scale 1:5000000, 970729

The boundary between *Graham Land* in the north and *Palmer Land* in the south is marked by a line from *Cape Jeremy* located at (69° 24'S/68° 51'W), the northwestern tip of land (about (-7700.000 N/500.000 E)) south of *Wordie Ice Shelf* and the northeastern entrance to *George VI Sound*, to *Cape Agassiz*, located at (68° 29'S/62° 56'W), tip of Hollick-Kenyon Peninsula, at (-7640.000 N/650.000 E). Graham Land is named after James Graham, First Lord of the Admiralty at the time of John Biscoe's exploration of the west side of Graham Land in 1832. Palmer Land is named for Nathaniel Palmer, an American sealer who explored the Antarctic Peninsula area south of Deception Island in the *Hero* in November 1820.

The western coastline of Graham Land (already described for map m297e289-305n63-68 Antarctic Peninsula (Graham Land)) is generally rugged and

steep, with steep glaciers tumbling into the sea and several islands offshore. Parts of the coastline seen on this map are, from north to south, *Graham Coast* (Cape Renard (65° 01'S/63° 47'W)(N) to Cape Bellue (66° 18'S/65° 53'W), north side of Darbel Bay), *Loubet Coast* (Cape Bellue (N) to Bourgeois Fjord (67° 40'S/67° 05'W), between Pourquoi Pas and Blaiklock Islands), and *Fallieres Coast* (Bourgeois Fjord (N) to Cape Jeremy (69° 24'S/68° 51'W)). Cape Jeremy marks the boundary between Graham Land and Palmer Land, and also the north entrance of George VI Sound. The east coast of Graham Land is bounded by *Larsen Ice Shelf*, which extends from Jason Peninsula (at ≈ 7360.000 S) to Palmer Land and Hearst Island (≈ 7640.000 S) for 280 km N–S and about 200 km E–W.

Parts of the coastline on the eastern side of Graham Land on this map are *Oscar II Coast* (Cape Fairweather (65° 00'S/61° 01'W) to Cape Alexander (66° 44'S/62° 37'W), head of Churchill Peninsula, just south of Jason Peninsula in Larsen Ice Shelf, north of Cabinet Inlet), *Foyn Coast* (Cape Alexander (N) to Cape Northrop (67° 24'S/65° 16'W, 1160 m), north of Whirlwind Inlet; opposite Adelaide Island), and *Bowman Coast* (Cape Northrop (N) to Cape Agassiz (68° 29'S/62° 56'W), tip of Hollick-Kenyon Peninsula at the south end of Larsen Peninsula).

Many glaciers drain into Larsen Ice Shelf. The satellite USGS Satellite Image Map (Ferrigno et al. 1996, inset) marks inlet names rather than glacier names (Mill Inlet, Whirlwind Inlet, Seligman Inlet, Solberg Inlet, Mobiloil Inlet). Some glacier names are given in Swithinbank (1988). Whereas "the backbone of the peninsula", the landmass, is only 60 km wide in Graham Land, it is 200 km wide in Palmer Land. The interior of Palmer Land north of ≈ 71.5° S is *Dyer Plateau*, north of 70° S is *Wakefield Highland* and, east of that, *Eternity Range*, inland of Wilkins Coast. Further south are the *Welch Mountains* (near 71° S/65° W), *Batterbee Mountains* (near 71.5° S/67° W on Rymill Coast) and *Gutenko Mountains* (near 72° S/65° W). Mt. Jackson (3050 m) on Black Coast appears to be the highest mountain of the Antarctic Peninsula.

Coastline sections on the east coast of Palmer Land are *Wilkins Coast* (Hollick-Kenyon Peninsula, Cape Agassiz to Cape Boggs (70° 33'S/61° 23'W), headland of Eielson Peninsula, the south boundary of Smith Inlet with Clifford Glacier) with Hearst Island, a 60 km large ice-covered island offshore south of Larsen Ice Shelf, surrounded by shelf ice, *Black Coast* (Cape Boggs to Cape Mackintosh (72° 50'S/59° 54'W), the tip of Kemp Peninsula, Mason Inlet). Large glaciers on Wilkins Coast are Bingham Glacier opposite Hearst Island, Anthony Glacier and Clifford Glacier, on Black Coast Gruening and Beaumont Glaciers, entering Hilton Inlet south of Cape Knowles.

Western coasts of Palmer Land on this map are *Rymill Coast*, named for John Rymill, leader of the British Graham Land Expedition (1934–37), extending from Cape Jeremy to Buttress Nunataks (72° 22'S/66° 47'W), a group of coastal rock exposures up to 635 m high, 16 km WNW of the Seward Mountains, bordering George VI Sound and Ice Shelf, and *English Coast*, named after Robert English, captain of a ship of the Byrd Expedition 1933–35, extending from Seward Mountains to Spaatz Island, precisely from Buttress Nunataks to Rydberg Peninsula (73° 10'S/79° 45'W) between Fladerer Bay and Caroll Inlet in Ellsworth Land (see map m285e267-303n71-77 Ellsworth Land), bordering George VI Ice Shelf. Glaciers on Rymill Coast are Chapman Glacier, Meiklejohn Glacier, Bertram Glacier, Ryder Glacier, and Goodenough Glacier.

George VI Sound is a major fault depression, 500 km long in the shape of the letter J and 25–60 km wide, between Rymill Coast of the Antarctic Peninsula and English Coast of Ellsworth Land in the east and Alexander Island in the west. The sound is covered by ice, the George VI Ice Shelf, between Niznik Island, located about 50 km south of the sound's entrance between Cape Jeremy and Cape Brown on Alexander Island, to Ronne Entrance at the southwestern end of the sound (ice coverage quoted after Alberts 1995(!), p. 274).

George VI Ice Shelf flows in two directions, N and S/SW, from an ice divide with a 500 m thickness, ice thickness on the edges is ≈100 m (Swithinbank 1988, p. B113 and fig. 86). Swithinbank (1968) observed several peculiarities in field work on George VI Ice Shelf: meltwater channels on top of the ice show that it flows across the sound rather than along it, and the meltwater channel direction does coincide with the direction of movement. Most of the input is from Millet, Bertram, and Ryder Glaciers on Palmer Land, which causes east to west flow across the sound, but Grotto and Uranus Glaciers on Alexander Island yield influx that causes flow in the opposite direction. Smaller glaciers do not cause west-eastern flow of ice shelf parts. The ice then discharges to the north and south of the sound. Despite the melt ponds on top, the balance on top is positive, and most mass losses are through bottom melting. Wind-blown dust from rock outcrops decreases the albedo on the ice shelf and increases ablation. Ice coverage in George VI Sound and Ice Shelf has changed drastically in recent years.

A 1978 satellite image showing the many glaciers that drain into *Marguerite Bay* and *Lallemand Fjord* (the northernmost bay on the west coast of Graham Land on our map) is shown in Swithinbank 1988, fig. 81, p. B110). The glaciers are valley glaciers, most of them end in fjords and/or calve into the sea. Most of the bays are ice-covered (in December 1978; i.e. in mid-summer), but do not contain ice shelves. *Müller Ice Shelf* is a small ice shelf, a few kilometers in diameter, in Lallemand Fjord (near 70° S) and the northernmost ice shelf on the west side of the Antarctic Peninsula. *Jones Ice Shelf* near 67° 30'S is also a thin ice shelf, it is located between the north end of Bijourdau Fjord and Bourgeois Fjord north of Pourquois Pas Island in Marguerite Bay.

At the southeastern end of Marguerite Bay is *Wordie Ice Shelf*, which has become well-known as the example of ice-shelf break-up.

Wordie Ice Shelf was punctuated by ice rises and ice rumples in 1979 (satellite image fig. 82, p. B112, Swithinbank 1988), and the whole ice shelf did not appear very well connected. Fleming Glacier is the largest glacier entering Wordie Ice Shelf, flowing from the east, it has a thickness of $\approx$ 1000 m, and its flowlines can be traced easily across Wordie Ice Shelf for $\approx$ 50 km (1979), the thickness decreases to 150 m near the ice shelf edge. Seller Glacier and Airy Glacier enter north of Fleming Glacier, Seller Glacier appears to flow more strongly. Prospect Glacier and another small glacier enter south of Fleming Glacier. The ice mass of Fleming Glacier is bifurcated by Balfour Island. From the north, Harrot Glacier enters. The shelf ice edge receded in the southern part of the ice shelf between 1966 and 1974 about 25 km, and everywhere a few to ten kilometers between 1974 and 1979.

Swithinbank (1968) collected radio echosoundings in 1966, but horizontal brine penetration from fractures top to bottom of the ice shelf hindered soundings in the area that retreated later, however, such horizontal brine penetration is an indication of the disintegration of the ice shelf. Hence the ice shelf was showing signs of waning already in 1966, but the break-up and retreat occurred in 1972 or 1973 (585 km^2 calved), a further 250 km^2 calved between 1974 and 1979 (Doake 1982). According to Thomas et al. (1979) longitudinal stress in an ice shelf should be compressive upstream of

an ice rise, and hence disintegration was unlikely to happen before the ice rise disappeared. Such a situation — ice shelf upstream of ice rise — existed near Buffer Ice Rise in Wordie Ice Shelf. However, Wordie Ice Shelf had already retreated past several ice rises (contrary to the hypothesis of Thomas et al. (1979)). Possible reasons for thinning are accelerated bottom melting (Robin 1979) or surface meltwater penetration (Hughes 1982). Mercer (1978) already saw break-up of the Antarctic Peninsula ice shelves as a warning sign of CO_2-induced global warming which might lead to disintegration of the West Antarctic Ice Sheet. Buffer Ice Rise is only 3 km in diameter and so the conceptual borderline case between ice rise and ice rumple — most of the ice flows around the grounded area but some across it.

One of the first occurrences of catastrophic disintegration of an Antarctic ice shelf was the break-up of Wordie Ice Shelf (Vaughan 1993). Meanwhile, Wilkins Ice Shelf and ice in George VI Sound have shown signs of disintegration. Larsen Ice Shelf (B) broke up in 2002.

The northern edge of *George VI Ice Shelf*, to the contrary, is mapped to have remained stable between 1949 and 1974, according to Swithinbank (1988), but retreated in the 1930s and between 1974 and 1979. On George VI Ice Shelf, tidal sea lakes exist — the ice flows towards land of Alexander Island, and shortly before the land it thins to nothing melting into a lake (a similar situation exists in the Schirmacher Oasis). George VI Ice Shelf has started to disintegrate, 5% of its area were affected by 2001 (Vaughan et al. 2001).

In *Wilkins Ice Shelf*, west of Alexander Island, there are two sea lakes in the ice shelf, a rare occurrence and a sign that the ice shelf is at the climatic limit of existence. Wilkins Ice Shelf has also undergone major changes recently.

On *Alexander Island*, there are several north-south trending mountain ranges, reaching 3000 m in the Douglas Range, with deep glacier-filled trenches to 500–800 m below sea level.

(D.3) Latitude Row 71-77°S: Maps from ERS-1 Radar Altimeter Data

Map m333e315-351n71-77 Riiser-Larsen Ice Shelf

(Coordinates e315-351 correspond to w45-9)

This map shows a large section of the western coast of Queen Maud Land to the Weddell Sea (Caird Coast and Princess Martha Coast), which is bounded by the extensive Riiser-Larsen Ice Shelf, measuring 650 km along the coast, including Brunt Ice Shelf west of Stancomb-Wills Glacier. The section of the Riiser-Larsen Ice Shelf northeast of the Kraul Mountains is seen on the easterly adjacent map m357e339-15n71-77 New Schwabenland, a small section (Seal Bay) just west of Cape Norvegia is mapped on m15we23w-7wn67-721 Ekström Ice Shelf in the next northerly map row.

Caird Coast is the continuation of Coats Land between the terminus of the Stancomb-Wills Glacier (20°W) and the vicinity of Hayes Glacier (27° 54'W). Caird Coast was sighted by Shackleton in 1915 and named for the patron of his expedition (Alberts 1995).

Princess Martha Coast extends from 5°E westward to the terminus of Stancomb-Wills Glacier at 20°W. The name Princess Martha Coast was originally given only to the section of coast near Cape Norvegia, which was discovered by Capt. Hjalmar Riiser-Larsen and charted from the air during that expedition in 1930. To the east of Princess Martha Coast lies *Princess Astrid Coast*, which extends from 5°E to 20°E and is named for Princess Astrid of Norway, this part of the coastline was discovered by Capt. H. Halvorsen of the ship *Sevilla* in March 1931.

The coast is fairly mountainous. Several large glaciers extend into Riiser-Larsen Ice Shelf. The *Kraul Mountains* form the center of a peninsula in the northeast of the map, the *Heimefront Range* is at the eastern edge of the map (the higher land south of Veststraumen Glacier extends from the Heimefront Range). Heimefront range was originally named Heimefrontfjella (norwegian). The Tottan Hills (center coordinate (75°02'S/12°25' W)) are located south of Heimefront Range, inland and between Veststraumen Glacier and Stancomb-Wills Glacier (National Geographic Atlas map, National Geographic Society 1992, p. 102). Ac-

cording to Alberts (1995), the Tottan Hills are a 30 km large group of rocky hills, which form the southwestern part of Heimefront Range (for geology, see Tingey 1991).

The section of ice shelf southwest of Stancomb-Wills Glacier, off Caird Coast, is the Brunt Ice Shelf (with the British overwintering station Halley). The Brunt Ice Shelf in Halley Bay has an unusually high ice edge, which stands 60 m above the water. In Halley Bay is a polynia, a warm-water anomaly that permits landing a research vessel at the ice edge for a larger part of the year than in surrounding ice areas (observation by the author during the German Antarctic Expedition ANTARKTIS VI, 1987/88, with R/V *Polarstern*). For comparison, the ice edge in the Atka Bay is only 20–30 m high above the sea ice or the water.

Several smaller glaciers drain into Riiser-Larsen Ice Shelf north of the Kraul Mountains, which is deduced from the fact that higher contour lines extend into the ice shelf. The two large glaciers in the area are *Veststraumen Glacier* south of the Kraul Mountains and north of Heimefront Range and *Stancomb-Wills Glacier* further south. The part "straumen" means ice stream, hence the word "glacier" in "Veststraumen Glacier" is superfluous (but a consequence of naming all Antarctic geographic features in English). The drainage basin of Veststraumen Glacier extends about 175 km east, however, it is likely that the glacier drains ice from above the Heimefront Range and Kirwan Escarpment.

The drainage basin of *Stancomb-Wills Glacier* extends at least to the 2200 m contour line (see description of m357e339-15n75-80 Western Queen Maud Land (North)) 250 km inland from its terminus. The general drainage basin extends about 500 km further east into Queen Maud Land (3200 m contour on map m357e339-15n75-80 Queen Maud Land (North) and map m357e339-15n71-77 New Schwabenland). Of course, observations of ice flow would be needed to mark in particular the smaller ice divides clearly; here we

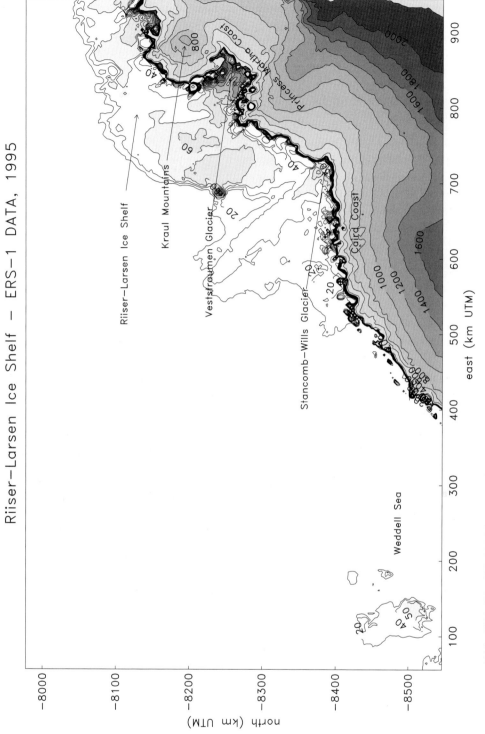

Riiser-Larsen Ice Shelf - ERS-1 DATA, 1995

e315-351n71-77, WGS84, Gaussian variog., central mer. 333, slope corrected, scale 1:5000000, 980112

173

draw conclusions based on the principle that gravitational flow generally follows the surface topography, which permits deduction of iceflow and location of drainage basins and ice divides from the morphology displayed in the elevation maps.

Stancomb-Wills Glacier (see also detail map 2, section (F.2)) has a long tongue which extends 230 km northwest (about 10 m above surrounding areas on our map, 30 m in the areas close to the terminus). The ice tongue is 30 km wide near its end and up to 50 km wide in various places close to the coast. The glacier tongue contour is surrounded by another contour, which indicates that the ice surrounding the tongue is elevated. Further east is the Riiser-Larsen Ice Shelf with elevations of over 60 m, in comparison, we conclude that the Stancomb-Wills Glacier Tongue extends into sea ice (and not into shelf ice) to the west of the Riiser-Larsen Ice Shelf. A comparison with satellite imagery will show that the ice area surrounding the glacier tongue consists of so-called "low ice shelf", which is a thicker form of sea ice.

The Stancomb-Wills Glacier Tongue was discovered already in January of 1915 by Shackleton (1919), who saw its seaward terminus as an ice-shelf promontory (Stancomb-Wills Promontory) (Swithinbank 1988, p. B97). The ice stream itself was discovered much later, in 1957 by Sir Vivian Fuchs. Stancomb-Wills Glacier is the largest glacier between Jutulstraumen (m3we11w-5n67-721 Fimbul Ice Shelf) and Slessor Glacier (m333o315-351n78-815 Filchner Ice Shelf), according to McIntyre (after Swithinbank 1988), it drains 35.000 km^2 which is less than the area of the drainage basin as described here.

Flowlines and crevasses of (part of) Stancomb-Wills Glacier are visible in an annotated LANDSAT MSS image (from 1981) (Swithinbank 1988, fig. 73, p. B98). The glacier has a wide shear margin to the slow-moving shelf ice of Riiser-Larsen Ice Shelf (to its northeast) and Brunt Ice Shelf (to its southwest). Noticeably, the centerline of the offshore part of the glacier has a smooth low for some part of its length reminiscent of the center of Jakobshavns Isbræ, a fast-moving ice stream in western Greenland (Mayer and Herzfeld 2001). The strain between the glacier tongue and the Brunt Ice Shelf is so large that it cannot be accomodated by deformation anymore, consequently, large icebergs calve from the ice sheet at or near the grounding line and are pulled along by the fast-moving ice stream, which forces them into a counterclockwise rotation (Swithinbank 1988, p. B97). The gaps are filled by so-called "low" ice shelf (sea ice which grows thicker over time). To the contrary, slowly-moving ice shelves usually have a vertical stratification, they consist of a thinning wedge of ice of land origin overlain by a thickening wedge of locally accumulated snow (and sea ice from to the bottom).

Lyddan Island is an ice rise between Stancomb-Wills Glacier and Riiser-Larsen Ice Shelf, it obstructs the westward flow of the ice shelf. Velocities of 1300 m a^{-1} on the right bank of the ice stream (Thomas 1973) and of 4000 m a^{-1} at the ice front (Orheim 1982) have been reported; from these observations and an estimate of 200 m ice thickness, Swithinbank (1988) estimates an ice discharge of 40 km^3 a^{-1} .

Map m357e339-15n71-77 New Schwabenland

New Schwabenland is an extensive mountainous area in northwestern Queen Maud Land ($\approx$ 800 km E to W). New Schwabenland, originally called Neuschwabenland, was first discovered by the German Antarctic Expedition 1938–39, led by A. Ritscher. The map gives a good impression of the topographic regions — the ice of central Queen Maud Land slopes down towards the northern coast (Atlantic sector of the circum-Antarctic ocean) and the western coast (Weddell Sea).

Queen Maud Land was originally called Dronning Maud Land after the Queen of Norway by the Norwegian Antarctic Expedition, led by Hjalmar Riiser-Larsen in 1930. Queen Maud Land extends from the terminus of Stancomb-Wills Glacier (20° W; see map m333e315-351n71-77 Riiser-Larsen Ice Shelf) to Shinnan Glacier (44°38'E) on Prince Olav Coast between Shirase Glacier and Rayner Glacier (see map m45e37-53n67-721 Prince Olav Coast). A southerly limit of this vast area is not explicitly given, nor are there other place names, so we use the name Queen Maud Land throughout 81.5°S.

In the higher elevation parts (3400–3000 m) the slope is very gentle ($\approx$ 200 m in 100–150 km), below 2800 m the regional topographic features start to dominate. The upper drainage basin of *Jutulstraumen Glacier* is one of the largest features, it extends 300 km from the terminus of Jutulstraumen Glacier directly south (to the 2800 m contour), near the 1800 m contour the main trough is joined by Penck Trough from the southwest. Jutulstraumen is the largest outlet glacier between 15°E and 20°W, draining an area of 124.000 km² (Swithinbank 1988, p. B92).

The nunatak area between *Penck Trough* and the main, southerly trough is called *Neumayer Cliffs*, it is part of the *Kirwan Escarpment*. The use of the name "Jutulstraumen" for the upper reaches varies from map to map. In the USGS Satellite Image Map (Ferrigno et al. 1996) and in Swithinbank (1988, p. B94, fig. 69), Jutulstraumen follows the southerly-easterly branch and is joined by ice from the more westerly Penck Trough; west of Borg Massif is Schytt Glacier, named after Swedish glaciologist Valter Schytt. In the text (Swithinbank 1988, p. B93), however, it is said that the main branch of Jutulstraumen Glacier

flows through Penck Trough and then follows a northerly trough, and is joined by an ice stream that flows in from east of the Neumayer Cliffs. We suggest to use "Western Upper Jutulstraumen Glacier" for the branch in Penck Trough and "Eastern Upper Jutulstraumen Glacier" for the branch that comes in from the east (see also fig. 70 in Swithinbank 1988).

Jutulstraumen Glacier has been studied extensively, several references are given in Swithinbank (1988), e.g. Orheim (1979). The geology and gravity field of the Jutulstraumen area have been described by Decleir and van Autenboer (1982). The bottom of Penck Trough reaches a few hundred meters below sea level. Ice velocities have been measured at 390 ma^{-1} at 72°15'S (discharge 12.5 km³a^{-1} at 72°15'S), and 1000 ma^{-1} at the ice front. In 1967, a 53x104 km ice tongue calved off, caused by another large iceberg floating by and impacting the front of Jutulstraumen Glacier.

The Kirwan Escarpment (see Helferich and Kleinschmidt 1998) is the prominent escarpment that runs in a really straight line SW to NE for 350 km in extension from the Heimefront Range (Heimefrontfjella) at point (260.000 E/-8320.000 N) to point (530.000 E/-8120.000 N) near the junction of Penck Trough and the southerly trough leading to Jutulstraumen Glacier. The Kirwan Escarpment is marked by a steep step descending from 2800 m to almost 2000 m. The escarpment appears to continue, at least morphologically, northeast of Jutulstraumen Glacier for another 120 km (but step only 2400 m to 2000 m). Further east are the Sverdrup Mountains and the Gielsvik Mountains, which run N–S on the National Geographic Atlas map (National Geographic Society 1992, p. 102) and are parts of the Mühlig-Hofmann Mountains. On the satellite image map (annotated NOAA 7 AVHRR mosaic of Queen Maud Land, produced by IfAG (1982) and reproduced in Swithinbank (1988, p. B93, fig. 68); AVHRR – advanced very high resolution radiometer; IfAG – Institut für Angewandte Geodäsie) the escarpment also appears to continue northeast of the southern branch of Jutulstraumen Glacier and further into the Mühlig-Hofmann Mountains.

Only part of the escarpment is actually called Kirwan Escarpment. North of Kirwan Escarp-

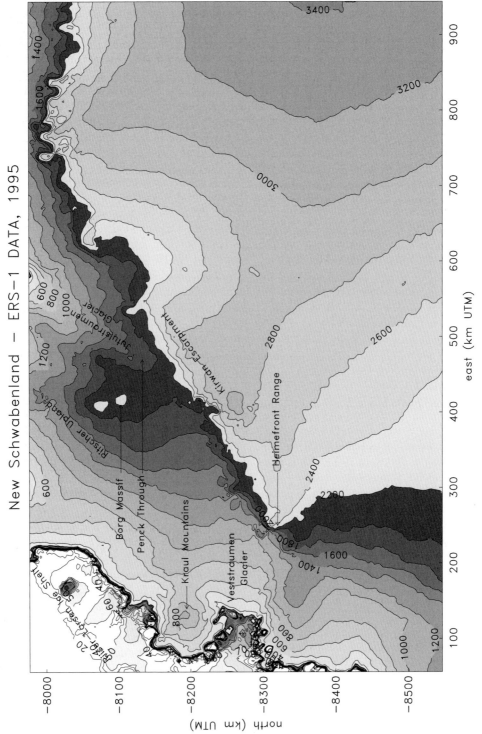

New Schwabenland – ERS–1 DATA, 1995

e339-15n71–77, WGS84, Gaussian variog., central mer. 357, slope corrected, scale 1:5000000, 980112

east (km UTM)

north (km UTM)

ment (where labeled on the map) and west of Jutulstraumen Glacier, the Ritscher Upland extends northwards, the mountains immediately west of Jutulstraumen Glacier include Borg Massif (2157 m), Mt. Hallgren (2337 m), and other mountains of 2546 m and 2579 m.

The *Heimefront Range* is about 80 km southwest of the Kirwan Escarpment and about 100 km long, it consists of three mountain groups (including *Kottas Berge*) and was discovered by air by the Norwegian-British-Swedish Antarctic Expedition in 1952. The Heimefront Range has exposed rocky escarpment steps of 400 m elevation difference.

Notably, Heimefrontfjella contains the "famous" Scharffenbergbotnen, the first glacier whose ice surface, ice thickness and bottom topography were mapped using geostatistics during the German Antarctic Expedition ANTARKTIS VI 1987/88, from radio-echosounding data collected under the (Swedish) SWEDARP program (Herzfeld and Holmlund 1988, 1990). The Swedish SWEA sta-

tion was built in the Heimefrontfjella. Scharffenbergbotnen is an interesting glacier in that it has an inflow that opposes its main flow direction.

Coastward of the Heimefront Range and *Ritscher Upland*, the ice descends with a steeper slope towards the Princess Martha Coast and the extensive Riiser-Larsen Ice Shelf (described on map m333e315-351n71-77 Riiser-Larsen Ice Shelf). The ice northwest of Ritscher Upland drains to *Ekström Ice Shelf* (convex 600 m contour), separated by *Cape Norvegia* (extension of the northwestern-most land on this map) from the Riiser-Larsen Ice Shelf (between Ekström Ice Shelf and Cape Norvegia in the small Quar Ice Shelf).

The *Kraul Mountains* are located close to the coast. *Veststraumen Glacier* (notice that "straumen" already means "ice stream") occupies a large trough that extends E–W about 100 km from Heimefront Range to Princess Martha Coast south of the Kraul Mountains.

Map m21e3-39n71-77 Sør Rondane Mountains

This map shows the *Wohlthat Mountains* and the *Sør Rondane Mountains* of Northeastern Queen Maud Land. The Sør Rondane Mountains are about 160 km east to west, which is their largest extension, and direct the flow of the inland ice to the east and west, creating an unusually steep slope to the north towards Princess Ragnhild Coast (see map m21e13-29n67-721 Erskine Iceport). The Wohlthat Mountains are in a range with the *Petermann Ranges, Weyprecht Mountains, Payer Mountains, Humboldt Mountains, Drygalski Mountains, Conrad Mountains, Kune Mountains, Filchner Mountains* (south of Princess Astrid Coast), which continue into the *Mühlig-Hofmann Mountains* (Luz Range, Gablenz Range, Gielsvik Mountains, Sverdrup Mountains) south of Fimbul Ice Shelf (to 0° E/W).

The Sør Rondane Mountains, a 150 km long mountain range, have been studied extensively since the time of the Norwegian Antarctic Expedition 1930–31. The Sør Rondane Mountains are named after the Rondane Mountains in Norway. The highest peak is *Isachsen Mountain* (72° 11'S/26° 15'E), with 3425 m, named after Gunnar Isachsen, leader with Hjalmar Riiser-Larsen of the 1930–31 Norwegian expedition. Both the Wohlthat Mountains and the Sør Rondane Mountains have been the objective of several Antarctic expeditions, and have been studied for their role in the geologic history of the Gondwanaland breakup, the former large continent in the southern hemisphere (parts of Africa, South America, Antarctica, Australia, and other landmasses) (Kleinschmidt et al. 1995, Kleinschmidt and Buggisch 1993, Buggisch and Kleinschmidt 1999). The Weddell Sea opened 70 million years ago.

The Japanese Asuka Station is located in the northern Sør Rondane Mountains near Mt. Romnaes (1500 m).

To the south and southeast, the land of Queen Maud Land increases to above 3600 m. The gradient lessens considerably above 3200 m elevation (200 m in over 100 km) as opposed to 200 m in 50 km (3200–3000 m) and 200 m in 25 km (3000–2400 m) in the area between the Sør Rondanes and the Wohlthat Mountains and even steeper towards the coast (1000 m in 75 km, corresponding to 0.76°), north of the Sør Rondane Mountains.

Named glaciers in the Sør Rondane Mountains are Byrdbreen, Mjell Glacier, Gjel Glacier, Gunnestad Glacier, which together discharge only 0.65 km^3a^{-1} (after Swithinbank 1988, p. B83; Van Autenboer and Decleir 1978). Surface velocities are less than 40 ma^{-1} and ice thicknesses of valley glaciers are 1000–1500m. Field expeditions have found small dry valleys that were ice-filled at earlier times, and other signs of strong lowering of the ice surface. The glacial level was several hundred meters above today's ice level.

No melting occurred (it is too cold there) but the ice was wasted by sublimation. Bare ice areas are visible in satellite images (1973, 1976). Precipitation in this area is very low, and snow accumulation, if any, occurs mostly by redistribution caused by strong winds. This fact makes observations of precipitation and accumulation in Antarctica very difficult to observe, but precipitation is an important variable needed to understand climate and glacier changes. (A satellite image of the Sør Rondane Mountains is given in Swithinbank (1988, fig. 61, p. B81).)

The Wohlthat and *Orvin Mountains* further west, also shown on a satellite image (Swithinbank 1988, fig. 62, p. B85), deflect the flow of the inland ice to the east and west, but have many glaciers that transect them (e.g. Humboldt Glacier, Somov Glacier, Glop Glacier, Sand Glacier). This area has been studied since the German Antarctic Expedition led by Ritscher. As in the Sør Rondane Mountains, dry valleys are found here, and signs of downwasting of several hundred meters of ice from the former to the present glacial level (Hermichen et al. 1985).

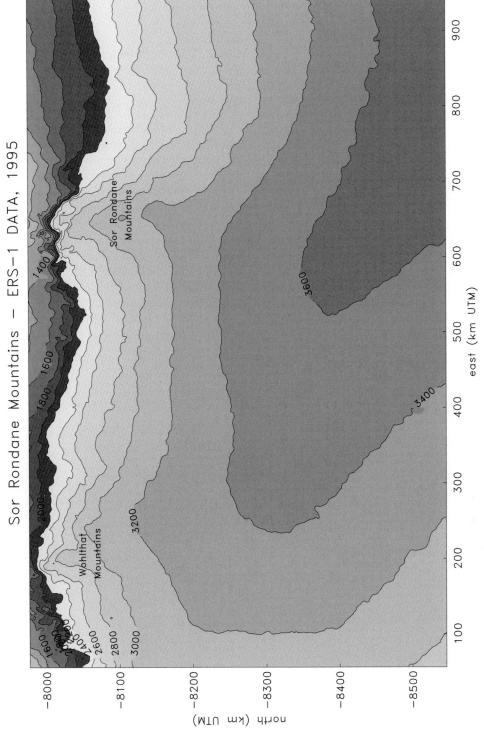

Sor Rondane Mountains — ERS—1 DATA, 1995

east (km UTM)

north (km UTM)

e3—39n71—77, WGS84, Gaussian variog., central mer. 21, slope corrected, scale 1:5000000, 970811

Map m45e27-63n71-77 Belgica Mountains

This map covers the area of eastern Queen Maud Land between Lambert Glacier to the east, *Enderby Land, Prince Olav Coast, Riiser-Larsen Peninsula* and the eastern part of *Princess Ragnhild Coast* to the north, and southern parts of Queen Maud Land to the south. Notably, the coast and coastal area of Enderby Land between Shinnan Glacier at 44° 38'E and William Scoresby Bay 59° 34'E does not have a special name (Alberts 1995, p. 221).

The topography here is fairly simple — in the northwestern part of the map are the *Belgica Mountains*, the elevation rises towards the south-southeast (from 2200 m to above 3600 m), towards *Valkyrie Dome* (77° 30'S/37° 30'E)(see map m45e27-63n75-80 Valkyrie Dome, to the south of this map). In the eastern part of the map, the surface slopes down towards the *Lambert Glacier system*. The divide between the interior of Queen Maud Land and the Lambert Glacier drainage basin is clearly determined and dominant in this map sheet.

The *Belgica Mountains* are an isolated, about 15 km long chain of mountains located southerly of and, in an east-westerly direction, between the *Sør Rondane Mountains* about 100 km to the west (see m21e3-39n71-77 Sør Rondane Mountains) and the *Queen Fabiola Mountains* in the east (see m33e25-41n67-721 Riiser-Larsen Peninsula). The highest point in the Belgica Mountains is Mt. Victor (2588 m). The mountain range was discovered by the Belgian Antarctic Expedition in 1957–58 under G. de Gerlache and named after the ship Belgica, commanded by his father, A. de Gerlache, leader of the 1897–99 Belgian Antarctic Expedition (Alberts 1995, p. 56).

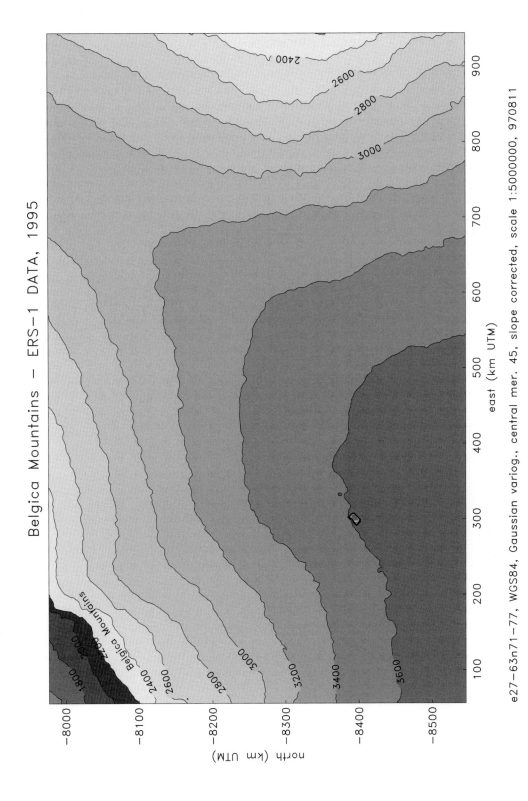

Belgica Mountains – ERS-1 DATA, 1995

e27-63n71-77, WGS84, Gaussian variog., central mer. 45, slope corrected, scale 1:5000000, 970811

Map m69e51-87n71-77 Upper Lambert Glacier

Upper Lambert Glacier is, of course, only visible on the ERS-1 map because of its southerly location. This map is the southern continuation of map m69e61-77n67-721 Lambert Glacier (and maps E and W of it). The most prominent feature is the evenness of the Lambert Glacier drainage basin, with U-shaped contour lines of almost equal spacing continuing throughout this map (and beyond, see map m69e51-87n75-80 South of Lambert Glacier, as the drainage basin of Lambert Glacier is much larger). The spacing of the contour lines is narrower in the western side of Lambert Glacier, in the area of the rugged Prince Charles Mountains, than in the eastern, American Highland, area, where the land is less rugged and less steep. (Upper) Lambert Glacier is formed by the confluence of (upper) *Lambert Glacier* (proper) and *Mellor Glacier*. Lambert Glacier comes in from directly near Mawson Escarpment and has traceable flowlines (on LANDSAT images) for about 100 km farther upstream from Mawson Escarpment; Mellor Glacier flows approximately 20 km farther west, between Mellor and Lambert Glaciers lie (from N to S) *Cumpston Massif, Mt. Maguire, the Blake Nunataks*, and *Wilson Bluff*; ice of (upper) Lambert Glacier and Mellor Glacier appears to also flow around Mt. Twigg and Mt. Borland; all these are nunataks which stick out of the flowing ice (Swithinbank 1988). The distinction between Mellor Glacier and Lambert Glacier ice does not seem to be unambiguous — on the USGS Satellite Image Map of Antarctica (Ferrigno et al. 1996), all the ice described here is only Lambert Glacier, whereas the name "Mellor Glacier" is attributed to a smaller and subordinate glacier coming in from farther west. So, the main stream of flow into Lambert Glacier is actually termed Lambert Glacier, which makes the most sense. The third branch that joins Mellor Glacier and upper Lambert Glacier to form the main trunk of Lambert Glacier is *Fisher Glacier*. It is about as wide as Mellor Glacier and upper Lambert Glacier and flows in from the west. Fisher Glacier has been hypothesized to surge (see discussion in section (E.7)).

Meltwater lakes have been reported near Cumpston Massif, upper Lambert Glacier.

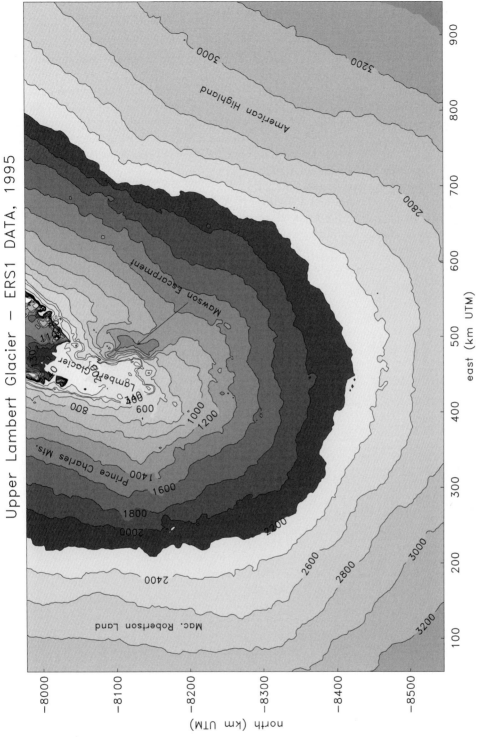

Upper Lambert Glacier – ERS1 DATA, 1995

e51–87n71–77, WGS84, Gaussian variog., central mer. 69, slope corrected, scale 1:5000000, 970801

183

Map m93e75-111n71-77 American Highland

The dominant feature on this map is a broad ridge, with elevations of 3200 m to above 3600 m; this constitutes the divide between the *Lambert Glacier* drainage on its western side and the *Denman Glacier* drainage on its eastern side. Whereas on the western side, ice clearly drains to the Lambert Glacier/Amery Ice Shelf system, ice on the eastern side reaches glaciers on the coast of *Wilkes Land*. *American Highland* appears as an area of rough surface ice or snow with features of 3–5 kilometers characteristic spacing, possibly induced by subglacial mountains or changes in surface slope.

These features have a larger characteristic spacing and are offset at shorter distances than the so-called "megadunes" described by Swithinbank (1988) for an area in map 117e99-135n75-80 East Antarctica (Vostok), but they could also have causes of subglacial morphology, surface morphology, or snow albedo changes.

The *Grove Mountains* are a group of nunataks near (72° 45'S/75° E), scattered over an area of 60 km by 30 km; close by (to the east) is the *Gala Escarpment*. Otherwise the surface of American Highland consists of snow and ice.

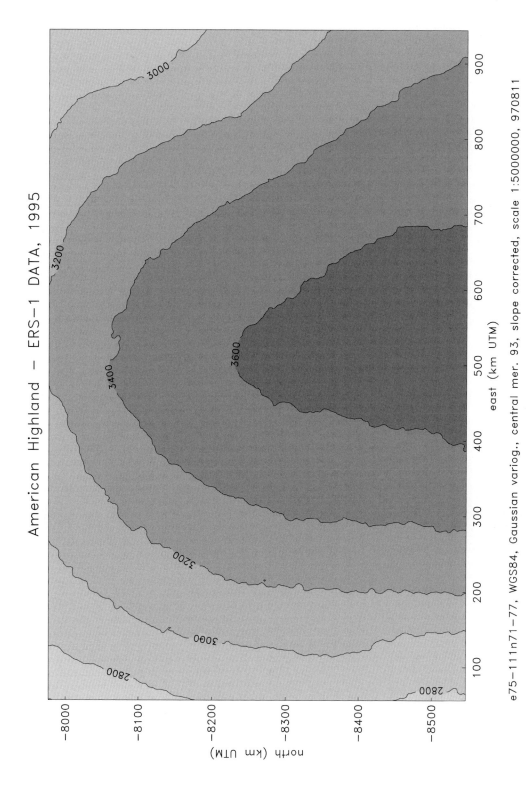

American Highland — ERS-1 DATA, 1995

north (km UTM)

east (km UTM)

e75–111n71–77, WGS84, Gaussian variog., central mer. 93, slope corrected, scale 1:5000000, 970811

Map m117e99-135n71-77 Dome Charlie

The ice on the northern and central part of this map, which shows part of *Wilkes Land*, drains northwest towards *Denman Glacier*, and more generally, *Queen Mary Coast* and *Knox Coast*. The ice below the concave part of the 2800 m and 3000 m contour lines in the NW corner of the map appears to flow towards the Denman Glacier Basin, which we deduce from comparison with the northerly and westerly adjacent maps.

Dome Charlie (center coordinates (75° S/125° E)) is a morphologic feature of less than 200 m elevation difference to the surrounding area (about 100 m above the connecting "ridge" to the southwest, judging from the general slope in its neighbourhood). A subtle ice divide extends northeast from Dome Charlie. The ice dome was termed "Dome C" or "Dome Charlie" (Charlie is the communication code word for the letter c) or "Dome Circe", and "Dome Charlie" became the official name. Dome Charlie was the site of ice core drilling by several nations in the 1970's. The U.S. Naval Support Force suffered damage to three LC-130 Hercules aircraft during attempted takeoff from Dome Charlie, which were later repaired and flown out (Alberts 1995, p. 130).

It may be worth noting that the elevation (less than 3200 m) reported elsewhere, likely by field teams (Alberts 1995, p. 130), coincides with the elevation mapped by altimetery, at this level of accuracy.

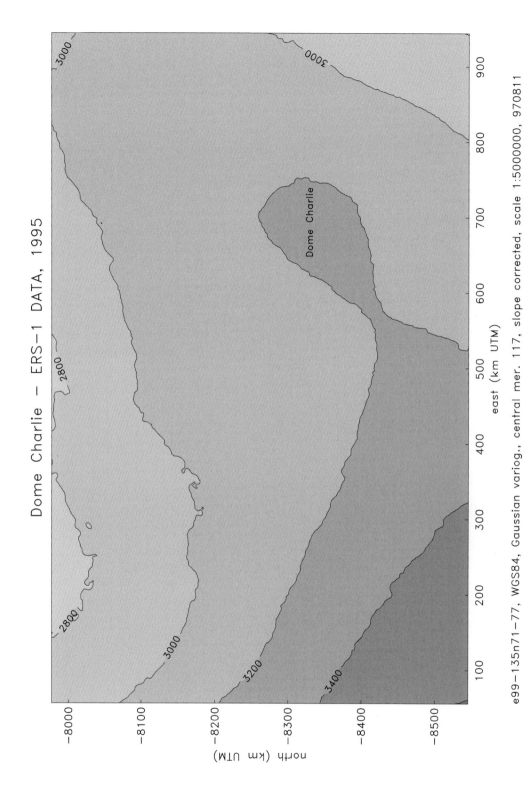

Dome Charlie — ERS–1 DATA, 1995

north (km UTM)

east (km UTM)

e99–135n71–77, WGS84, Gaussian variog., central mer. 117, slope corrected, scale 1:5000000, 970811

Map m141e123-159n71-77 Southern Wilkes Land (e123-159)

Map m141e123-159n71-77 Southern Wilkes Land (e123-159) shows that part of Wilkes Land that reaches from near Dome Charlie to Victoria Land. *Dome Charlie* is recognized by the small patch of dark gray above the 3200 m contour (near the label -8400 on the left map margin), Dome Charlie is an ice dome rising to above 3100 m (see map m117e99-135n71-77 Dome Charlie, the westerly adjacent map). With a very low gradient (lowest in the range 2400–2600 m, 200 m in 200 km; $0.1\% = 0.057°$ gradient) the ice surface descends form the interior to the large mountainous headlands of Victoria Land which separate the circum-Antarctic oceans from the Ross Sea to the east. Below 2200 m elevation, the gradient to the east steepens significantly (to 200 m in 40 km; $0.5\% = 0.29°$), leading towards Terra Nova Bay and Scott Coast (see maps m165e147-183n71-77 Victoria Land, m153e145-161n67-721 Cook Ice Shelf, m165e157-173n67-721 Pennell Coast).

Centered at $(73° \, S/158° \, E)$ is *Talos Dome*, a large ice dome that rises to 2262 m southwest of the Usarp Mountains. In our map, this is in the area of the white nose between 2200 m and 2400 m. Ice thickeness at Talos Dome is 2476 m. The dome overlies the eastern margin of Wilkes Subglacial Basin. (Talos Dome is named after Talos in the Greek mythology who assisted Minos in the defense of Crete (Alberts 1995, p. 732).)

Wilkes Subglacial Basin is a large subglacial basin situated generally south of George V Coast and westward of Prince Albert Mountains, centered at $(75° \, S/145° \, E)$, it was mapped by U.S. seismic groups in 1958-1960 (Alberts 1995, p. 813).

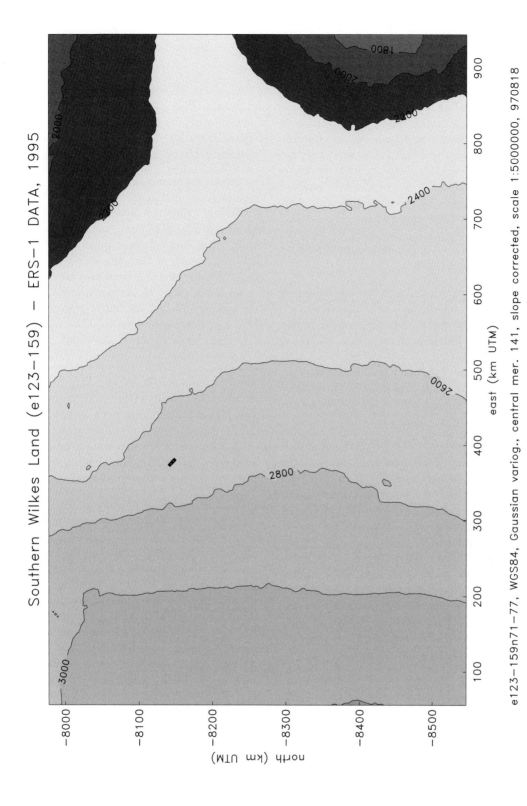

Map m165e147-183n71-77 Victoria Land

This map shows *Victoria Land* and its east coast, south of Cape Adare (71° 17'S/170° 14'E). *Cape Adare* is the north tip of Adare Peninsula, it is formed of black basalt and, as such, in contrast to the surrounding snow and ice covered sea and coast. The cape was discovered in 1841 by Captain James Ross and the site of the first landing in 1895 by Borchgrevink (see description of map m165e157-173n67-721 Pennell Coast). Other than the coast of Wilkes Land, the coast of Victoria Land has been an area of intensive exploration by many research expeditions.

The light area above the 2200 m contour enhances the location of the ice divide between the Ross Sea to the east and the circum-Antarctic ocean to the north, this divide branches off at 2200 m, an extension leads towards Cape Adare – of course locally, only some glaciers transcend the Transantarctic Mountains and so are outlet glaciers of the Antarctic inland ice, while many others are of local origin in the mountains.

Talos Dome is centered at (73° S/158° E), about in the bifurcation part of the contour lines, this large ice dome rises to an elevation of 2262 m (see also map m141e123-159n71-77 Southern Wilkes Land (e123-159)). At the scale of 1:5.000.000, the area to the east of the 2200 m contour appears as a large, shallow bowl, centered around the bay of David Glacier on Scott Coast. There are coastal mountain ranges in its south and north. The north-south trending valley descending to below 1000 m in the north of the map at 450.000 E is the valley of Rennick Glacier (see description of map m165e157-173n67-721 Pennell Coast), to the west of this is the southern part of the Usarp Mountains.

Borchgrevink Coast, named after Carstens E. Borchgrevink who landed on Cape Adare in 1895 and was the first overwinterer in Antarctica (1898–1900) as leader of the British Antarctic Expedition (1898–1900), extends from Cape Adare to *Cape Washington* (74° 39'S/165° 25'E); the latter is a 275 m high cape below Mt. Melbourne at the southern end of the peninsula that separates *Wood Bay* and *Terra Nova Bay* (named for Capt. Washington, secretary of the Royal Geographic Society 1836–1840). *Scott Coast*, named after British Antarctic Explorer Robert Scott, lies to the south

of Borchgrevink Coast, between Cape Washington and *Minna Bluff* (78° 31'S/166° 25'E), a narrow, bold peninsula, 40 km long and 25 km wide, that projects from Mt. Discovery into Ross Ice Shelf (on map m165e147-183n75-80 Scott Coast).

The mountain ranges and some of the glaciers of Victoria Land inland of Borchgrevink Coast are Mt. Sabine (3720 m) directly above *Cape Adare* and *Mt. Minto* (4165 m) further east, the *Admiralty Mountains*, further inland is Mt. Aorangi (3135 m), south of the Admiralty Mountains is *Tucker Glacier*, flowing southeast to east. Tucker Glacier is an about 15 km wide and 170 km long local glacier (i.e. not an outlet glacier), it has a 60 km long floating tongue (Swithinbank 1988, fig. 38, p. B48) and ends in *Tucker Inlet*.

South of Tucker Glacier are the *Victory Mountains* with Mt. Riddolls (3295 m), Mt. Northampton (2467 m) and Mt. Brewster (2027 m). South of the Victory Mountains is Mariner Glacier, it descends from the *Evans Neve*, a glaciated high plateau east of upper *Rennick Glacier* and northwest of Mt. Supernal (3655 m) to the coast, Mariner Glacier has a tongue and flows through a small ice shelf.

To the east of the ice shelf is Mt. Murchison (3600 m), and further south *Aviator Glacier* (140 km long and 5–10 km wide), which takes a turn to the east and then has a long ice tongue in Wood Bay. The ice tongues of Mariner Glacier and Aviator Glacier are likely the ones on our map, as they are the longest tongues. Even further south is *Mt. Melbourne* (2730 m), a prominent mountain near the coast (a stratovolcano with hot ground near its summit), south of it *Campbell Glacier* (140 km long, 9–10 km wide; already on Scott Coast, as Mt. Melbourne is above Cape Washington, the southern limit of Borchgrevink Coast). Between this glacier and Terra Nova Bay are the German Gondwana Station and the Italian Terra Nova Station (both summer stations).

In the area near Mt. Melbourne every glacier that reaches the coast extends into the sea as a glacier tongue (as seen on the LANDSAT image of the area, Swithinbank 1988, fig. 37, p. B46). The floating tongues tend to hold the fast ice in place until late in the summer and sometimes all year. *Campbell Glacier* is a local glacier, its val-

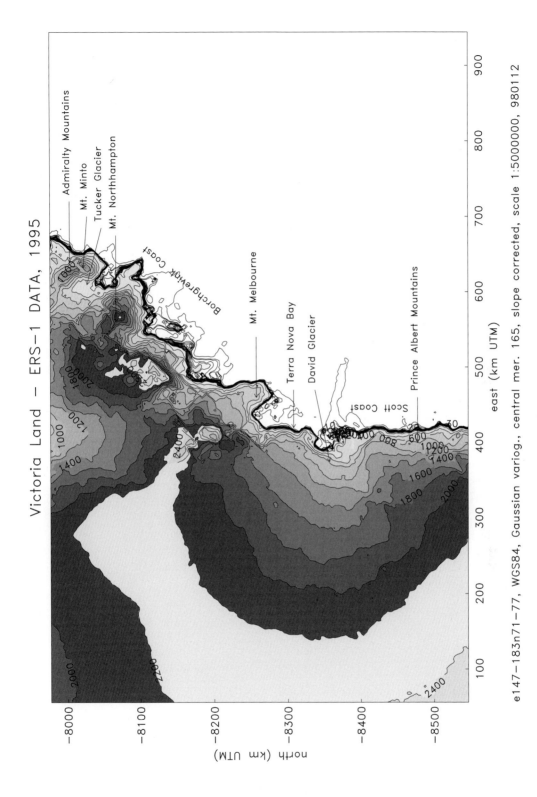

Victoria Land – ERS–1 DATA, 1995

e147–183n71–77, WGS84, Gaussian variog., central mer. 165, slope corrected, scale 1:5000000, 980112

ley is discernable trending in the direction of the peninsula with Mt. Melbourne. The next valley south of Campbell Glacier is occupied by Priestley Glacier, which shows distinct flowlines on a satellite image and may be moving faster. *Priestley Glacier* turns south, following a turn in the mountain valley, flows through *Nansen Ice Sheet* and terminates near Inexpressible Island. *Reeves Glacier* is a shorter glacier entering Nansen Ice Sheet. Priestley and Reeves Glaciers are both outlet glaciers of the East Antarctic Plateau, all other glaciers in the area are local. Sublimation due to dry, warmer wind coming from the plateau may explain why extensive ablation areas are confined to the two outlet glaciers (Swithinbank 1988, p. B45). In Terra Nova Bay is a *polynia* (an open water anomaly), whose existence is also attributed to the relatively warm katabatic winds coming down Reeves Glacier. Reeves Glacier has significant characteristic shear patterns on its floating tongue, which flows around a rock obstacle, while Priestley Glacier shows only flowlines.

The mountains inland of Scott Coast are all the *Prince Albert Mountains*, they extend from inland of Terra Nova Bay (the northern limit of Scott Coast) to 76.5° S. Between 76.5° S and 72° S is Convoy Range, with Mawson Glacier in the north and Mackay Glacier in the south (see map m165e147-183n75-80 Scott Coast).

South of the small headland (-8320.000 N/410.000 E) south of Terra Nova Bay is *David Glacier* that terminates in *Drygalski Ice Tongue*. This is the largest glacier (over 100 km long) in the region and drains from the inland ice. As concluded from the elevation contours on our map, David Glacier extends 300 km inland. This matches observations of a subglacial trench (Steed and Drewry 1982). The drainage basin is 224.000 km^2 (McIntyre after Swithinbank 1988). David Glacier is also the largest glacier in the bowl-shaped area of the Antarctic Plateau west of Victoria Land. Because Priestley and Reeves Glaciers are also outlet glaciers, David Glacier does not drain the entire bowl, but likely a segment delineated by a diagonal line from the northern headland northwest and a diagonal line from the southern headland southwest, the latter being indicated by topography.

David Glacier/Drygalski Ice Tongue are described in detail with the detail map David Glacier/Drygalski Ice Tongue (section (F.12)).

Mawson Glacier, an outlet glacier further south at (76° 13'S/162° 05'E), also has an ice tongue with a different name, *Nordenskjöld Ice Tongue*.

As more mountainous regions lie close to the coast, more small glaciers enter the sea.

Map m213e195-231n71-77 Ruppert Coast

(Coordinates e195-231 correspond to w165-129)

This map shows part of the northwestern coast of the Marie Byrd Land, West Antarctica, Saunders Coast, Ruppert Coast, and Hobbs Coast.

The Phillips Mountains are at (-8500.000 N/ 640.000 E), and Brennan Point is at (-8430.000 N/ 530.000 E) marking the limit of *Saunders Coast* to the south and west and *Ruppert Coast* to the east. Ruppert Coast extends eastward to Cape Burks (74° 45'S/136° 50'W), which is a prominent rock cape, the northwest extension of McDonald Heights (Mt. Gray) and the east side of Hull Bay. On Cape Burks is Russkaya, a Russian research station.

Further south are the *Ford Ranges*, to the south of those the *Clark Mountains, Mackay Mountains*, and the *Allegheny Mountains*. South of Hull Glacier and *Hobbs Coast* (which is east of Hull Bay), are the *Flood Ranges*, from west to east Mt. Berlin (3498 m), Mt. Moultan (3069 m), Mt. Bursey (2779 m). *Mt. Berlin* interrupts a northward-flowing ice sheet and has an exposed height of 1300 m on the south side and 2100 m on the N side (Swithinbank 1988, p. B127), it is a late Cenozoic stratovolcano (LeMasurier 1972) with a 1500 m diameter summit crater. Ice thickness is 1500 m. These mountains are part of the *Marie Byrd Land volcanic province*, a volcanic chain that parallels the coast of West Antarctica from the Jones Mountains at longitude 94° W (266° E) in the Ellsworth Mountains (see map m285e267-303n78-815 Ellsworth Mountains) to the Fosdick Mountains at 145° W (215° E). Based on volcanic evidence, it can be shown that glaciation in Marie Byrd Land began in the Eocene, but it is not clear whether the glaciation remained uninterrupted to the present (LeMasurier and Rex 1982).

Land Glacier is a prominent glacier between the Phillips Mountains and McCuddin Mountains, it drains from the interior ice. *Hull Glacier* is a much smaller glacier. Land Glacier is broad, 55 km long and heavily crevassed. According to Alberts (1995), Hull Glacier is also 55km long, but it is narrower. On satellite maps, Hull Glacier appears much smaller than Land Glacier (Ferrigno et al. 1996; see also Swithinbank 1988, fig. 98, p. B132, for Hull Glacier). Between Land Glacier and Hull Glacier are no significant glaciers (Swithinbank 1988, p. B128). There is little channeling of the glaciers, including Hull Glacier, which is consistent with the immature volcanic landscape of Pleistocene age. Instead, there is an active ice wall coastline which discharges icebergs into the sea. This area of Antarctica is known (from weather satellite observations) as being one of the oldest sea ice areas. Although the individual icebergs continually move north, the pattern of icebergs stays about the same, since new icebergs calve off. A question is what forces icebergs to separate and move apart. It has been hypothesized that freezing sea water in tidal cracks causes the initial separation of icebergs (Debenham 1948; Swithinbank 1957a) or that continued minor calving from adjacent icebergs forces them apart (Nichols 1960, after Swithinbank 1988, p. B129).

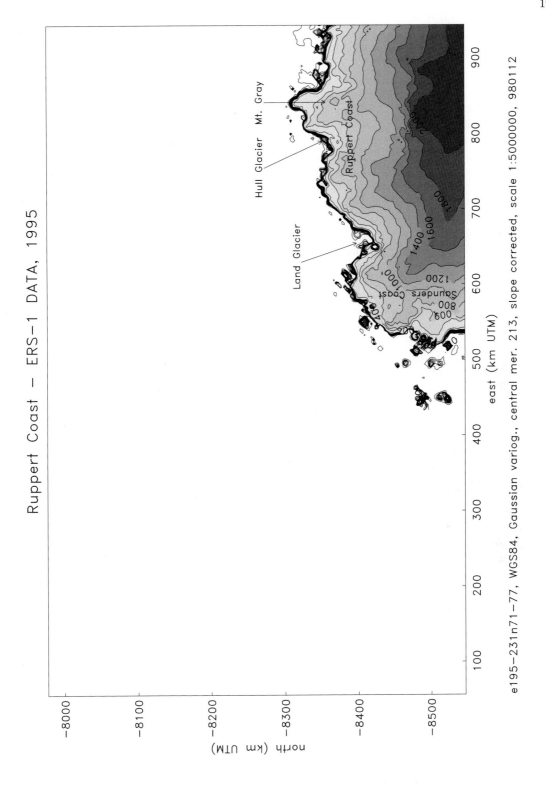

Ruppert Coast – ERS-1 DATA, 1995

e195-231n71-77, WGS84, Gaussian variog., central mer. 213, slope corrected, scale 1:5000000, 980112

Map m237e219-255n71-77 Bakutis Coast

(Coordinates e219-255 correspond to w141-105)

This map shows almost 900 km of a coast of Marie Byrd Land in West Antarctica. Approximately paralleling the coast is a volcanic province, i.e. the mountains rising above the ice sheet between Jones Range (94° W) in Ellsworth Land and Fosbick Range (145° W) are volcanoes. The volcanic province is described in the text to maps m237e219-255n75-80 Northern Marie Byrd Land and m213e195-231n71-77 Ruppert Coast. Executive Committee Range belongs to the volcanic ranges.

The coasts are Hobbs Coast (from the east side of Hull Bay, Cape Burks to a point on the coast at (74° 42'S/127° 05'W) opposite Dean Island, named for geography Professor W. Hobbs of University of Michigan). Bakutis Coast (from there to Martin Peninsula, ice-covered Cape Herlacher (73° 52'S/114° 12'W)) and Walgreen Coast (east of Martin Peninsula).

The entire length of Hobbs Coast and Bakutis Coast is bordered by Getz Ice Shelf, more than 500 km long and 30–100 km wide. Getz Ice Shelf is protected by Grant Island, Siple Island with Mt. Siple and Cape Dart, Carney Island (Russell Bay is between Siple and Carney Islands), Duncan Peninsula and Wright Island. Because of the presence of the islands, Getz Ice Shelf should be stable. Mt. Siple is also volcanic, it has the greatest exposed height (3110 m) and the greatest exposed diameter (50 km) of any volcano in West Antarctica (Swithinbank 1988, p. B127). The eastern end of Getz Ice Shelf is Martin Peninsula with Cape Herlacher (73° 52'S/114° 12'W).

South of Bakutis Coast and Hobbs Coast is the Executive Committee Range.

East of Martin Peninsula is Dotson Ice Shelf, bordered on the east by Bear Peninsula with Moore Dome and on the south by Kohler Range. Kohler Glacier flows past the northwest side of Kohler Range and into Dotson Ice Shelf at (-8370.000 N/750.000 E). Richmond Peak (3595 m) and Toney Mountain, also volcanic, are just south of Kohler Range. Mt. Takahe is south of Bear Peninsula (see map m237e219-255n75-80 Northern Marie Byrd Land). Smith Glacier flows southeast of Bear Peninsula and north of Mt. Murphy, it extends into the Amundsen Sea at (-8300.000 N/805.000 E).

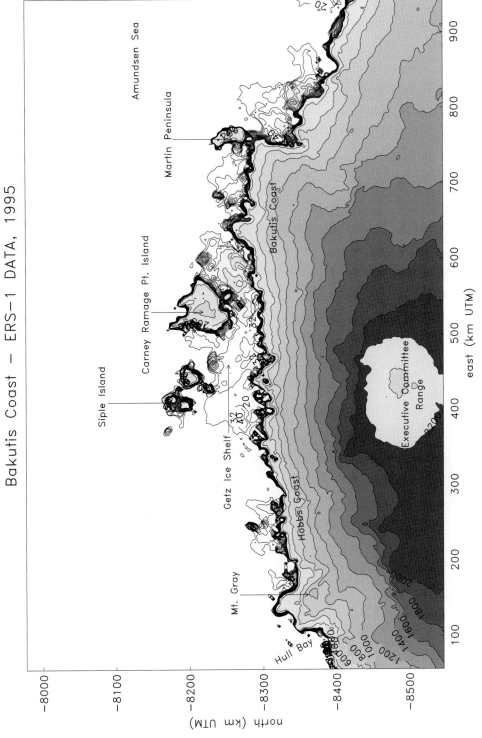

Bakutis Coast – ERS–1 DATA, 1995

e219–255n71–77, WGS84, Gaussian variog., central mer. 237, slope corrected, scale 1:5000000, 980112

Map m261e243-279n71-77 Walgreen Coast

(Coordinates e243-279 correspond to w117-81)

The map *Walgreen Coast* shows northeastern Marie Byrd Land, bordered here by Walgreen Coast, and northwestern Ellsworth Land, east of the longitude of *Pine Island Bay*. Walgreen Coast extends from Martin Peninsula (Cape Herlacher (73° 52'S/114° 12'W) on map m231e219-255n71-77 Bakutis Coast) to King Peninsula (Cape Waite (72° 44'S/103° 16'W)). Walgreen Coast was discovered in 1940 by the Byrd Expedition in flights from the ship *Bear* and named after Charles Walgreen of Walgreen Drug Company of Chicago, who supported the 1933–35 and 1939–41 expeditions financially.

Cape Waite on King Peninsula is the northeastern tip of Ellsworth Land and the northeastern continental land of Amundsen Sea. *Amundsen Sea* is named after Norwegian explorer Roald Amundsen who first reached the South Pole in 1911. Beyond Peacock Bay, south and north of Cape Waite, and further east beyond Abbot Ice Shelf is Thurston Island with the Walker Mountains. In the small ice shelf between Thurston Island and Eights Coast is Sherman Island. On the ERS-1 altimetry map, these islands appear connected to King Peninsula, and the costal topography contains small-scale errors. Canisteo Peninsula is correctly mapped, however.

Eights Coast and *Bryan Coast* form the coasts of Ellsworth Land to Bellingshausen Sea in the north. Thaddeus Bellingshausen was leader of the Russian Antarctic Expedition 1819–1821. *Eights Coast* extends from Cape Waite, King Peninsula, east to Pfrogner Point (72° 37'S/89° 35'W), Fletcher Peninsula. James Eights counts as the first American scientist in the Antarctic region, as he carried out geologic investigations in the South Shetland Islands (at 62° N, northwest of the tip of the Antarctic Peninsula).

To the east of Fletcher Peninsula lies *Bryan Coast*, its eastern limit is the northern tip of Rydberg Peninsula. George Bryan was a hydrographer, 1938–1946. The coast on this map is largely bordered by ice shelves, but those have different properties. On the Marie Byrd Land section of Walgreen Coast are *Dotson Ice Shelf* at

8300.000–8380.000 S at the western map margin, *Thwaites Glacier Tongue* and ice in *Pine Island Bay*. Thwaites Bay does not contain an ice shelf, but Pine Island Bay does (see discussion on detail maps 15 Thwaites Glacier (F.13) and 16 Pine Island Glacier (F.14)). On the west coast of Ellsworth Land, also Walgreen Coast, are parts of *Pine Island Bay*, and, north of Canisteo Peninsula, *Cosgrove Ice Shelf*. Inland and south of Canisteo Peninsula are the *Hudson Mountains*. Dotson Ice Shelf is described with map m237e219-255n71-77 Bakutis Coast.

East of Fletcher Peninsula and west of Allison Peninsula on Bryan Coast is a small ice shelf, *Venables Ice Shelf*. Siple Station (USA) is near (76° S/85° W).

Pine Island Glacier and *Thwaites Glacier* are the two biggest ice streams draining the northern margin of the West Antarctic Ice Sheet between 90° W and 160° W, that is, in all of Ellsworth Land and Marie Byrd Land, and they are located only 200 km apart.

Thwaites Glacier extends further out into the sea than any other glacier, it has a large floating glacier tongue and a grounded iceberg tongue (see detail map in (F.13)). The tongue was discovered in 1946–47 and has been existing since then. In 1966, the glacier tongue was 63 km long, and the iceberg tongue was 110 km long. Ship travel was hindered by the Thwaites Glacier and Iceberg Tongue. In a LANDSAT1MSS image of 1972–73 (Swithinbank 1988, p. B126, fig. 93), Thwaites Iceberg Tongue is broken off from Thwaites Glacier Tongue and rotated about 40° westward, but otherwise it appears to have stayed in place, it is grounded on shoals (Holdsworth 1985). The main part of Thwaites Glacier occupies a deep trench that extends inland well below sea level. It has been hypothesized by Hughes (1977) that Thwaites Glacier surges, because the iceberg tongue increased by 60% between 1965 and 1974, as noted by Southard and MacDonald (1974) and MacDonald (1976), which could be the result of a recent surge. (A glacier surge is a sudden acceleration of a glacier to up to 100 times its nor-

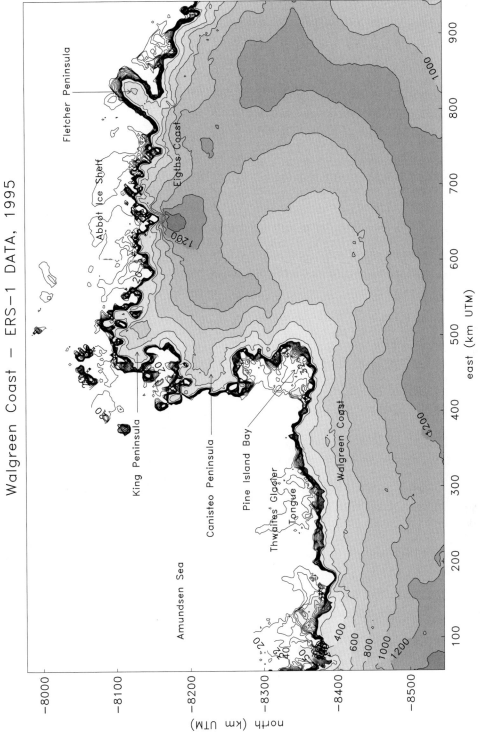

Walgreen Coast − ERS−1 DATA, 1995

east (km UTM)

north (km UTM)

Fletcher Peninsula

Eights Coast

Abbot Ice Shelf

King Peninsula

Canisteo Peninsula

Pine Island Bay

Thwaites' Glacier Tongue

Walgreen Coast

Amundsen Sea

e243−279n71−77, WGS84, Gaussian variog., central mer. 261, slope corrected, scale 1:5000000, 980112

199

mal velocity. Only some glaciers surge. Surges occur quasi-periodically, with a short surge phase of rapid movement during which the glacier advances, and a long quiescent, normal phase.)

Thomas et al. (1979) quote a photogrammetric velocity estimate of 2 km a^{-1} after R. Allen and consider that equilibrium velocity. According to Lindstrom and Tyler (1985), the *Thwaites Glacier Tongue* moves 3.5 km a^{-1}. The glacier drains an area of 121.000 km^2 and should have a balance discharge of 47 km^3 a^{-1}. On the ERS-1 Atlas map, Thwaites Glacier and Iceberg Tongue appear to extend 100 km into Amundsen Sea. On morphological grounds, the drainage basin may extend south to the area of the Hollick-Kenyon Plateau, to 350 km further south (measured to the 2000 m contour). It is not clear, however, how much of the area drains to Thwaites Glacier.

Pine Island Glacier appears to drain the entire area south of the coastal volcanic ranges that border the northern coast of Ellsworth Land. The high elevation area at (-8180.000 S/660.000 E) indicates the Jones Mountains, the easternmost range of the volcanic province that extends to the Fosbick Mountains at 145° W (see other maps on the volcanic province). *Cosgrove Ice Shelf* (further north) has a much smaller drainage basin which extends inland about 50 km, whereas the drainage basin of Pine Island Glacier extends 350 km inland (eastward). From the ice divide, that is mapped by the 1200 m contour, ice drains eastward to the Ronne Ice Shelf (Rutford Ice Stream, glacier in Carlson Inlet, Evans Ice Stream, see maps m285e267-303n71-77 Ellsworth Land and m285e267-303n75-80 Zumberge Coast).

According to Crabtree and Doake (1982), the drainage basin of Pine Island Glacier is 214.000 km^2 +/− 20.000 km^2 and mass flux at the ice front is 25 +/− 6 gigatons, equal to 28 km^3, per year. As concluded from a LANDSAT1MSS image (1973), Pine Island Glacier is afloat for only about 80 km before it calves into Pine Island Bay. The grounding line is close to a 1400 m thickness measurement. Icefront velocity from satellite images (1973, 1975) is 2.4 km a^{-1}, and comparison to 1966 aerial photographs indicates 10 km retreat or calving.

The *West Antarctic Ice Sheet* is called a *marine ice sheet*, because the ice sheet lies on a rock basement that is well below sea level. In scenarios of a break-up of the West Antarctic Ice Sheet (see Introduction, chapter (A)), Pine Island Glacier and Thwaites Glacier play a special role, because they are large ice streams that are not protected from the sea by ice shelves. This presents a rarity among West Antarctic ice streams.

In a disintegration model by T. Hughes (after Swithinbank 1988, p. B124), this area would most likely start a collapse of the West Antarctic Ice Sheet: "Surging could produce a basal water layer that would uncouple the ice from its bed and thus draw down the surface level of the ice sheet. In the absence of a high bedrock sill to prevent it, the grounding line could migrate inland until ultimately the whole marine portion of the ice sheet is converted into an ice shelf." Hughes (1973) and Thomas et al. (1979) suggested that today the northern part of the ice sheet could already be collapsing. The exceptionally low ice surface gradient, which is obvious east of Pine Island Bay, might support this hypothesis. In contrast, Crabtree and Doake (1982) modeled the longitudinal profile of Pine Island Glacier using steady-state assumptions and found no evidence of instability.

Map m285e267-303n71-77 Ellsworth Land

(Coordinates e267-303 correspond to w93-57)

The "backbone" of the Antarctic Peninsula swings around to the west and continues in the mountains of Ellsworth Land. The large geographic units on this map are *Palmer Land* (the southern part of the Antarctic Peninsula, see also map m292e284-300n67-721), *Alexander Island, Ellsworth Land*, the *Bellingshausen Sea* to the north and the *Ronne Ice Shelf* and the *Weddell Sea* to the east. Palmer Land is very rugged, whereas Ellsworth Land has a smoother surface topography. The large basin in the center of the map drains to *Evans Ice Stream*, which enters Ronne Ice Shelf north of Fowler Ice Rise, an ice-covered peninsula. The ice masses of Evans Ice Stream are seen to elevate the surface of Ronne Ice Shelf by 90 m near the coast, and extending for 400 km across Ronne Ice Shelf (see also southerly adjacent map m285e267-303n75-80 Zumberge Coast). The drainage basin extends 230 km inland to an ice divide between Ronne Ice Shelf and the Bellingshausen Sea (-8270.000 N/410.000 E), the width of the basin is about 400 km. Ice thickness measurements of Evans Ice Stream are 800 m to 1240 m (Swithinbank 1988, p. B121, fig. 89).

The east coast of Palmer Land at this latitude is *Black Coast* (Cape Boggs (70° 33'S/61° 23'W), the headland of Eielson Peninsula, to Cape Mackintosh (72° 50'S/59° 54'W), the tip of Kemp Peninsula, near Mason Inlet) and *Lassiter Coast* (Cape Mackintosh, Kemp Peninsula, to Cape Adams (75° 04'S/62° 20'W), the southern tip of Bowman Peninsula and northern entrance to Gardner Inlet). Both Black Coast and Lassiter Coast in this map row 71-77 are better seen and described on the easterly adjacent map, m309e291-327n71-77 Black Coast.

The ice area at (-8300.000- -8350.000 N/810.000-860.000 E), where the Ross Ice Shelf borders the Weddell Sea, is receiving ice from several glaciers north and south of Bowman Peninsula (Nantucket Inlet to its north, Westmore Glacier, Ketchum Glacier in Gardner Inlet south of the Peninsula, Ueda Glacier, Hansen Inlet further south). The northern edge of the *Ronne Ice Shelf* coincides roughly with *Bowman Peninsula*, i.e. Lassiter Coast/Palmer Land borders the Weddell Sea and Orville Coast and Zumberge Coast on

Ellsworth Land border Ronne Ice Shelf. Precisely, *Orville Coast* extends from Cape Adams (Bowman Peninsula) to Cape Zumberge (76° 14'S/69° 40'W), a rocky cape. James Zumberge was an American glaciologist who worked in Antarctica, Orville a naval meteorologist who designed part of the RARE Program. Cape Zumberge is located between the Hauberg Mountains to its NE, and the Behrendt Mountains, to its NW. (John Behrendt is a geophysicist who has worked in Antarctica repeatedly in 6 decades: 1957–2002.) Only small glaciers drain from Orville Coast.

Zumberge Coast is the southwestern coast of Ronne Ice Shelf, it extends to Skytrain Ice Rise and the coast near the Ellsworth Mountains (see map m285e267-303n78-815 Ellsworth Mountains). Cape Zumberge (at -8500.000 N/610.000 E) marks a point between rugged topography (north and east) and smooth topography with ice-covered peninsulas and large ice streams, the northernmost of which is Evans Ice Stream.

In the northern part of the map, we also distinguish a coastal and inland area of higher relief in Palmer Land (English Coast and surroundings) and an area of less relief on the northern coast of Ellsworth Land, but this is not as smooth as Zumberge Coast. In the northeast, the southern continuation of George VI Sound is seen (for the northern part, see the northerly adjacent map m292e284-300n67-721 Antarctic Peninsula (Palmer Land)). The sound separates Alexander Island and Palmer Land and is occupied by George VI Ice Shelf (except for a small area in the north).

In a January 1973 LANDSAT1 MSS image, the shelf ice of George VI Ice Shelf extends past the small Eklund Islands and DeAtley Island to Spaatz Island, but there are many areas of ice rumples and large crevassed zones which are weakness zones that may facilitate disintegration of the ice shelf. George VI Ice Shelf has started to disintegrate, about 5% of its area were affected by 2001 (Vaughan et al. 2001). North of Spaatz Island is Ronne Entrance of George VI Sound, which is ice-free, Beethoven Peninsula is also part of Alexan-

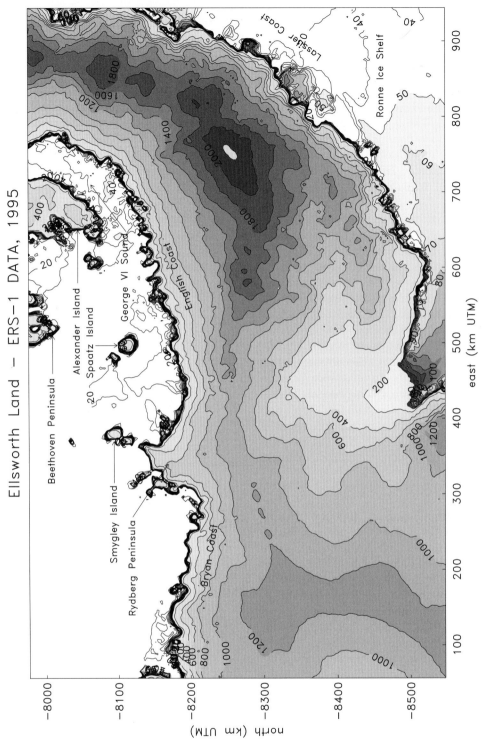

Ellsworth Land – ERS–1 DATA, 1995

e267–303n71–77, WGS84, Gaussian variog., central mer. 285, slope corrected, scale 1:5000000, 980112

der Island. Several small glaciers enter the ice shelf from the east.

West of Spaatz Island is ice-covered Stange Sound, protected by Smygley Island, on its far side. The coastal sections are (NE to SW to W): *Rymill Coast* (Cape Jeremy (69° 24'S/68° 51'W), the northern limit of Palmer Land, to Buttress Nunataks (72° 22'S/66° 47'W)) and *English Coast* (Buttress Nunataks to Rydberg Peninsula (73° 10'S/79° 45'W)). English Coast borders both the southern part of Palmer Land along the southern part of George VI Sound and a part of Ellsworth Land, so the geographic limits are a bit oddly chosen. West of Rydberg Peninsula lies *Bryan Coast*, its western limit is Pfrogner Point (72° 37'S/89° 35'W) at Fletcher Peninsula. West of Pfrogner Point is *Eights Coast*, which extends all the way to the northwesternmost headland of Ellsworth Land, Cape Waite on Kings Peninsula (see westerly adjacent map m261e243-279n71-77 Walgreen Coast). In the western part of Bryan Coast is *Venable Ice Shelf*, $\approx$ 60 km long by 40 km wide. Off Eights Coast is the large *Abbott Ice Shelf*, partly protected by small islands and by 150 km long Thurston Island in the west (see map m261e243-279n71-77 Walgreen Coast). The interior of Ellsworth Land rises above 1200 m.

Map m309e291-327n71-77 Black Coast

(Coordinates e291-327 correspond to w69-33)

Most of the area on this map is covered by the *Weddell Sea*. The land in the western part of the map is the southeastern part of the Antarctic Peninsula (*Palmer Land, Black Coast* and *Lassiter Coast*) with many outlet glaciers that protrude into the sea ice, the sea, and the *Ronne Ice Shelf*. In the southeastern part of the map area, the Weddell sea is bordered by the Ronne Ice Shelf. The large Ronne Ice Shelf is also seen and described on maps m285e267-303n71-77 Ellsworth Land, m285e267-303n75-80 Zumberge Coast, m285e267-303n78-815 Ellsworth Mountains, m309e291-327n75-80 Ronne Ice Shelf, and m309e291-327n78-815 Berkner Island; good descriptions of the ice shelf are given with the latter two maps.

The eastern coast of the Peninsula is bordered by many small ice shelves, fed by outlet glaciers of the mountainous Peninsula. The area near the edge of the Ronne Ice Shelf is the area near Nantucket Inlet. For an image map of the Peninsula, see Ferrigno et al. (1996), inset map Antarctic Peninsula, but no elevations are given there.

At (-8400.000 to -8500.000 N/700.000 to 800.000 E), large icebergs are floating.

Wetmorc Glacier and *Ketchum Glacier* are joining to enter *Gardner Inlet* just south of the north-ern limit of Ronne Ice Shelf and to the south of *Nantucket Inlet*, separated by *Bowman Peninsula* (the boundary cape between *Lassiter Coast* (N) and *Orville Coast* (S)). In the same area are *Ueda Glacier* and *Hansen Inlet*. All those glaciers add to the ice mass mapped at (-8320.000 to -8380.000 N/110.000-140.000 E) near 74.5–75°S in Ronne Ice Shelf. Approximately, Lassiter Coast borders the Weddell Sea, and Orville Coast borders Ronne Ice Shelf. A glacier whose ice can he followed inland for 150 km enters New Bedford Inlet near 73.5°S, it passes south of Mt. Axworthy (1639 m) (Lassiter Coast).

Cape Mackintosh (72°50'S/59°54'W) at *Kemp Peninsula* (73°08'S/60°15'W) is the boundary of *Black Coast* (N) and *Lassiter Coast* (S). Large ice shelves are seaward of Black Coast, (1) offshore of Cape Knowles (71°48'S/60°50'W) to Kemp Peninsula is one large ice shelf of about 130 km, this ice shelf is seen between -8060.000 N and -8190.000 N on our map; it contains Gruening Glacier and Beaumont Glacier in Hilton Inlet, Butler Island, Cape Christmas, Violante Inlet, and Mason Inlet. (2) Just north of Cape Knowles is Odom Inlet, a large glacier that passes between Mt. Jackson (3050 m) and Rowley Massif enters here, its ice mass is seen offshore of Black Coast at (-8020.000 N/100.000 E).

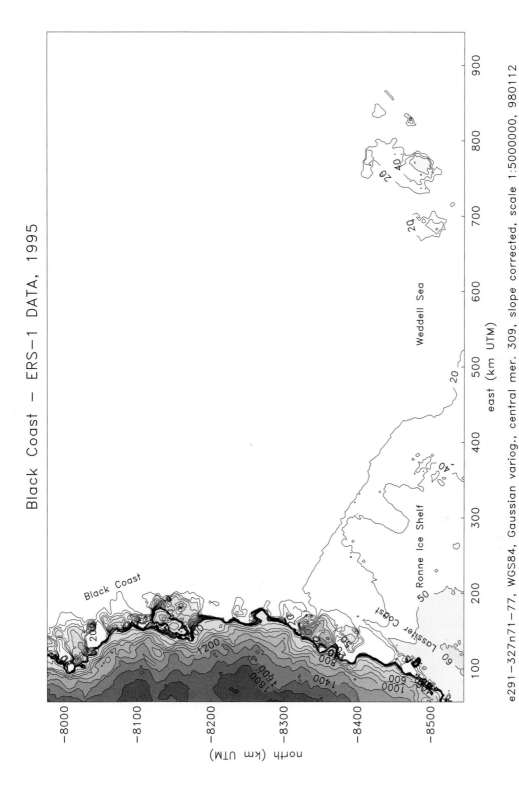

Black Coast – ERS–1 DATA, 1995

e291–327n71–77, WGS84, Gaussian variog., central mer. 309, slope corrected, scale 1:5000000, 980112

(D.4) Latitude Row 75-80°S: Maps from ERS-1 Radar Altimeter Data

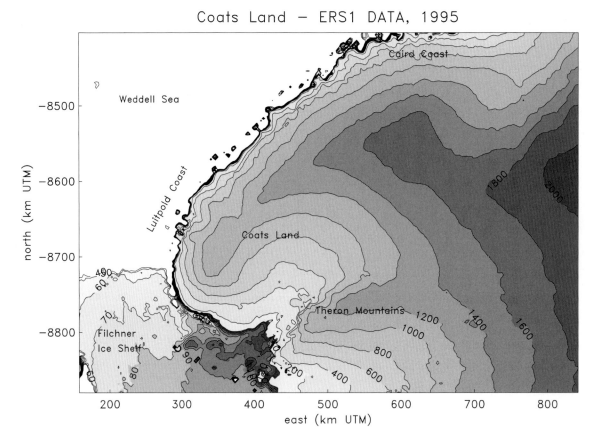

Coats Land — ERS1 DATA, 1995

e315–351n75–80, WGS84, Gaussian variog., central mer. 333, slope corrected, scale 1:5000000, 971105

Map m333e315-351n75-80 Coats Land

(Coordinates e315-351 correspond to w45-9)

Coats Land forms the westernmost part of southern Queen Maud Land, along the Luitpold Coast (southerly part of the west side of Coats Land) and Caird Coast (northerly part), which borders the Weddell Sea. Caird Coast extends from the terminus of the Stancomb-Wills Glacier (20° W) to the vicinity of Hayes Glacier (27° 54'W). Caird Coast was sighted by Shackleton in 1915 and named for the patron of his expedition (Alberts 1995). Luitpold Coast is westerly adjacent to Caird Coast, it extends from the vicinity of Hayes Glacier to 36°

W, the geographic eastern limit of the Filchner Ice Shelf. Luitpold Coast was discovered by W. Filchner during the German Antarctic Expedition 1911–12 and named for Prince Regent Luitpold of Bavaria. The southern coast of Coats Land borders *Filchner Ice Shelf*.

The whole land mass of Coats Land follows a wide arc, trending southwest to west to west-northwest to northwest to west-southwest and finally to south-southwest, throughout the area of this map, with elevations decreasing from above

2000 m to the coast. This arc forms a divide between the northern areas and the southerly basins which drain into Filchner Ice Shelf. The divide continues east into Queen Maud Land (see description to map m357e339-15n75-80 Western Queen Maud Land (North)). Small glaciers must drain off the northwest coast into the Weddell Sea, but are not distinguishable here. The gradient of the northwesterly coast increases from 200 m in 30 km (1600 m to 1400 m; 0.67% = 0.38°) to 200 m in 20 km (1000 m to 800 m; 1% = 0.57°) and to 200 m in less than 5 km between 600 m and 400 m elevation (4% slope = 2.3°). The 800 m contour and lower contours indicate rugged terrain along most of the coast. The slope towards Filchner Ice Shelf is less (200 m in 10–15 km, 1–2%), and the terrain is smooth. A glacier system drains into Filchner Shelf Ice north of the Theron Mountains. The *Theron Mountains* peak at 1175 m above sea level and have an extension of 40 km. The Argentinian Belgrano Station is located on southern Luitpold Coast. The southernmost drainage basin seen on this map is that of the large *Slessor Glacier* (see map m333e315-351n78-815 Filchner Ice Shelf and detail map Slessor Glacier, section (F.1)).

Western Queen Maud Land (North) — ERS1 DATA, 1995

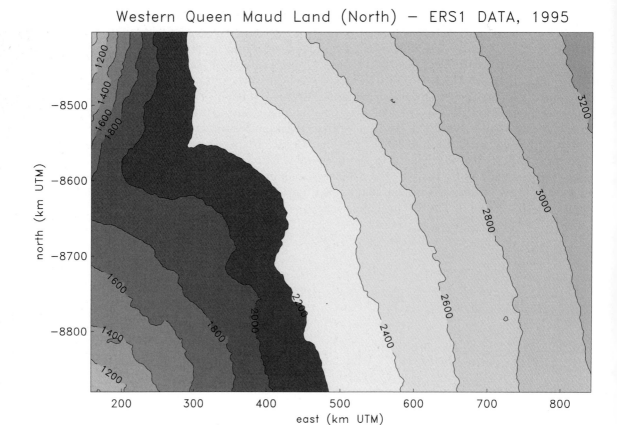

e339−15n75−80, WGS84, Gaussian variog., central mer. 357, slope corrected, scale 1:5000000, 971106

Map m357e339-15n75-80 Western Queen Maud Land (North)

The Western Queen Maud Land (North) map shows the ice divides between the drainage basin of *Stancomb-Wills Glacier* (which terminates between Brunt Ice Shelf and Riiser-Larsen Ice Shelf in the eastern Weddell Sea, see map m333e315-351n71-77 Riiser-Larsen Ice Shelf) and Bailey Ice Stream (which flows through Coats Land and drains into Filchner Ice Shelf north of the Theron Mountains); this divide is marked by the "nose" in the 2200 m contour line at location (-8570.000 N/300.000 E) and in the 2000 m contour line at (-8600.000 N/200.000 E) and continues on map m333e315-351n75-80 Coats Land in a large northerly-westerly-southerly arc to the head of Coats Land.

Further south is a more subtle divide (mapped at (-8720.000 N/420.000 E) in the 2200 m contour

and also in the 2000 m contour) that separates the Bailey Ice Stream drainage from the drainage system of *Slessor Glacier* (see maps m333e315-351n75-80 Coats Land and m333e315-351n78-815 Filchner Ice Shelf). The gradient in the upper Stancomb-Wills Glacier drainage is much steeper (800 m from 1800 m to 1000 m elevation in 65 km, corresponding to 200 m in about 16 km, 1.23% = 0.71°) than the gradient in the upper basin of the glaciers that flow to the Filchner Ice Shelf (Bailey Ice Stream and Slessor Glacier) (600 m from 1800 m to 1200 m in 180 km, corresponding to 200 m in 60 km, 0.3% = 0.19°). Above the 2200 m contour, the gradient is typical of the slope of interior Queen Maud Land (200 m in about 100 km, 0.2% = 0.1°), increasing slightly to 3200 m elevation (which is the maximum elevation on this map).

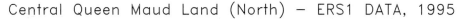

Central Queen Maud Land (North) — ERS1 DATA, 1995

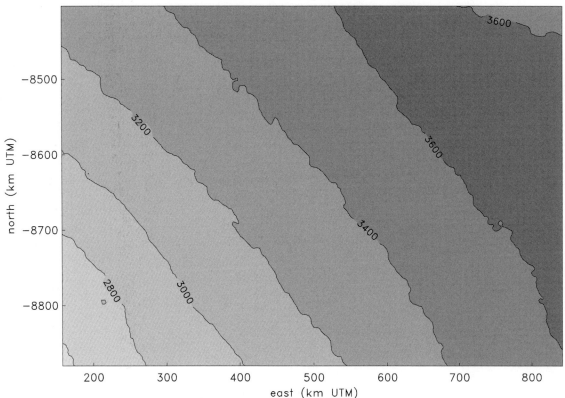

e3–39n75–80, WGS84, Gaussian variog., central mer. 21, slope corrected, scale 1:5000000, 971106

Map m21e3-39n75-80 Central Queen Maud Land (North)

This map shows Central Queen Maud Land (North), south of *New Schwabenland* (first called Neuschwabenland by the German Antarctic Expedition 1938–39, led by Alfred Ritscher), a mountainous upland that extends 800 km E to W in the northern part of Queen Maud Land. The mountains are all north of the area covered by this map.

In the northeastern part of the map, there is a dark grey area bordered by two 3600 m contours, this is a western-to-northwestern continuation of the divide on which *Valkyrie Dome* lies. Drainage to the north is towards the northern coast of the Atlantic sector of the circum-Antarctic sea; drainage to the east (of the Valkyrie Dome divide) is to *Lambert Glacier*, drainage to the west is to the Weddell Sea. The surface slopes down in a southwesterly direction toward the *Coats Land* sector of the eastern coast of the Weddell Sea (more than 600 km distant).

210

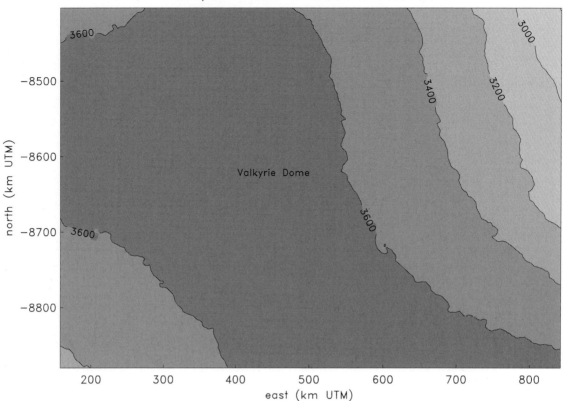

e27-63n75-80, WGS84, Gaussian variog., central mer. 45, slope corrected, scale
1:5000000, 971105

Map m45e27-63n75-80 Valkyrie Dome

Valkyrie Dome is an ice dome that rises to about 3700 m at center coordinates (77° 30'S/37° 30'E) in eastern Queen Maud Land — and it is the only distinguished feature on this map. Queen Maud Land in the area is fairly flat, as seen from the wide spacing of contour lines — the slope is only 200 m in 100–150 km.

In 1963–64 an oversnow traverse of the Soviet Antarctic Expedition crossed the northern part of Valkyrie Dome (first called Valkyriedomen), in 1967–1969 the dome was surveyed by airborne radio-echosounding by members of the Scott Polar Research Institute (Cambridge, UK) and the Technical University of Denmark (Lyngby) (after Alberts 1995, p. 775). On Valkyrie Dome, the permanent Japanese Station Dome Fuji was built. The historic U.S. American Plateau Station was located 2 degrees due south. From the wide area of Valkyrie Dome, the surface slopes gently to the southwest, the northnorthwest, and in the direction of the Lambert Glacier drainage basin, to the east and northeast. The upper elevations of the Lambert Glacier drainage basin are on this map, and Valkyrie Dome is part of the divide between Lambert Glacier drainage basin and interior Queen Maud Land.

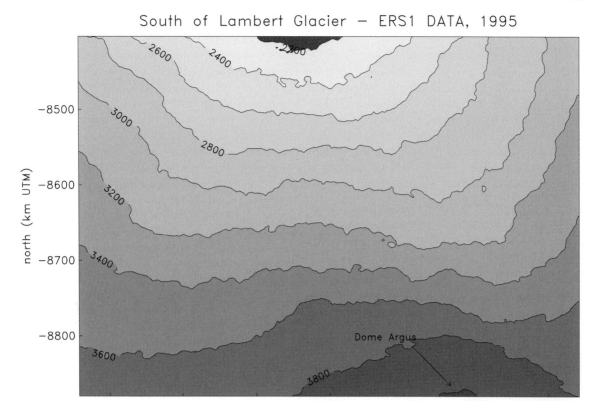

South of Lambert Glacier – ERS1 DATA, 1995

e51–87n75–80, WGS84, Gaussian variog., central mer. 69, slope corrected, scale 1:5000000, 971105

Map m69e51-87n75-80 South of Lambert Glacier

This map shows the area south of Lambert Glacier. The topography here is dominated by one main feature — the southern part of the fairly evenly shaped drainage basin of Lambert Glacier (see m69e61-77n67-721 Lambert Glacier, m69e51-87n71-77 Upper Lambert Glacier and adjacent maps to the east and west). Elevations in the area range from 2400 m (in the north center) to just above 4000 m in the area of Dome Argus at the southeastern map edge.

The drainage basin appears to have two arms — a westerly and an easterly one. The westerly one appears to turn into *Mellor Glacier* and *Fisher Glacier*, which is immediately west of Mellor Glacier, while the easterly one becomes *(upper) Lambert Glacier*. At the confluence, those glaciers are actually close together. Fisher Glacier, Mellor

Glacier and upper Lambert Glacier form the main trunk of Lambert Glacier.

Dome Argus is a fairly flat ice dome (see map m69e51-87n78-815 Dome Argus), but the highest ice feature in Antarctica (of course not the highest mountain — that is Mt. Vinson in the Ellsworth Mountains at 5140 m) with just over 4000 m elevation. Dome Argus (first called "Dome A") was mapped by the SPRI-NSF-TUD airborne radio-echosounding program (1967–1979) (Alberts 1995, p. 26); the abbreviations stand for Scott Polar Research Institute, Cambridge, UK (SPRI), National Science Foundation, U.S.A. (NSF), Technical University of Denmark, Lyngby, Denmark (TUD).

The area of the *Gamburtsev Subglacial Mountains* (≈70–80° E/ 77–79° S on the USGS satellite image map (Ferrigno et al. 1996)) is on this map, north to northwest of Dome Argus. A center coordinate of (80° 30'S/ 76° 0'E) is given in Alberts (1995, p. 267). The Gamburtsev Subglacial Mountains were detected by a Soviet seismic party in 1958 (named after geophysicist Grigory A. Gamburtsev (1903–1955)) and are said to extend beyond the area of Dome Argus. The discrepancies in coordinates indicating the extension of the subglacial mountains in different literature sources attest to the state of knowledge in the interior of Antarctica.

A LANDSAT1 MSS image of the ice surface above the Gamburtsev Subglacial Mountains reproduced in Swithinbank (1988, p. B64, fig. 49) indicates a surface topography which descends in offset steps to the northwest, in the direction of Lambert Glacier. Subdued escarpment slopes cause shadows in the (visual) satellite image, which indicate the presence of subglacial ridges or steps. Information on the thickness of the ice above the subglacial mountains is not given, or possibly in the original Russian field report. There are subregular snow dunes at the foot of many of the escarpments. A field party of the Soviet Antarctic Expedition during the International Geophysical Year made seismic measurements, the ice thickness above the rugged and elevated subglacial mountain range is in places only 600 m to 800 m. Ice surface elevation is about 3500 m. The existence of subglacial lakes has been postulated from strong bottom reflections in seismic records in the neighbourhood (Robin et al. 1970; after Swithinbank 1988, p. B65), as long as 180 km.

East Antarctica (Sovetskaya) – ERS1 DATA, 1995

e75–111n75–80, WGS84, Gaussian variog., central mer. 93, slope corrected, scale 1:5000000, 971105

Map m93e75-111n75-80 East Antarctica (Sovetskaya)

Map m93e75-111n75-80 East Antarctica (Sovetskaya) shows the southern extension of the broad ice divide that is seen on the (northerly adjacent) map m93e75-111n71-77 American Highland and that separates the *Lambert Glacier* drainage basin from the interior of *Wilkes Land* which drains to the Indian Ocean sector of the circum-Antarctic ocean, and, in its eastern part, to the Ross Sea. At elevations above 3600 m, the ridge is 300–350 km wide. The ridge reaches 3800 m in its southern center, hence the E–W slope is 200 m in 180 km, but in the SW corner of the map is a steep section with a 3600–3800 m N–S slope over 30 km.

Further southwest is *Dome Argus* (see map m69e51-87n78-815 Dome Argus). The Russian Sovetskaya Station is at approximately (78° S/88° E).

214

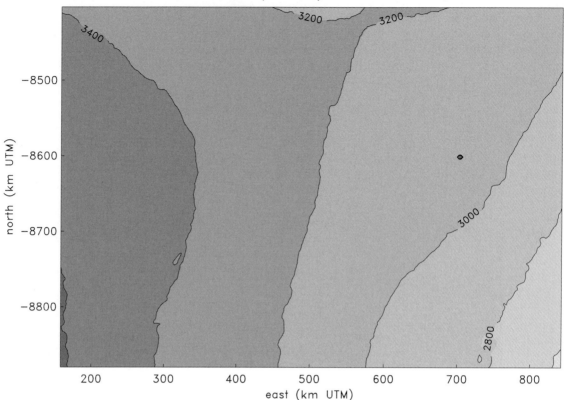

e99−135n75−80, WGS84, Gaussian variog., central mer. 117, slope corrected, scale 1:5000000, 970723

Map m117e99-135n75-80 East Antarctica (Vostok)

In the part of the interior of East Antarctica that is mapped here, the ice surface has elevations from 2600 m (SE map corner) to 3600 m above WGS 84 (SW map corner) and generally slopes down to the east with a gradient of 200 m in 180 km (3400 to 3200 m) and to the north with a very low gradient. The narrow section of ice above 3200 m in the center north of the map is the shallow ridge that extends to Dome Charlie, further northeast (see map m117e99-135n71-77 Dome Charlie).

Drainage to the east is towards the Ross Sea (as may be concluded from looking also at the easterly adjacent map m141e123-159n75-80 East Antarctica (Mt. Longhurst). Mt. Longhurst is on the Ross Coast!

Swithinbank (1988) concludes the existence of snow megadunes at elevations of roughly 3000 m in the region (126°–132° E/76°–77° S) from a LAND-SAT 1MSS image (Swithinbank 1988, fig. 45, p. B57–58). The features are subparallel, slightly wavy, a bit offset to each other and have a characteristic distance of 2.5 km from one maximum to the next. Swithinbank (1988) attributes the existence of such "megadunes" most likely to the redistribution of newly fallen snow by wind, with lighter bands representing newer snow than darker bands, or to morphologic structures, where slight change in slope may result in change in albedo. Snow megadunes are likely to be physically similar to sand megadunes, as described by Cornish (1914) and Bagnold (1941) for eastern Iran and the

Namib desert; the height of sand dunes is several tens of meters, the height of snow dunes 1–2 m, so they may not be noticeable to the surface traveler. These megadunes appear to be different patterns than the structures seen in the USGS Landsat map for American Highland — more elongated and of shorter characteristic length.

Vostok Station, a Russian research station, is located at approximately (78.5° S/107° E), at an elevation of 3488 m. Here an ice thickness of 3700 m was measured (National Geographic Atlas, National Geographic Society 1992, p. 102) which means that the Antarctic Ice Shield extends below sea level.

If all the ice would melt, however, the top of the then exposed rock surface would not be below sea level anymore, because weight of the ice sheet pushes the rock layer downward (into the Earth's mantle, simply speaking, which is more viscous; it may be imagined as a thick fluid), and as the ice melts, the weight is taken off and the rock layer (the crust) rises, with some time lag. This is called isostatic compensation, since there is a time lag, the land mass may still be rising after the ice is long gone. Only after the land stopped rising, one speaks of isostatic equilibrium (also, when the ice is not growing in thickness and the land is neither sinking nor rising, ie. when the forces are balanced). For instance, land in Scandanavia, around the northern end of the Baltic Sea, is still rising several millimeters per year, after the Fennoskandian Ice Shield melted after the end of the last ice age, over 10,000 years ago, and hence, the area is not in isostatic equilibrium.

Vostok Station bears the distinction of being the coldest place on Earth (or, the coldest place with temperature measurements), as a record low temperature of -89.2° C (-128.5° F) was measured here on the 21st of July, 1983.

Not far away as is the *Geomagnetic South Pole*. The Geomagnetic South and North Poles mark the axis of the Earth's magnetic field; they wander with time (and so the compass declination changes slowly at every place on Earth).

East Antarctica (Mt. Longhurst) — ERS1 DATA, 1995

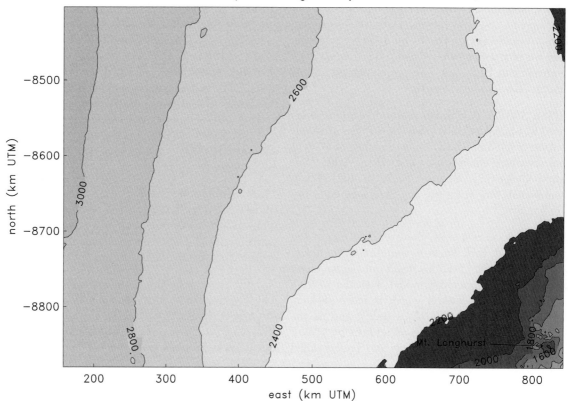

e123–159n75–80, WGS84, Gaussian variog., central mer. 141, slope corrected, scale 1:5000000, 971105

Map m141e123-159n75-80 East Antarctica (Mt. Longhurst)

In this area of East Antarctica, the ice surface descends from the interior (Highpoint at *Dome Charlie*, northwest of the map area above the 3000 m contour) in an eastward direction towards the Ross Sea in the east (3000 m to 2200 m contours). Between the interior and the coast lies the Victoria Land section of the Transantarctic Mountains. In the southeastern corner of the map is *Mount Longhurst* (79° 26'S/157° 18'E) (2846 m) near *Hillary Coast*, which borders Ross Ice Shelf. Mt. Longhurst is the highest point of Festive Plateau in the Cook Mountains (discovered by the British National Antarctic Expedition 1901–1904, led by Robert Scott). In the mountain ranges, the elevation is not correct, because areas of high vertical relief cannot be mapped accurately with satellite radar altimetry. Mt. Longhurst appears as between 2000 m and 2200 m high). But Mt. Longhurst is a useful land mark that can be identified on three maps: (1) this map, (2) m165e147-183n75-80 Scott Coast, and (3) m165e147-183n78-815 Hillary Coast, but not on m141e123-159n78-815 Byrd Glacier, giving a good idea of the overlap and relative location of adjacent maps.

Scott Coast — ERS1 DATA, 1995

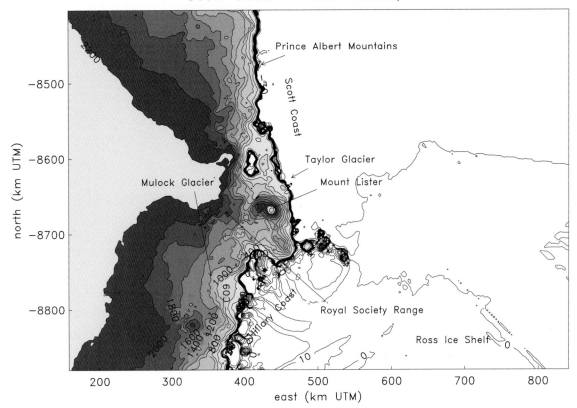

e147−183n75−80, WGS84, Gaussian variog., central mer. 165, slope corrected, scale 1:5000000, 971105

Map m165e147-183n75-80 Scott Coast

This map shows southern Victoria Land, Scott Coast and Hillary Coast. Scott Coast borders the Ross Sea largely, Hillary Coast borders the Ross Ice Shelf. *Scott Coast* extends from Cape Washington (74° 39'S/165° 25'E) between Wood Bay and Terra Nova Bay to Minna Bluff (78° 31'S/166° 25'S). The northern part of Scott Coast is described in the text of map m165e147-183n71-77 Victoria Land, including Mawson Glacier and Mackay Glacier.

The area south of Mackay Glacier contains the *"Dry Valleys"* of Victoria Land, and from north to south, Olympus Range, Asgard Range and Royal Society Range. A good image and map at scale 1:1.000.000 of the area is given in the insert "Mc-Murdo Sound Area" of the USGS Satellite Image Map (Ferrigno et al. 1996). The area has seen a lot of research and exploration because of the location of the American McMurdo Research Station, the largest station in Antarctica, and Scott Station (New Zealand).

The Dry Valley region is a cold desert, about 2000 km² large, that is essentially free of snow. The major valleys are *Taylor*, *Wright* and *Victoria Valleys*. They separate east-west trending mountain ranges. Victoria Valley is north of *Olympus Range*, Wright Valley is south of Olympus Range and north of Asgard Range, Taylor Valley is south of Asgard Range and north of *Kukni Hills*. Ferrar Glacier south of Kukni Hills is not dry but a real glacier. There are also small subordinate dry valleys. Outlet glaciers from the inland ice sheet enter the valleys from the west, but the central valleys are dry. In the coastal areas the precipitation is higher again, and this explains why glaciers form in the lower valleys (e.g. Victoria Lower Glacier), and piedmont glaciers form.

Large glaciers are *Debenham Glacier*, *Victoria Lower Glacier*, *Wright Lower Glacier* (Olympus and Asgard Range), all ending in *Wilson Piedmont Glacier*; *Ferrar Glacier*, ending in New Harbor (a piedmont glacier), *Blue Glacier* in the Royal Society Range, and *Bowman Piedmont Glacier*. Small mountain glaciers exist at higher elevations, they flow about normally toward the dry valleys and end above the valley floor — they dry out. This phenomenon is really unique and has made the "Dry Valleys" famous.

Because of the very cold (-20°C mean) and arid climate, the glaciers in the valleys are slow and inactive. Here, a glacier advance would indicate a warming climate (the reverse of the usual relationship of climatic warming and glacier retreat). Summer precipitation is higher than winter precipitation, because the air can carry more moisture.

The alpine glaciers are all cold-based and cause very little erosion, they leave the bed largely unchanged and stand above the surface. Accumulation is mostly by wind redeposition. Ablation is mostly by sublimation. Velocities are typically 1 ma^{-1}, so the glaciers are very slow. (The description of the Dry Valley region follows Chinn, in Swithinbank 1988, p. B39–41.)

South of Royal Society Range is Koettlitz Glacier, on the west and southwest bordered by Brown Peninsula on which Mt. Discovery (2861 m) sits (Minna Bluff extends from here).

Ross Island, seen off the coast on the map, with the volcanoes Mt. Terror (3262 m), Mt. Terra Nova, Mt. Erebus (3794 m) is located at (77°30'S/168° E). The stations are in the southwest. Ross Island is 68 km large from Cape Bird in the north to Cape Armitage in the south, and from Cape Royal in the west to Cape Crozier in the east. *Mt. Erebus* is an active volcano. Erebus Glacier Tongue extends into the sea between the main part of the island and Hut Point Peninsula.

Ross Island and Scott Coast form *McMurdo Sound*. Minna Bluff, the boundary point between Scott Coast and Hillary Coast, is only a short distance to the south. *McMurdo Ice Shelf* (between Minna Bluff and Ross Island) has one of the largest ablation areas in Antarctica, indicative of a warming climate. The 25 km long floating tongue of Koettlitz Glacier is also wasting away (Swithinbank 1988, p. B32).

Major glaciers on Hillary Coast are Skelton Glacier and Mulock Glacier. *Hillary Coast* has a "white" appearance, distinctly different from the dry valley region to the north. *Skelton Glacier* is a local glacier, draining from the Skelton Neve, not from the interior ice sheet, whereas *Mulock Glacier* is a major outlet glacier. Between both is Worcester Range with 2760 m Mt. Harmsworth.

Mulock Glacier is 12 km wide, and has characteristic crevasse patterns with at least three main directions forming polygons and organized in large flowline-parallel stripes (Swithinbank 1988, fig. 26, p. B31). The surface falls from 1400 m to 100 m over 80 km length of the glacier. Mulock Glacier moves 390 ma^{-1} near its mouth (Swithinbank 1963) and discharges 5–6 km^{3}a^{-1} into the Ross Sea. The contour map shows an ice tongue or floe.

Mt. Longhurst (2846 m) is also visible on this map, at about (-8820.000 N/330.000 E). North of Mt. Longhurst is Carlyon Glacier, south of it is Darwin Glacier with a tributary, Hatherton Glacier (south of that is Byrd Glacier, described on map m165e147-183n78-815 Hillary Coast). Darwin Glacier and Hatherton Glacier have high rock barriers at their heads, barring ice from the interior (Swithinbank 1988, fig. 23, p. B26–27).

The "nose" in the 2200 m contour in the center of the map marks another ice divide, extending south of Scott Coast.

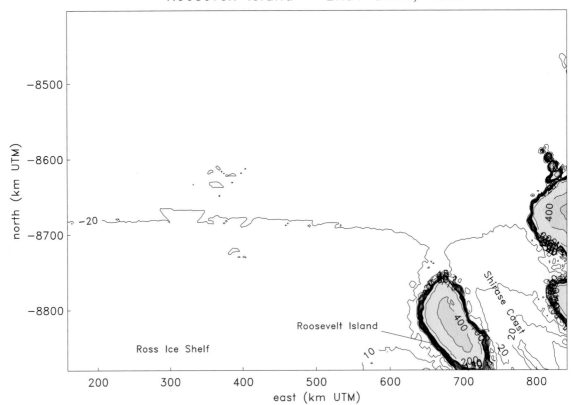

Roosevelt Island — ERS1 DATA, 1995

e171−207n75−80, WGS84, Gaussian variog., central mer. 189, slope corrected, scale 1:5000000, 971105

Map m189e171-207n75-80 Roosevelt Island

(Coordinates e171-207 correspond to w189-153)

This map shows *Roosevelt Island*, parts of *Shirase Coast* in the east and part of *Ross Ice Shelf*. The shelf ice edge should be on this map, but is not contained in the ERS-1 radar altimeter data set (see description of map m189e171-201n78-815 Ross Ice Shelf). Ice from the West Antarctic Ice Streams flows north through Ross Ice Shelf and around Roosevelt Island, as indicated by the elevated ice surface contours. The ice streams are labeled A, B, C, D, E (Ice Stream B is now called Whillans Ice Stream, after American glaciologist Ian Whillans); the ice between Roosevelt Island and Shirase Coast stems from ice stream E, the northernmost one. The ice streams are further described in the text of map m213e195-213n78-815 Shirase Coast.

Roosevelt Island is an ice rise about 140 km long in northwest-southeast direction and 60 km wide. An ice rise is ice that does not flow in the direction of the surrounding ice but rests on a seafloor shoal on rocks, i.e. it is an ice-covered island. According to Alberts (1995, p. 629) its northern tip is only 5 km from the ice edge at the Bay of Wales, and its elevation is 550 m, which matches the elevation on the ERS-1 map (approximately, by extrapolation of the gradient). Ice thickness is 750 m, according to the National Geographic Atlas map (National Geographic Society 1992, p. 102). The island was discovered in 1934 by the Antarctic Expedition led by Richard Byrd and named after the contemporary president of the United States of America, Franklin Roosevelt.

On Shirase Coast, we see the area north and south of *Prestrud Inlet*, the bay at -8750.000 N on the eastern map margin. To the north of the inlet are the Rockefeller Mountains. The coastline is more completely mapped and with the inland on maps m213e195-231n78-815 Shirase Coast and m213e195-231n75-80 Saunders Coast.

Saunders Coast − ERS1 DATA, 1995

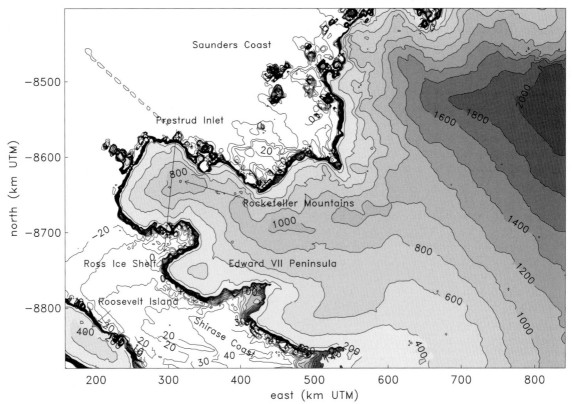

e195−231n75−80, WGS84, Gaussian variog., central mer. 213, slope corrected, scale 1:5000000, 971105

Map m213e195-231n75-80 Saunders Coast

(Coordinates e195-231 correspond to w165-129)

This map shows the northern part of the eastern coast of *Ross Ice Shelf* (*Shirase Coast*), part of the Ross Ice Shelf between that coast and *Roosevelt Island*, and part of the northern coast of Marie Byrd Land (Saunders Coast). *Shirase Coast* extends from (80° 10'S/151° W) to Cape Colbeck (77° 07'S/157° 54'W), located at the northwest tip of *Edward VII Peninsula*, and hence at the extreme northwest end of Marie Byrd Land. *Saunders Coast* is limited by Cape Colbeck and Brennan Point (76° 05'S/146° 31'W), the east side of the entrance to Block Bay, in which Balchen Glacier enters. Saunders Coast was explored from the air in 1929 by the Byrd Antarctic Expedition (1928–1930) and mapped from those photographs

by Harold Saunders. It was as late as 1959–65 that Saunders Coast was mapped completely, this was carried out by the US Geological Survey, using ground surveys and aerial photography. East of Saunders Coast is *Ruppert Coast* which is mapped on m213e195-231n71-77 Ruppert Coast.

Ice from *West Antarctic Ice Streams D, E, and F* flows northwest through Ross Ice Shelf between Roosevelt Island and the coast (for the ice streams, see m213e195-231n78-815 Shirase Coast). The mouth of Ice Stream E is just at the southern edge of the map (at 500.000 E). The smaller Ice Stream F is located in the bay at (-8800.000 S/420.000 E). The northern margin of Marie Byrd Land is mountainous, with several ranges and nunataks rising above the ice, whereas the higher interior is ice-covered. There are numerous glaciers.

Prestrud Inlet with Kiel Glacier is located on Shirase Coast across from the northern tip of Roosevelt Island and south of Rockefeller Mountains. *Rockefeller Mountains* are the head of *Edward VII Peninsula*. On Saunders Coast are (east to west): Bartlett Inlet, inland of that McKinley Peak (77° 54'S/148° 18'W) in the Ford Ranges (the smaller brother and namesake of Mt. McKinley, Alaska, the highest mountain of North America, but named after Grace McKinley, the wife of the photographer on the Byrd Expedition (1929)), Scott Nunataks, Alexandra Mountains, *Sulzberger Ice Shelf* with Sulzberger Bay offshore at (-8600.000 N/450.000 E), Boyd Glacier, Saunders Mountains, Crevasse Valley Glacier, Fosdick Mountains, extending coastward to Guest Peninsula (-8500.000 N/560.000 E), Balchen Glacier, Phillips Mountains (-8500.000 N/640.000 E), extending to Brennan Point and seaward in Nickerson Ice Shelf, further inland are the *Ford Ranges*, south of these the *Clark Mountains*, *Mackay Mountains*, and *Allegheny Mountains* and further east the *Flood Ranges*. *Nickerson Ice Shelf* extends east to a headland that borders *Land Glacier* on its eastern side (see map m213e195-231n71-77 Ruppert Coast). On Saunders Coast there are significant differences in the National Geographic Atlas Map (National Geographic Society 1992, p. 102) and in the USGS Satellite Image Map (Ferrigno et al. 1996). Features identified here with coordinates match the USGS Satellite Image Map.

Northern Marie Byrd Land — ERS1 DATA, 1995

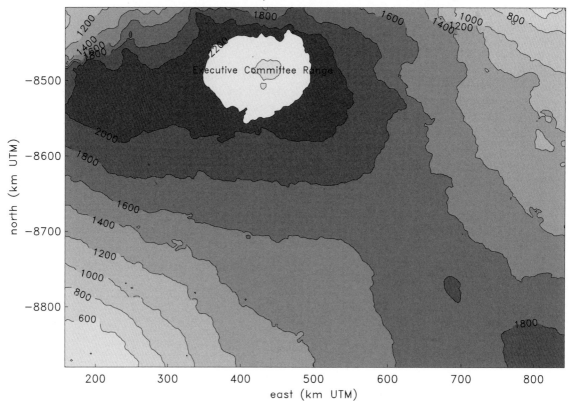

e219−255n75−80, WGS84, Gaussian variog., central mer. 237, slope corrected, scale 1:5000000, 971105

Map m237e219-255n75-80 Northern Marie Byrd Land

(Coordinates e219-255 correspond to w141-105)

In the northern center of the map is *Executive Committee Range*, part of the Marie Byrd Land volcanic province. Marie Byrd Land volcanic province is a volcanic chain that parallels the coast of West Antarctica from the *Jones Mountains* at longitude 94° W (266° E) in the Ellsworth Mountains (see map m285e267-303n78-815 Ellsworth Mountains) to the *Fosdick Mountains* at 145° W (215° E). Based on volcanic evidence, it can be shown that glaciation in Marie Byrd Land began in the Eocene, but it is not clear whether the glaciation remained uninterrupted to the present (Le Masurier and Rex 1982).

Executive Committee Range blocks a southward-flowing ice sheet that drains into the Ross Ice Shelf (Swithinbank 1988, p. B127). An elevated ridge above 1600 m extends southeast from Executive Committee Range to the interior of Marie Byrd Land (see also the southern adjacent map m237e219-255n78-815 Southern Marie Byrd Land). Ice to the northeast of this map drains to Walgreen Coast (map m261e243-279n71-77 Walgreen Coast) and the Amundsen Sea. North of Executive Committee Range is *Bakutis Coast.*

Mt. Sidley (north face) in the Executive Committee Range rises 2000 m above the ice sheet, Mt. Waesche (south face) 2200 m, Mt. Sidley has a 5 km diameter caldera open to the south, its rim (4181 m) is the highest peak in the range. In our maps, the sumit appears only above 2200 m, because elevations of peaks cannot be mapped accurately with radar altimetry (the signal returns from a large area, and elevation is averaged out). Contemporary fumarolic activity has been observed (Mt. Hampton 3325 m), i.e. the volcanoes are still active. Further east and a little lower are the Crary Mountains, the highest peak is Mt. Frakes (3677 m).

The volcano with the largest caldera (8 km diameter) is *Mt. Takahe* (3398 m, (76° 17'S/112° 05'W)), a partially buried shield volcano which rises about 2000 m above the ice sheet.

226

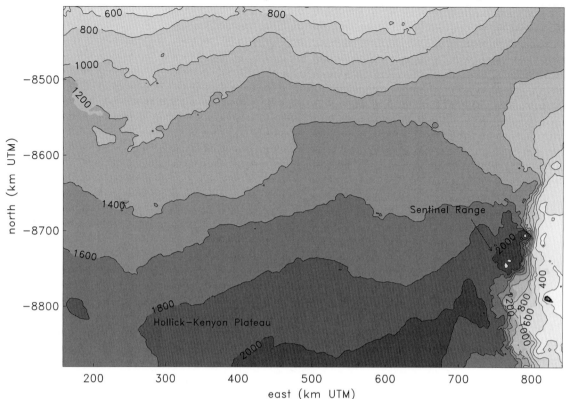

e243−279n75−80, WGS84, Gaussian variog., central mer. 261, slope corrected, scale 1:5000000, 971105

Map m261e243-279n75-80 Northern Hollick-Kenyon Plateau

(Coordinates e243-279 correspond to w117-81)

This map shows eastern central *Marie Byrd Land* and southern *Ellsworth Land*, West Antarctica. The map is named after *Hollick-Kenyon Plateau*, the only larger named geographic feature in the region. Drainage to the north is to *Thwaites Glacier* (concave contour lines 600 m to 1800 m between 200.000 E and 400.000 E on this map) and to *Pine Island Glacier* (concave contour lines 800 m to 1200 m at 500.000–800.000 E and from the top of the map to -8500.000 N) on Walgreen Coast. The glaciers and their key role in scenarios of instability of the West Antarctic Ice Sheet is discussed in the description of map m261e243-279n71-77 Walgreen Coast and detail maps 15 Thwaites Glacier (F.13) and 16 Pine Island Glacier (F.14). Drainage

to the east is to the Ronne Ice Shelf (see also description of map m285e267-303n75-80 Zumberge Coast).

In the southeastern part of the map are the *Ellsworth Mountains*, which border the Ronne Ice Shelf and consist of the *Sentinel Range* (77.3–79° S/≈85° W) in the north, trending N–S, and the *Heritage Mountains* (79–80.5° S/≈85° W) in the south, trending NW–SE. The Sentinel Range contains the Antarctic continent's highest mountain, *Vinson Massif* (78° 35'S/85° 25'W) peaking at 5140 m above sea level. A geologic bibliography of the Ellsworth Mountains is given by Webers and Splettstoesser (1982). For a general geologic history of the Ellsworth Mountains, see Tingey (1991).

Hollick-Kenyon Plateau is a large, relatively featureless snow plateau, centered at (78° S/105° W), which was discovered by Lincoln Ellsworth on his transantarctic airplane flight in 1935 and named for his pilot, Herbert Hollick-Kenyon.

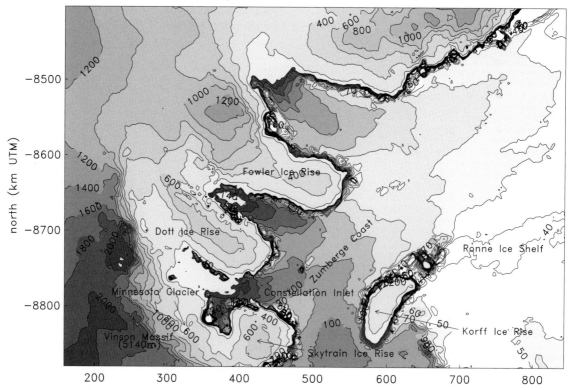

Zumberge Coast – ERS1 DATA, 1995

e267–303n75–80, WGS84, Gaussian variog., central mer. 285, slope corrected, scale 1:5000000, 971105

Map m285e267-303n75-80 Zumberge Coast

(Coordinates e267-303 correspond to w93-57)

The map m285e267-303n75-80 Zumberge Coast contains a complex array of geographic features in West Antarctica. Running diagonally and then south across the map is the western coastline of *Ronne Ice Shelf*. In the western half of the map, eastern *Ellsworth Land* and the *Ellsworth Mountains* are shown. In the center are ice rises along *Zumberge Coast*, from north to south, *Fowler Ice Rise*, *Fletcher Ice Rise* (*Dott Ice Rise*) and *Skytrain Ice Rise*, between which large glaciers and ice streams drain into the Ronne Ice Shelf, increasing the offshore ice elevation for one hundred to several hundred kilometers by initially 100 m (40–140 m). *Korff Ice Rise* is completely surrounded by Ronne Ice Shelf. In the northeastern part of the map, the southern extension of the arc of the

Antarctic Peninsula is seen, bordered by *Orville Coast*.

The northern edge of the Ronne Ice Shelf coincides roughly with *Bowman Peninsula*, i.e. *Lassiter Coast*/*Palmer Land* borders the Weddell Sea and *Orville Coast* and *Zumberge Coast* on Ellsworth Land border Ronne Ice Shelf. *Zumberge Coast* is the southwestern coast of Ronne Ice Shelf, it extends to Skytrain Ice Rise and the coast near the Ellsworth Mountains. Cape Zumberge (at -8500.000 N/610.000 E) marks a point between rugged topography (north and east) and smooth topography with ice-covered peninsulas and large ice streams, the northernmost of which is *Evans Ice Stream*. *Orville Coast* extends from *Cape Adams* (75° 04'S/62° 20'W), the rocky tip of Bowman Peninsula and the north side of the entrance to Gardner Inlet, to Cape Zumberge (76° 14'S/69° 40'W), a rocky cape. James Zumberge was an American glaciologist who worked in Antarctica, Orville a naval meteorologist who designed part of the Ronne Antarctic Research Expedition 1947–48 (United States Navy Reserve) led by Ronne. Cape Zumberge is located between the Hauberg Mountains to its NE, and the *Behrendt Mountains*, to its NW. (John Behrendt is a geophysicist who has worked in Antarctica repeatedly in 6 decades: 1957–2002.) Only small glaciers drain from Orville Coast.

The upland in the northwest corner of the map is the region of the *Jones Mountains*, the easternmost part of the volcanic province in Ellsworth Land and Marie Byrd Land. In the southwestern part of the map are the *Ellsworth Mountains*, which border the Ronne Ice Shelf and consist of the *Sentinel Range* (77.3–79° S/≈85° W) in the north, trending N–S, and the *Heritage Mountains* (79–80.5° S/≈85° W) in the south, trending NW–SE. A geologic bibliography of the Ellsworth Mountains is given by Webers and Splettstoesser (1982). For the geology of Antarctica, see Tingey (1991).

The *Transantarctic Mountains* contain old rocks from the Paleozoic Era. Sedimentation occurred during the Cambrian Period throughout most of the Transantarctic Mountains. The sedimentation phase was terminated by the Ross Orogeny, which lasted a long time and centered around 520–450 Ma (million years before present), i.e. in the Late Cambrian and Early Ordovician Periods. An orogeny is a phase of mountain building, the Ross Orogeny involved folding, metamorphism of the sediments, and uplift, accompanied and followed by widespread intrusion of granitoid batholiths, sheets and dykes.

In the Shackleton Range, the Ross Orogeny was completed before deposition of the Blaiklock Glacier Group in the Ordovician. In the Pensacola Mountains, rhyolite flows were embedded around 510 Ma. In the Thiel, Horlick and Queen Maud Mountains, the Ross Orogeny was marked by granitoid intrusions (collectively called Queen Maud Batholith). In the Ellsworth Mountains, 3000 m thick sediments, the upper layers of which are dated Devonian (408–360 Ma), overlay the Cambrian Heritage Group (Laird 1991). Permian sediments are also found in the Ellsworth Mountains (Barrett 1991).

With 5140 m, *Vinson Massif* in the Sentinel Range of the Ellsworth Mountains is the highest mountain in Antarctica, it was first climbed in 1966–67 by members of an American expedition sponsored by the National Geographic Society, the National Science Foundation and the American Alpine Club. Its neighbour Mt. Tyree (4965 m) is the second highest mountain in Antarctica, first climbed by New Zealand mountaineer Gary Ball and friends in the 1990's. The position of Vinson Massif is marked exactly on the map. Hence it is apparent that the steep eastern margin of Ellsworth Mountains is most prominent, mapped as a 1400 m drop-off. While the elevations in this mountainous terrain are incorrect, as expected, since satellite altimetry cannot map mountain ranges exactly, it is correct that the eastern drop-off is much larger and steeper than the western slope. Only 60 km to the east of Vinson Massif is the thickest floating ice sheet in the world (2000 m thick), so the peak-to-trough relief is 7000 m, the largest on any continent, but some oceanic trenches have greater relief. The ice plateau to the west is at 1500–2000 m elevation (mapped correctly), i.e. Vinson appears to be cut out of the data set or not recorded.

The Ellsworth Mountains dam the natural drainage of the ice sheet and divert the flow to the north and south from an ice divide at about -8550.000 N that is seen on map m261e243-279n75-80 Northern Hollick-Kenyon Plateau. *Rutford Ice Stream* drains all the ice that flows around the northern end of the Sentinel Range. Rutford Ice Stream (79° S/81° W) is a 284 km long and 24 km wide ice stream, which drains southeastward between the Ellsworth Mountains and Fletcher Ice Rise into the southwestern part of Ronne Ice Shelf. It is named after geologist Robert Rutford, member of the U.S. Antarctic Research Program (USARP) expedition to Antarctica and leader of the University of Minnesota Ellsworth Mountains Party (1963–1964) (Alberts 1995, p. 639).

Rutford Ice Stream is a fast-moving ice stream. Its central part has been researched intensively by field parties, consequently the grounding line position is known, the grounding line starts from a peninsula across from Vinson Massif and curves sinuously across the glacier, which is attributed to the existence of pinning points. Ice thickness near the grounding line is 1670, 1880, 2000 m (spot soundings).

The velocity near the center is fairly uniform, 400 m a^{-1} to 380 m a^{-1}, measured in a 40 km network of stakes near the glacier center. The drainage basin of Rutford Ice Stream is 40.500 km^2 +/- 4000 km^2 and mass flux at the grounding line is 18.5 +/- 2 Gt a^{-1} (20.3 km^3 a^{-1}) (Crabtree and Doake 1982). Rutford Ice Stream was the first glacier for which velocity was investigated with SAR interferometry (Goldstein et al. 1993), a technique that uses phase differences of two SAR data sets to study change or movement. Mountain glaciers flow out of the Sentinel Range to join Rutford Ice Stream at almost normal angle.

"Minnesota Glacier" is labeled slightly incorrectly on the ERS-1 map, in the position of this label should be a label "Rutford Ice Stream". Rutford Ice Stream flows southeastward east of Vinson Massif, which trends NW–SE, and Nimitz Glacier and Minnesota Glacier are two smaller glaciers, flowing southeastward (Nimitz Glacier) and southeastward-to-eastward (Minnesota Glacier) to the southwest of Vinson Massif. Nimitz Glacier ends near the tip of the arrow "Constellation Inlet", Minnesota Glacier ends a bit further south (according to the USGS Satellite Image Map of Antarctica (Ferrigno et al. 1996); labels were written according to the National Geographic Atlas map (National Geographic Society 1992, p. 102), on which Rutford Ice Stream is not labeled and Minnesota Glacier appears to be the largest glacier in the area).

On a satellite image of *Fletcher Ice Rise*, lower Rutford Ice Stream, the Sentinel Range and glaciers south of it (Swithinbank 1988, fig. 90, p. B122), three glaciers are distinguished and labeled southwest of Vinson Massif. *Nimitz Glacier* flows along the western edge of Vinson Massif for 60 km, where flowlines are clearly visible, its valley extends at least another 100 km northward, just south of Vinson is a step in the glacier bed and ice surface from a surface elevation of 1380 m downward. Ice thickness below the step is 720 m, 25 km downstream it is 1400 m. *Minnesota Glacier* parallels Nimitz Glacier 15 km further south and south of a small mountain range, it swings around the southern end of those mountains to join Nimitz Glacier. Minnesota Glacier occupies a 150 km long trench. Both glaciers are about 5 km wide

and drain ice from the polar west of the mountains. The joint Nimitz and Minnesota Glaciers are joined by *Splettstoesser Glacier*, which flows SW–NE and does not drain inland ice. All three glaciers pass the southern end of the Sentinel Range and then join Rutford Ice Stream, the location where all glaciers join is located approximately where "Constellation Inlet" is marked on the ERS-1 map; the joint ice stream passes between Fletcher Ice Rise and Skytrain Ice Rise and flows into Ronne Ice Shelf.

Fletcher Ice Rise (center coordinate (78° 20'S/81° W)) is 160 km long and 60 km wide and completely ice covered, it lies between Rutford Ice Stream and Carlson Inlet. On the satellite image (Swithinbank 1988, p. B122, fig. 90), Fletcher Ice Rise appears not to be connected to the mainland, but on the USGS Satellite Image Map of Antarctica (Ferrigno et al. 1996) it is called "Fletcher Promontory" and connected to the mainland. Fletcher Ice Rise is labeled *"Dott Ice Rise"* on the National Geographic Atlas map (National Geographic Society 1992, p. 102) and on our map. According to Alberts (1995, p. 196), Dott Ice Rise (79° 18'S/81° 48'W) is a much smaller, 30 km long peninsula-like feature that is ice-drowned except for Barrett Nunataks, extending eastward from the Heritage Range and terminating at Constellation Inlet. This may demonstrate to the reader that features described by field parties or mapped by expeditions are often hard to identify in other maps or satellite data — these difficulties are shared by our maps and other maps.

Korff Ice Rise is 160 km long, 40 km wide and about 500 m high (Swithinbank 1988, p. B103), here it appears to be over 200 m high. Ice thicknesses of up to 980 m have been measured. As noted for Berkner Island, the land mass of Korff Ice Rise does not rise above sea level, but reaches up to 115 m below sea level (Behrendt 1962; Swithinbank 1988, p. B103, also p. 119ff). Hence Korff Ice Rise is not truly an island, because its landmass does not reach above sea level.

Not much ice flows from the inland into Carlson Inlet (Swithinbank 1988, p. B119). *Evans Ice Stream* enters Ronne Ice Shelf north of Fowler Ice Rise; *Fowler Ice Rise* is a peninsula covered by ice. Several small glaciers enter Ronne Ice Shelf on Orville Coast, which is a mountainous, high-relief coastline (see map m285e267-303n71-77 Ellsworth Land).

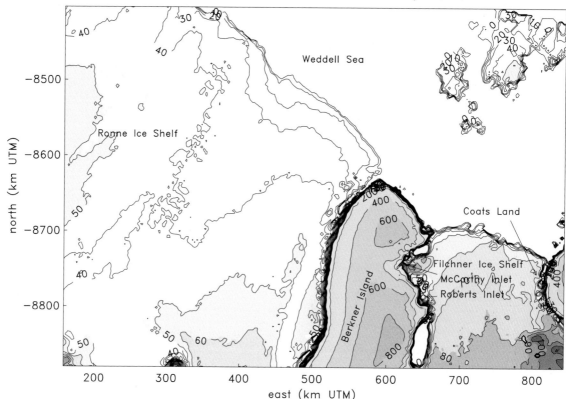

Ronne Ice Shelf – ERS1 DATA, 1995

e291–327n75–80, WGS84, Gaussian variog., central mer. 309, slope corrected, scale 1:5000000, 971105

Map m309e291-327n75-80 Ronne Ice Shelf

(Coordinates e291-327 correspond to w69-33)

This map shows the northern parts of the *Ronne Ice Shelf*, the *Filchner Ice Shelf*, and of *Berkner Island* — separating the two ice shelves in the north, which are connected south of Berkner Island — and the western nose of *Coats Land* (see m333e315-351n75-80 Coats Land). The features offshore are large icebergs. For general geographic information on the Filchner and Ronne Ice Shelves, see description to map m309e291-327n78-815 Berkner Island.

The northern part of the fast-moving Filchner Ice Shelf, which is shown on this map, is also shown on a LANDSAT 1 MSS satellite image of 1973 (Swithinbank 1988, fig. 76, p. B102). Our map shows the ice-surface elevation, which decreases from 80 m (above the WGS84 ellipsoid) in the southern part of the map to 50 m near the ice front with a gradient of about 10 m in 10 km (0.1% = 0.057°) to 10 m in 50 km (0.02% = 0.011°), steeper near the ice front. The ice front appears to be 40–50 m high between Berkner Island and Coats Land. Ice surface elevation increases to above 90 m further south and above 100 m south of Berkner Island (see map m305e291-327n78-815 Berkner Island).

High surface elevations (to above 100 m) in the southeastern corner of the map are caused by inflowing glacier ice that descends from an area of southern Coats Land and north and west of the

Theron Mountains (see map m333e315-351 n75-80 Coats Land). The LANDSAT imagery complements our elevation mapping, in as far as it shows the flowline pattern in the Filchner Ice Shelf. The ice flow diverges north of the line of narrowest distance between Berkner Island and Coats Land (see map m333e315-351n75-80; this area of the ice shelf is shown on map m309e291-327n75-80 Ronne Ice Shelf). The fast flow causes the so-called "Grand Chasms", a large rupture 100 km long and 400 m to 5 km wide, and 53 m deep from top to top of sea ice (in 1957; Neuburg et al. 1959), 115 km long and 11 km wide in 1973 (Swithinbank 1988, p. B101), and 19 km wide in 1985, when it reached its maximum width. In 1986, the area north of the chasm had separated and formed into icebergs. When such a rift reaches an area of diverging flow – as is the case here in the northern part – then the rift widens because parts of the ice that are closest to the flanking land hinges, and rifts perpendicular to the ice front form. Seismic measurements undertaken by Behrendt during the International Geophysical Year 1957–58 (see Behrendt 1962) indicate thicknesses of 290 m to 640 m in the area, about 400 m near the ice front (330–420 m), up to 520 m north of the Grand Chasms, and 500–610 m immediately south of the Grand Chasms, increasing to 610–640 m further upstream on the ice shelf (Swithinbank 1988, fig. 76, p. 102).

Ice-surface elevations in the *Ronne Ice Shelf* are generally lower than those in the *Filchner Ice Shelf*, they decrease from above 60 m west of Berkner Island and in the southern part of this map to 20 m near the ice front (above WGS84). Higher ice surface elevations in the Ronne Ice Shelf are seen on maps m309e231-327n78-815 Berkner Island (adjacent to the south), increasing to above 100 m south of Berkner Island and Henry Ice Rise, m280e267-303n78-815 Ellsworth Mountains, reaching above 100 m between Henry Ice Rise and Korff Ice Rise, and south and west of Korff Ice Rise, and on map m285e267-303n75-80 Zumberge Coast, where outlet glaciers contribute land ice. The ice front of Ronne Ice Shelf appears lower than that of Filchner Ice Shelf, however, it is only contoured down to 0 m and may actually be higher. (This depends on the geoid in the area; cf. section (C.4). Elevation information is still contained in the ATLAS DTM.) The Ronne Ice Shelf has a lower surface gradient than the Filchner Ice Shelf.

There are large areas of ice rumples in Ronne Ice Shelf. Actually the largest known area of ice rumples is shown in a satellite image in Swithinbank (1988, fig. 77, p. 104), it is located in Ronne Ice Shelf. *Rumples* form where the ice shelf flows over large seafloor shoals, and the ice thickness and surface elevation increases upstream of the shoal. Ice thicknesses of 1000 m were measured by Smith (1986) between Korff Ice Rise and Henry Ice Rise further south.

Berkner Island is the world's largest ice rise (378 km long, 150 km wide, and up to 1000 m above sea level (Swithinbank 1988, p. 103); max elevation on our map is above 800 m; which is about the same). The *ice rise* is an independent ice cap built on a shoal on the continental shelf; the highest bedrock elevation is 80 m below sea level (Behrendt 1962) — so the land mass of the "island" does not rise above sea level, which means, it is not truly an island.

The northern part of Berkner Island is mapped on m309e291-327n75-80 Ronne Ice Shelf. We can see outlet glaciers in McCarthy Inlet and further south at (-8500.000 N/650.000 E) UTM (near the label "Filchner Ice Shelf"). Roberts Inlet can also be distinguished. The large glacier draining from the mainland into Filchner Ice Shelf is Recovery Glacier (in the southeastern corner of the map).

Henry Ice Rise is wedge-shaped with a 135 km N–S extension in its eastern part and a 120 km E–W extension in its southern part; elevations reach above 200 m in the satellite-altimetry-derived map (notice that actual elevations are usually higher in reality, due to effects explained in chapter (C)). Ice surface elevations in the Ronne Ice Shelf between Henry Ice Rise and southern Berkner Island decrease from 100 m (above WGS 84) to 50 m with a fairly even gradient of 10 m in 50 km ($0.02\% = 0.011°$; 100 m to 90 m) to 10 m in 70 km ($0.014\% = 0.008°$; 60 to 50 m) and with a lesser gradient to 40 m.

(D.5) Latitude Row 78-81.5°S: Maps from ERS-1 Radar Altimeter Data

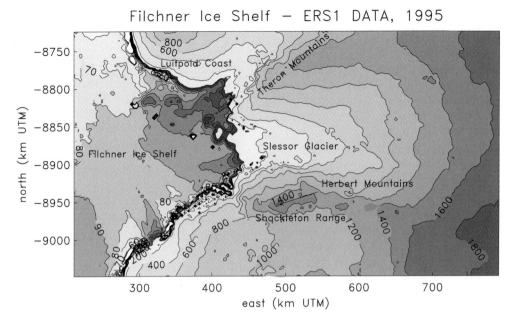

Filchner Ice Shelf — ERS1 DATA, 1995

e315–351n78–815, WGS84, Gaussian variog., central mer. 333, slope corrected, scale 1:5000000, 970724

Map m333e315-351n78-815 Filchner Ice Shelf

(Coordinates e315-351 correspond to w45-9)

The *Filchner Ice Shelf* map shows a lot of detail in many geographic features, and hence it is an excellent example of the capabilities of the Atlas mapping using geostatistical analysis of radar altimeter data. Enlarging from the Atlas scale of 1:5.000.000, even more details become apparent. A detail map of *Slessor Glacier* is given in section (F.1), since Slessor Glacier is the largest glacier in this area of Antarctica and geographically very interesting.

The map shows a part of *Filchner Ice Shelf* between Luitpold Coast (Coats Land, see m333e315 n75-80 Coats Land) in the north, the *Theron Mountains*, Slessor Glacier and its drainage basin, the area of the *Transantarctic Mountains* with major ice streams draining from it in the south. Filchner Ice Shelf is a fast-flowing Ice Shelf, the flowline pattern is in some regions visible in satellite im-

agery. Parts of *Filchner* and *Ronne Ice Shelves* are also seen on maps m309e291-327n78-815 Berkner Island and m285e267-303n78-815 Ellsworth Mountains.

Slessor Glacier is one of the largest coherent ice streams known on Earth. It measures about 500 km from Parry Point upstream (the corner point formed by the northern margin of Slessor Glacier and the "coast"-line inland ice/ Filchner Ice Shelf) as determined by flowlines visible in satellite imagery (Swithinbank 1988, fig. 75, p. B100), plus 350 km from Parry Point to the ice front (Crabtree and Doake 1980), so the glacier is at least 850 km long. The drainage basin is 575.000 km² (McIntyre after Swithinbank 1988). Slessor Glacier, at its narrowest point, is 30 km wide. It descends west with a constant slope of 0.32° to a flatter area before joining the Filchner

Ice Shelf. Surface elevations 10–30 m higher than the surrounding ice shelf, mark its protrusion into the ice shelf (at (-8900.000 N/400.000 E); Herzfeld et al. 2000b). Annual discharge is estimated at 34 km^3. No velocity or ice-thickness measurements have been made.

In the south, the glacier is bordered by the *Shackleton Range*. A 1974 LANDSAT image (Swithinbank 1988, fig. 75, p. B100) of Slessor Glacier and the Shackleton Range shows flowlines indicative of rapid motion on the glacier and on the part that is in the Filchner Ice Shelf. Isolated smooth undulation between Parry Point and Blaiklock Glacier may be indicative of isolated grounded areas in otherwise floating ice, so this area may be near the grounding line. Comparing this location to the Atlas map, the grounding line would coincide with the 140 m contour line and coincide with the location of the extension of the line between the inland ice and Filchner Ice Shelf.

Taking calculations of slopes and breaks in slope from the detail map into account also, the most likely candidate for an approximation of the grounding line is the 130 m elevation line. However, the "line" is probably included in a grounding zone, that extends a bit downward and possibly some distance upstream.

Ice surface characteristics in the upper reaches of Slessor Glacier (at 1000–1500 m a.s.l.) are typical of those of a large ice stream, with series of escarpments trending generally across the flowline, spaced about 10 km apart and attributed to subglacial ridges (Swithinbank 1988, fig. 74, p. B99). The subglacial topography, as indicated by ice-surface features seen in satellite imagery (Swithinbank 1988, fig. 74, p. B99) indicates that the *Shackleton Range* extends eastward under Slessor Glacier (Marsh 1985, after Swithinbank 1988). The subglacial features are different than those of the *Gamburtsev Subglacial Mountains*.

On some of the few other discovered subglacial features in Antarctica, see map m69e51-87n78-815 Dome Argus.

Several small glaciers join Slessor Glacier, both from the Theron Mountains region and from the Shackleton Range, the latter include *Stratton Glacier* and *Blaiklock Glacier* located at the tip of the Filchner Ice Shelf. "Blaiklock Glacier" (Group) is also a type locality for a geologic unit

mapped during the "German Expedition into the Shackleton Range" (GEISHA) in 1987/88 (see Kleinschmidt et al. 1995).

The Shackleton Range is a northern part and the "Filchner part" of the Transantarctic Mountains, which extend all the way across the Antarctic continent, along the Filchner-Ronne Ice Shelf area, close to the South Pole, along the eastern side of the Ross Ice Shelf region and back north to Victoria Land at the South Pacific Ocean side of the Antarctic continent. The Transantarctic Mountains form the boundary between the area called "West Antarctica" (Filchner-Ronne Ice Shelf, Ross Ice Shelf, Antarctic Penninsula, Ellsworth Land, Marie Byrd Land) and the area called "East Antarctica" (a geologically old shield, which includes Queen Maud Land, the Lambert Glacier region, Wilkes Land, and the Polar Plateau). The terms "East" and "West" are, of course, misnomers. East Antarctica is largely covered by a connected, thick ice sheet, whereas West Antarctica is more geographically featured, and its ice "sheet" is far more susceptible to disintegration (see Introduction, chapter (A)).

The highest peak of the *Herbert Mountains*, in the central part of the Shackleton Range, is Mount Absalom (1642 m a.s.l.). The Shackleton Range rises to about 1800 m a.s.l. (Alberts 1995). The map kriged from altimeter data reaches over 1600 m in the Herbert Mountains, so the averaging effect that must be expected from both satellite altimetry and kriging is not severe here. The Shackleton Range and the Herbert Mountains were the destination of the "German Expedition into the Shackleton Range" (GEISHA) in 1987/88 and have been studied for their role in the break-up of Gondwanaland, an earlier large continent of the Earth (Kleinschmidt and Buggisch 1993; Flöttmann and Kleinschmidt 1993; Kleinschmidt et al. 1995; Helferich and Kleinschmidt 1998; Millar and Talarico 1999; Tessensohn et al. 1999; Buggisch and Kleinschmidt 1999).

At (-9000.000 N/320.000 E) another ice tongue enters Filchner Ice Shelf, this is the ice of *Recovery Glacier* (center coordinate (81°10'S/28°W)), a 100 km long and 60 km wide glacier that flows west south of the Shackleton Range and turns north into Filchner Ice Shelf. Recovery Glacier was studied by the Commonwealth Transantarctic Expedition in 1957, the name stems from the recovery of vehicles that broke into crevasses.

Many more large glaciers drain from the Transantarctic Mountains into Filchner Ice Shelf but are located outside of the region of radar altimeter data coverage. One of those is *Support Force Glacier* (center coordinate (82° 45'S/46° 30'W) in the Pensacola Mountains. Unfortunately, coverage by ERS-1 satellite radar altimeter data (and thus by any radar altimeter data so far) ends at 81.5° southern latitude, so detailed elevation maps cannot be constructed for the southern central part of Antarctica.

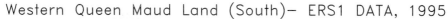

Western Queen Maud Land (South)– ERS1 DATA, 1995

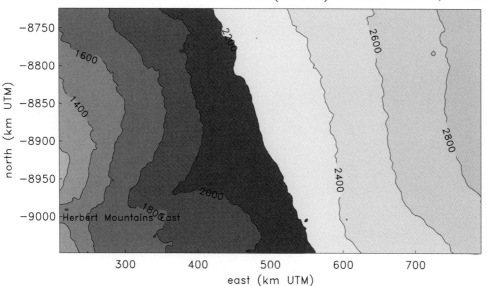

e339–15n78–815, WGS84, Gaussian variog., central mer. 357, slope corrected, scale 1:5000000, 990302

Map m357e339-15n78-815 Western Queen Maud Land (South)

This map of the Western Queen Maud Land (South) may be divided at about the 2200 m contour into an eastern part, where the surface slopes gently down to the west and which belongs to the interior of Queen Maud Land, and a western part, where the ice surface topography is regionally influenced by glacial drainage systems and the eastern part of the *Herbert Mountains*.

The basin to the north of the Herbert Mountains is the upper part of the Slessor Glacier drainage

basin. The basin to the south is not as ideally shaped as the Slessor Glacier basin and feeds several smaller glaciers south and west of the Shackleton Range. Ice from both basins drains into the Filchner Ice Shelf. The Herbert Mountains are a group of rocky summits east and slightly north of the Shackleton Range. They have been studied by geologists in connection with the Gondwana break-up (Kleinschmidt and Buggisch 1993; Kleinschmidt et al. 1995; Buggisch and Kleinschmidt 1999).

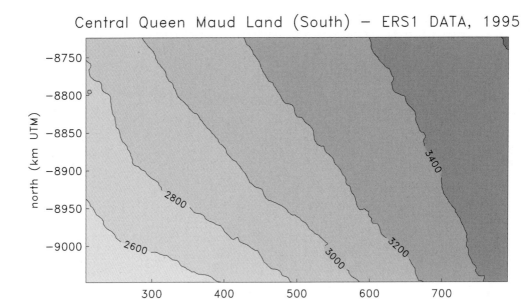

Central Queen Maud Land (South) – ERS1 DATA, 1995

e3–39n78–815, WGS84, Gaussian variog., central mer. 21, slope corrected, scale 1:5000000, 990219

Map m21e3-39n78-815 Central Queen Maud Land (South)

This map shows Central Queen Maud Land (South), in general, from the divide between Lambert Glacier drainage basin (on which Valkyrie Dome is located, see map m45e27-63n75-80 Valkyrie Dome), and drainage to the eastern coast of the Weddell Sea.

The surface in the mapped part of Queen Maud Land slopes gently (200 m elevation difference over 200–300 km) from above 3400 m in the east to below 2600 m in the west, with a westsouthwesterly gradient.

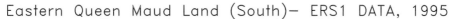

Eastern Queen Maud Land (South)— ERS1 DATA, 1995

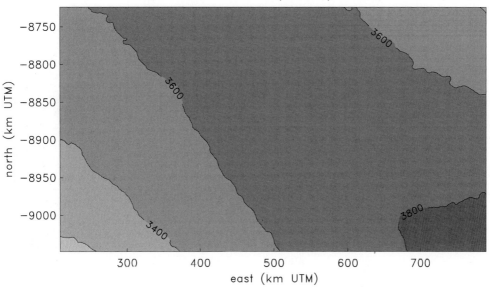

e27 63n78 815, WGS84, Gaussian variog., central mer. 45, slope corrected, scale 1:5000000, 990302

Map m45e27-63n78-815 Eastern Queen Maud Land (South)

This map shows the southern part of Eastern Queen Maud Land, the highest elevation in the center of the map (above 3600 m and increasing to above 3800 m in the southeastern corner of the map) are the continuation of the divide between interior Queen Maud Land (to the west) and the *Lambert Glacier drainage basin* to the northeast, as described on the maps m45e27-63n75-80 *Valkyrie Dome* and m45e27-63n71-77 Belgica Mountains. The divide rises to the northwest slightly, to the height of Valkyrie Dome, located at (77° 30'S/37° 30'E), just off this map.

240

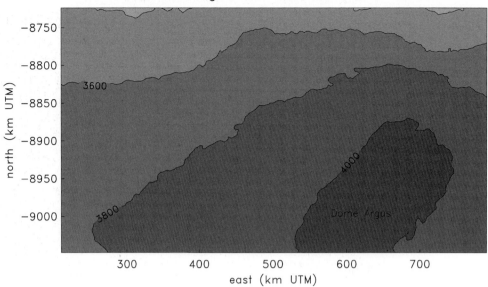

e51−87/n78−815, WGS84, Gaussian variog., central mer. 69, slope corrected, scale 1:5000000 990218

Map m69e51-87n78-815 Dome Argus

This map is located south of the Lambert Glacier maps. *Dome Argus* is the central feature on the map, it is a dome-and-oval-shaped highland or rise in elevation to above 4000 m (in our maps). Because of the sheer size of Dome Argus, its elevations should be accurate on our map. To the north and northwest extends the region of the Gamburtsev Subglacial Mountains (cf. map m69e51-87n75-80 South of Lambert Glacier), a large subglacial mountain range, which according to Alberts (1995) reaches as far as under Dome Argus.

For more information on the Gamburtsev Subglacial Mountains, the reader is referred to the description of map m69e51-87n75-80 South of Lambert Glacier.

Dome Argus is the highest ice feature in Antarctica (of course not the highest mountain — that is Mt. Vinson in the Ellsworth Mountains at 5140 m) with just over 4000 m elevation. Dome Argus (first called "Dome A") was mapped by the SPRI-NSF-TUD airborne radio-echosounding program (1967–1979) (Alberts 1995, p. 26).

East Antarctica (e75–111n78–815)– ERS1 DATA,1995

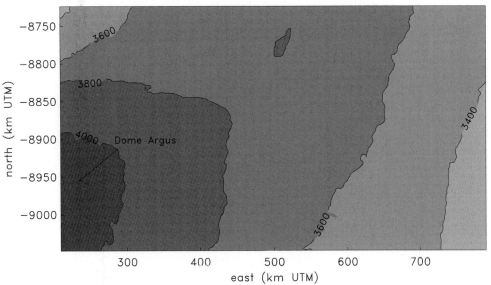

e75–111n78–815, WGS84, Gaussian variog., central mer. 93, slope corrected, scale 1:5000000, 990301

Map m93e75-111n78-815 East Antarctica (e75-111n78-815)

This map shows part of the interior of East Antarctica, the surface is all snow and ice. The high elevation area, above 4000 m, in the southwest corner of the map is part of *Dome Argus*, the center of which is seen on the westerly adjacent map m69e51-87n78-815 Dome Argus. From the corner of Dome Argus, a broad ridge continues above the 3600 m contour. This ridge continues on the two northerly adjacent maps m93e75-111n75-80 and m93e75-111n71-77 (the small area above 3800 m on this map is the same as the one on the next map, its position gives an idea of the overlap of maps in the two southerly rows), the ridge separates the Lambert Glacier drainage basin (to the west and northwest), and the Wilkes Land Coast and Ross Ice Shelf area to its east. At an elevation of 3200 m, drainage to the north (Wilkes Land, Coast of Indian-Ocean Sector of circum-Antarctic sea) and drainage to the east (Ross Ice Shelf) and northeast (Coast of Victoria Land) can be distinguished (on the easterly adjacent maps, m117e99-135n78-815 and m141e123-159n78-815; m117e99-135n75-80 and m141e123-159n75-80; m117e99-135n71-77 and m141e123-159n71-77). It is best to look at all the Wilkes Land maps together to get an idea of the general ice divides and drainage basins. (The main purpose of the Atlas mapping scheme was mapping the coastal features in greater detail.)

The *Vostok Subglacial Highlands* (center coordinate (80°S/102°E)) are a line of subglacial mountains which trend NNW–SSE and form an easterly extension of the Gamburtsev Subglacial Mountains. They were mapped by airborne radio-echosounding (Alberts 1995, p. 288).

242

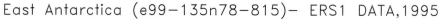

East Antarctica (e99−135n78−815)− ERS1 DATA,1995

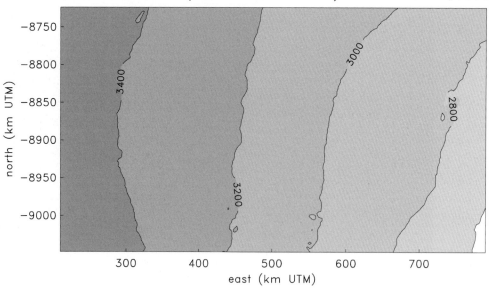

e99−135n78−815, WGS84, Gaussian variog., central mer. 117, slope
corrected, scale 1:5000000, 990219

Map m117e99-135n78-815 East Antarctica (e99-135n78-815)

In this part of East Antarctica, the surface slopes generally to the east, and the ice drains to the Ross Ice Shelf area, through the Transantarctic Mountain Range, as is seen on the easterly adjacent map (m141e123-159n78-815 Byrd Glacier). The elevation of the ice surface in the map area is above 3400 m (about 3500 m, as concluded from the slope) to 2600 m. The gradient is 200 m in 110 km ($0.18\% = 0.104°$; 3000–3200 m) to 200 m in 160 km ($0.125\% = 0.072°$; 3200–3400 m and 2800–3000 m).

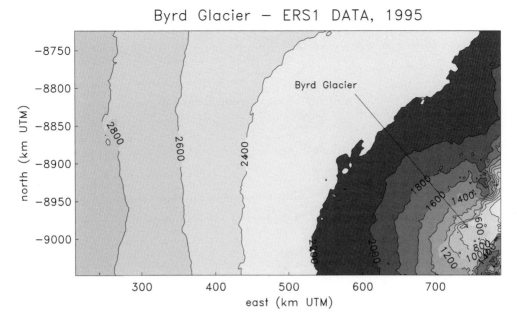

Byrd Glacier – ERS1 DATA, 1995

e123–159n78–815, WGS84, Gaussian variog., central mer. 141, slope corrected, scale 1:5000000, 980126

Map m141e123-159n78-815 Byrd Glacier

The area covered by this map extends from south (southeast) of *Dome Charlie* (75° S/125° E) in Southern Wilkes Land to the Transantarctic Mountains west of the Ross Ice Shelf. The Transantarctic Mountains are higher than the inland ice to their west and large glaciers such as Byrd Glacier cross through the mountain ranges from the inland ice to the sea or the Ross Ice Shelf.

For example, Mt. McClintock in the *Britannia Mountains* north of *Byrd Glacier* is 3492 m high, Mt. Hamilton south of Byrd Glacier is 1990 m high, further south are Mt. Wharton (2895 m) and Mt. Albert Markham (3207 m) in the Churchill Mountains (above small Starshot Glacier). North of Byrd Glacier is Hillary Coast, to the south is Shackleton Coast. Byrd Glacier is one of the largest glaciers in the area, it is about 135 km long and 24 km wide and drains an extensive area of the polar plateau. Byrd Glacier was used by Robert Scott and his party to ascend the Antarctic Inland Ice from the Ross Ice Shelf on their way to the South Pole (British Antarctic Expedition, 1910–1913).

244

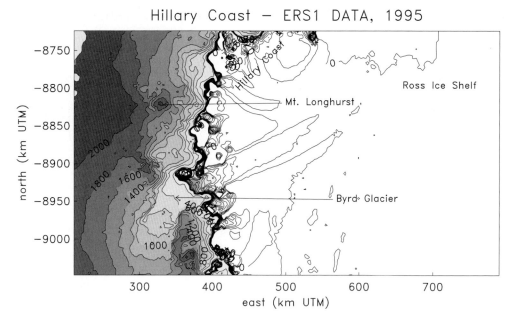

e147 183n78 815, WGS84, Gaussian variog., central mer. 165, slope corrected, scale 1:5000000, 980126

Map m165e147-183n78-815 Hillary Coast

This map shows the southernmost part of the *Transantarctic Mountain Range* and of the coast of the Ross Ice Shelf that is covered by satellite radar altimeter data. The northern part is a part of Hillary Coast, the southern part is a part of Shackleton Coast. *Hillary Coast* lies between Minna Bluff (see map m165e147-183n75-80 Scott Coast) and Cape Selborne, named after Sir Edmund Hillary, leader of the New Zealand party of the Commonwealth Transantarctic Expedition. *Cape Selborne* (80° 23'S/160° 45'E) is a high snow-covered cape at the south side of Byrd Glacier. South of Byrd Glacier is *Shackleton Coast*.

Mountain ranges and glaciers on Hillary Coast are (north to south) Royal Society Range, Skelton Glacier, Worcester Range, Mulock Glacier (all described on map m165e147-183n75-80 Scott Coast), Carlyon Glacier, Cook Mountains, Darwin Glacier, Britannia Range (Mt. McClintock 3490 m), and Byrd Glacier.

On the *Shackleton Coast* part of the map are (north to south) Byrd Glacier (the boundary between Hillary Coast and Shackleton Coast), Mt. Field (3010 m), the Churchill Mountains (extend-

ing ≈ 100 km south along the coast), Nursery Glacier, Mt. Nares and Swithinbank Range (off coverage!).

Ross Ice Shelf extends to 85.5° South (Amundsen Coast with Amundsen Glacier and Scott Glacier, among others).

Byrd Glacier is the most dominant glacier in this area. Its large drainage basin extends inland for several hundred kilometers. Lower Byrd Glacier is 23 km wide and shows significant flowlines indicative of high velocities in satellite images (Swithinbank 1988, fig. 23–25, p. B22–B30). (Byrd Glacier is at the southern extremity of LANDSAT coverage!)

Byrd Glacier is one of the largest valley glaciers of the Earth and possibly the most active outlet glacier of East Antarctica (Swithinbank 1988, p. B24–B25). The ice surface is heavily crevassed as seen in aerial photographs. According to McIntyre (after Swithinbank 1988), the drainage area is 1.012.000 km^2, greater than that of any other glacier in the world. Byrd Glacier follows a gently curving fjord trough with maximal vertical relief of

5000 m (peaks to bottom of trough). Over 180 km length the surface falls from 1400 m to 70 m, the valley through the Transantarctic Mountains is 100 km long. The glacier has been traced by radio-echosounding measurements for a length of 430 km (Shabtaie and Bentley 1982). The glacier is afloat downstream of the narrowest point of the valley, there still 22 km wide. The grounding line is considered to lie about 40–50 km south of Cape Selborne (seaward edge), there the glacier is 1300–1500 m thick. Ice thickness near Cape Selborne is 600–700 m, about 100 km upstream it is 3000 m. Bare ice 30–60 km up from the headlands is indicative of ablation due to katabatic winds.

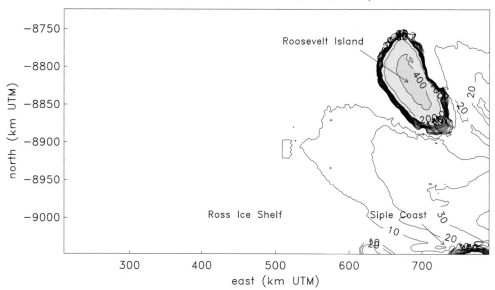

Ross Ice Shelf — ERS1 DATA, 1995

e171-207n78-815, WGS84, Gaussian variog., central mer. 189, slope corrected, scale 1:5000000, 990219

Map m189e171-207n78-815 Ross Ice Shelf

(Coordinates e171-207 correspond to w189-153)

This map is named "Ross Ice Shelf" because most of the map area is covered by shelf ice surface — white and no contours, as the ice shelf is flat at this scale, except for the area where ice streams enter, in the eastern part of the map (Ice Streams E (north) and D (south)). The ice edge is not mapped — the Ross Ice Shelf edge is not captured well in the radar altimeter data (compared to the Amery Ice Shelf and the Filchner and Ronne Ice Shelves). In the Ross area, the elevation of the geoid relative to the ellipsoid is much lower than in the Filchner-Ronne Area, but negative contour lines did not reveal the shelf ice edge — hence it is not included in the data set. The shelf ice edge is mapped best for the Amery Ice Shelf (see map m69e61-77n67-721 Lambert Glacier or detail map 5 Lambert Glacier, section (F.5)).

Roosevelt Island, the main geographic feature on the map, is an ice rise about 140 km long in northwest–southeast direction and 60 km wide. An ice rise is ice that does not flow in the direction of the surrounding ice but rests on a seafloor shoal

on rocks, i.e. it is an ice-covered island. According to Alberts (1995, p. 629) its northern tip is only 5 km from the ice edge at the Bay of Wales, and its elevation is 550 m, which matches the elevation on the ERS-1 map (approximately, by extrapolation of the gradient). Ice thickness is 750 m, according to the National Geographic Atlas map (National Geographic Society 1992, p. 102). The island was discovered in 1934 by the Antarctic Expedition led by Richard Byrd and named after the contemporary president of the United States of America, Franklin Roosevelt. Roosevelt Island is also partly seen on map m189e171-207n75-80 Roosevelt Island, m213e195-231n78-815 Shirase Coast and m213e195-231n75-80 Saunders Coast, which gives an impression of the overlap of adjacent maps.

The small section of coastline in the southeastern corner of the map is part of *Siple Coast* (see also map m213e195-231n78-815 Shirase Coast) of Byrd Land.

Shirase Coast – ERS1 DATA, 1995

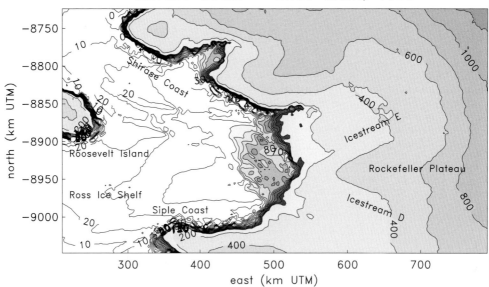

e195 231n78 815, WGS84, Gaussian variog., central mer. 213, slope
corrected, scale 1:5000000, 980126

Map m213e195-231n78-815 Shirase Coast

(Coordinates e195-231 correspond to w165-129)

Shirase Coast and *Siple Coast* form the eastern margin of the Ross Ice Shelf and are mapped here as far south as the coverage of ERS-1 radar altimeter data extends (to 81.5° S). *Siple Coast* extends from the north end of Gould Coast (83° 30'S/153° W) to the south end of Shirase Coast (80° 10'S/151° W). (Paul Siple was a member of the expedition led by Byrd.) The coastline at Siple Coast is relatively ill-defined.

Shirase Coast extends from the mentioned point to Cape Colbeck (77° 07'S/157° 54'W), the extreme northwest tip of Edward VII Peninsula and hence of Marie Byrd Land (seen on map m213e195-231n75-80 Saunders Coast, where the name "Edward VII Peninsula" is printed too far south). Shirase was leader of a 1912 Japanese Antarctic Expedition. Further south are Gould Coast, the Queen Maud Mountains and Dufek Coast (all south of the Ross Ice Shelf). The eastern margin of the Ross Ice Shelf is characteristically different from the southern and western margins. The southern and west-

ern margins are dominated by the Transantarctic Mountains with local glaciers that originate in the mountains and flow towards the ice shelf and with large outlet glaciers that drain ice from the polar plateau and southern Wilkes Land to the ice shelf. The ice surface elevations of Marie Byrd Land to the west of the Transantarctic Mountains and to the west of the Ross Ice Shelf are lower (approximately 1500–2000 m) than those of the polar plateau and Wilkes Land east of the Trantantarctic Mountains (2500–3000 m). The dominant feature of the western margin are the large West Antarctic Ice Streams, A, B (Whillans), C, D, E (from south to north), which drain ice from Marie Byrd Land. The ice streams are flowing faster than the surrounding ice. The ice streams have been the objective of intensive studies but still many questions are unanswered (e.g. Bindschadler 1991, Alley and Bindschadler 2001). Good satellite image maps of the West Antarctic Ice Streams are available through the U.S. Geological Survey.

In the classes of fast-moving ice (Clarke 1987), the fast Antarctic Ice Streams form their own class, sharing some properties with fast-moving glaciers because of reduced friction at their bottom but having absolutely much lower velocity than fast glaciers (e.g. Jakobshavns Isbræ, a continuously fast-flowing glacier in West Greenland, has a velocity of 19 km per year, whereas the West Antarctic Ice Streams flow 400–700 m per year).

Morphologically, the drainage basin of *Ice Stream E* extends 600 km eastnortheast into Marie Byrd Land north of Rockefeller Plateau, as seen on this map and on the eastward adjacent map m237e219-255n78-815. On map m237e219-255n75-80, the head of the drainage is seen which lies in an ice divide southeast of Executive Committee Range.

Rockefeller Plateau (center coordinate (80° S/135° W)) is an extensive, ice-covered plateau at about 1000–1500 m elevation, bordered by Siple Coast, Shirase Coast to its west and Ford Range, Flood Range, and Executive Committee Range to its north.

The ice surface of the ice shelf is elevated up to 50 m due to inflow from Ice Streams E and D. South of Ice Stream D, the surface rises to Siple Dome, a much-studied feature, an ice dome between Ice Streams D and C (to the Dome's south). The inlet on Shirase Coast on this map is the bay in which Ice Stream F terminates. Ice Stream F is much smaller and shorter than Ice Streams D and E. Roosevelt Island bifurcates the flow of Ice Stream E such that less than half the discharge from E flows to the east of the island and more than half of E and the total output of D flow to the west.

Ice Stream D drains an area of 170.000 km^2 (McIntyre after Swithinbank 1988) and should have a balance discharge of 27 km^{3}a^{-1}. Ice Stream E drains an area of 154.000 km^2 and has a balance discharge of 22 km^{3}a^{-1}. The margins of Ice Stream E are well-defined in a satellite image (Swithinbank 1988, p. B135–B136, fig. 101 and 102), the fact that the margins are so well-defined in satellite imagery is attributed to the fact that bottom melting occurs under Ice Stream E but not on either side (Rose 1979). Below the ice streams, bedrock is several hundred meters below sea level. The surfaces of the ice streams show flowlines, rumples, and crevasses whereas the surfaces of Siple Dome and other slow-moving ice are smooth at this scale.

Ice Stream B (Whillans) moves 827 m a^{-1}, Rutford Ice Stream flows at 400 m a^{-1} into the Ronne Ice Shelf (Doake et al. 1987). It has been observed that ice streams change speed — Ice Stream C dropped its velocity in ≈1730 A.D. from ≈700 m a^{-1} to its present 5 m a^{-1}. It is not understood, why ice streams switch between fast and slow movement, and how and where the transition between sheet flow of the Greenland or Antarctic Ice Sheets (≈30 m a^{-1}) and ice stream flow occurs, if bed-topographic boundaries are not the cause (i.e. for all the ice streams that do not follow a trough). Ice Stream B has been found to slide on an 8 m thick sediment layer, which may serve to explain fast flow despite low driving stress (Alley et al. 1986). Engelhardt et al. (1990) report that the base of Ice Stream B is at the melting point and the basal water pressure is close to the ice overburden pressure, which may explain the fast flow of the ice stream by basal sliding or shear deformation of sediments.

250

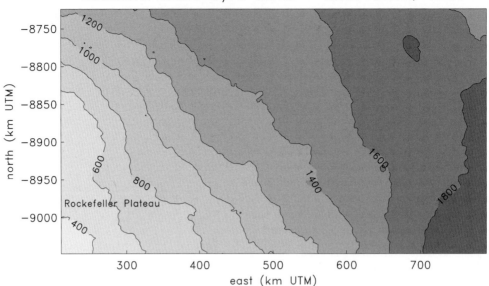

e219-255n78-815, WGS84, Gaussian variog., central mer. 237, slope corrected, scale 1:5000000, 980123

Map m237e219-255n78-815 Southern Marie Byrd Land

(Coordinates e219-255 correspond to w141-105)

This map shows part of *Marie Byrd Land* in West Antarctica, at an elevation of below 400 m to above 1800 m. Marie Byrd Land is lower than the *Polar Plateau* and *Wilkes Land* east of the Transantarctic Mountains. The ice thickness in the interior is 2500 m, so the ice bottom is approximately 1000 m below sea level in the central part of Marie Byrd Land. The ice drains to the west and westsouthwest, the 400 m contour marks the drainage of *Ice Stream D*; the 600 m part north of Rockefeller Plateau indicates the direction of the valley of *Ice Stream E* (on the westerly adjacent map, m213e195-231n78-815 Shirase Coast). The surface rises towards an ice divide above 1600 m, which runs from the Executive Committee Range to the southeast. Byrd Surface Camp (USA) is near (80° S/120° W), or (80° S/240° E). The surface gradient increases with lower elevations, from a very low gradient similar to those found in central East Antarctica (200 m over 120 km) between 1400 m and 1600 m to 200 m over 30 km between 600 m and 800 m elevation in the "valley" of Ice Stream E.

Southern Hollick—Kenyon Plateau — ERS1 DATA, 1995

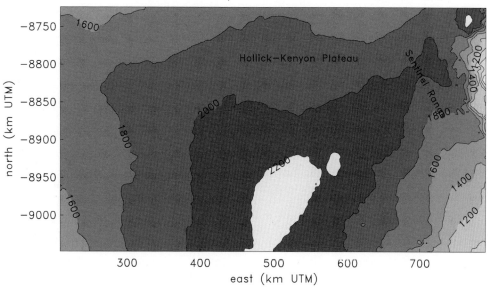

e243—279n78—815, WGS84, Gaussian variog., central mer. 261, slope corrected, scale 1:5000000, 980122

Map m261e243-279n78-815 Southern Hollick-Kenyon Plateau

(Coordinates e243-279 correspond to w117-81)

This map shows the interior of West Antarctica in the region of *southeastern Marie Byrd Land* containing the southern part of *Hollick-Kenyon Plateau*, and an area south of Ellsworth Land, west of the Ellsworth Mountains and east of Hollick-Kenyon Plateau that is not named. Hollick-Kenyon Plateau is a large, relatively featureless snow plateau, centered at (78° S/105° W), which was discovered by Lincoln Ellsworth on his transantarctic airplane flight in 1935 and named for his pilot, Herbert Hollick-Kenyon.

This map also contains the lowest known bedrock point in Antarctica, -2538 m at approximately (110° W/80° 30'S).

In the northeast of the map are parts of the *Ellsworth Mountains*, which border the Ronne Ice Shelf and consist of the *Sentinel Range* (77.3–79° S/≈85° W) in the north, trending N–S, and the *Heritage Mountains* (79–80.5° S/≈85° W) in the south, trending NW–SE. The Sentinel Range contains the Antarctic continent's highest mountain, *Vinson Massif* (78° 35'S/85° 25'W) peaking at 5140 m above sea level. A geologic bibliography of the Ellsworth Mountains is given by Webers and Splettstoesser (1982). Parts of the Ellsworth Mountains are also seen on maps m261e243-279n75-80 Northern Hollick-Kenyon Plateau and m285e267-303n78-815 Ellsworth Mountains, where a more detailed description is given. For the geology of these mountain ranges, see Tingey (1991).

Drainage to the north of this map is to Walgreen Coast, drainage to the east and to the southeast is to the Ronne Ice Shelf.

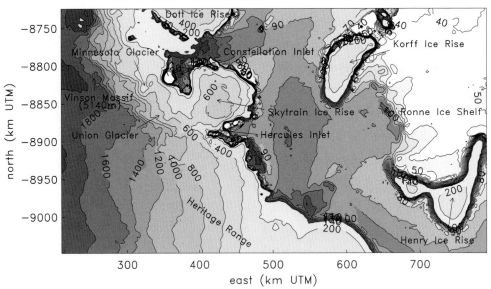

Ellsworth Mountains — ERS1 DATA, 1995

e267-303n78-815, WGS84, Gaussian variog., central mer. 285, slope corrected, scale 1:5000000, 990218

Map m285e267-303n78-815 Ellsworth Mountains

(Coordinates e267-303 correspond to w93-57)

The map m285e267-303n78-815 named after the *Ellsworth Mountains* shows the area of this mountain range, the southwestern part of the coast of Ronne Ice Shelf, and the ice shelf itself. The Ellsworth Mountains and the area of *Fletcher Ice Rise* (labeled Dott Ice Rise), and *Skytrain Ice Rise*, *Rutford Ice Stream* and neighbouring glaciers are described on the northerly adjacent map m285e267-303n75-80 Zumberge Coast. *Heritage Range*, the southeastern part of the Ellsworth Mountains, is shown on this map. It is lower than *Sentinel Range*.

Skytrain Ice Rise is an almost round ice rise, here it appears to be connected. Alberts (1995) lists Skytrain Ice Rise indeed as a large, flattish peninsula-like ice rise of 80 km extent (Skytrain is an airplane type). The USGS Satellite Image Map (Ferrigno et al. 1996) shows *Union Glacier* draining out of the central Heritage Range, and flowing east towards Skytrain Ice Rise, then north towards *Minnesota Glacier/Rutford Ice Stream*. The valley of

Union Glacier is seen in the ERS-1 map. *Patriot Hills* (80° 20'S/81° 25'W), an 8 km long line of rock hills just south of Heritage Range, are the staging location for research operations and mountaineering expeditions in the Ellsworth Mountains. The geology of the Ellsworth Mountains is treated in Tingey (1991). A short summary is given in the description of map m285e267-303n75-80 Zumberge Coast.

Hercules Inlet (80° 05'S/78° 30'W) is a large, narrow, ice-filled inlet which forms part of the southwest margin of Ronne Ice Shelf. It is bounded on the west by Heritage Range and in the north by Skytrain Ice Rise. (Hercules is an airplane type used for transport in Antarctica.) The ice streams entering Ronne Ice Shelf further south are not named.

In the area of the map are two of the larger ice rises in the interior of Ronne Ice Shelf, *Korff Ice Rise* and *Henry Ice Rise*. Korff Ice Rise is 160 km long, 40 km wide and, according to Swithinbank (1988,

p. B103), about 500 m high, here it appears to be over 200 m high. Ice thicknesses of up to 980 m have been measured. As noted for Berkner Island, the land mass of Korff Ice Rise does not rise above sea level, but reaches up to 115 m below sea level (Behrendt 1962; Swithinbank 1988, p. B103, also p. 119ff). Hence Korff Ice Rise is not truely an island, because its landmass does not reach above sea level.

Henry Ice Rise (80° 35'S/62° W) is a triangular shaped ice rise (described on map m309e291-327n78-815 Berkner Island) of over 200 m elevation, it has a northern tip and a western tip. (Captain Clifford Henry had been to Antarctica 14 times in support of the US Antarctic Program.) Between Korff and Henry Ice Rises are the Doake Ice Rumples (79° 45'S/67° W), an area of disturbed ice at the elevated area (> 100 m) (named after Christopher Doake, glaciologist at the British Antarctic Survey).

254

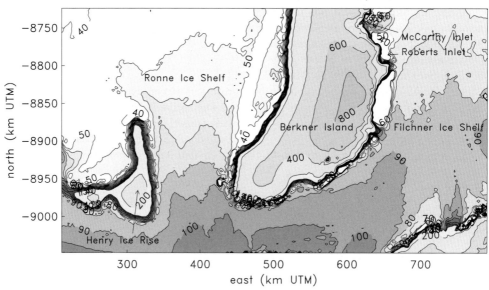

e291−327n78−815, WGS84, Gaussian variog., central mer. 309, slope corrected, scale 1:5000000, 990219

Map m309e291-327n78-815 Berkner Island

(Coordinates e291-327 correspond to w69-33)

This map shows the central part of the Filchner-Ronne Ice Shelf with *Berkner Island* and *Henry Ice Rise*.

Filchner-Ronne Ice Shelf is the large ice shelf separating the Antarctic Peninsula and Ellsworth Land from Queen Maud Land and the Transantarctic Mountains (here the Pensacola Mountains), located south of the Weddell Sea. Formally, there are two ice shelves, the (western) *Ronne Ice Shelf* and the (eastern) *Filchner Ice Shelf*, which are separated by the large Berkner Island in the north and connected south of Berkner Island. The Filchner-Ronne Ice Shelf and the *Ross Ice Shelf* are the two large ice shelves of the Antarctic continent; they are both part of so-called "West Antarctica" (see description of map m333e315-351n78-815 Filchner Ice Shelf); the Lambert Glacier/Amery Ice Shelf system is the largest ice-stream/ice-shelf system in East Antarctica and much smaller than the two West Antarctic ice shelves.

Filchner Ice Shelf is a fast-moving ice shelf (Swithinbank 1988, p. B101). Measured velocities vary around an average of 1240 ma^{-1} (Crabtree and Doake 1980) with a maximum of 2 km a^{-1} north of the Grand Chasms (Orheim 1979; on the Grand Chasms, see below).

The flowline pattern is visible in LANDSAT 1MSS satellite imagery (Swithinbank 1988, fig. 76, p. B102), the ice flow diverges north of the line of narrowest distance between Berkner Island and Coats Land (see map m333e315-351n75-80 Coats Land; this area of the ice shelf is shown on map m309e291-327n75-80 Ronne Ice Shelf). The fast flow causes the so-called "Grand Chasms", a large rupture 100 km long and 400 m to 5 km wide, and 53 m deep from top to top of sea ice (in 1957; Neuburg et al. 1959), 115 km long and 11 km wide in 1973 (Swithinbank 1988, p. B101), and 19 km wide in 1985, when it reached its maximum width. In 1986, the area north of the chasm had separated and formed into icebergs. When such a rift reaches an area of diverging flow – as is the case here in the northern part – then the rift widens because parts of the ice that are closest to the flanking land hinges, and rifts perpendicular to the ice front form.

Berkner Island is the world's largest ice rise (378 km long, 150 km wide, and up to 1000 m above sea level (Swithinbank 1988, p. B103); max elevation on our map is above 800 m; which is about the same). The *ice rise* is an independent ice cap built on a shoal on the continental shelf; the highest bedrock elevation is 80 m below sea level (Behrendt 1962) — so the land mass of the "island" does not rise above sea level, which means, it is not truly an island.

The northern part of Berkner Island is mapped on m309e291-327n75-80 Ronne Ice Shelf. We can see outlet glaciers in McCarthy Inlet and further south at (-8500.000 N/650.000 E) UTM (near the label "Filchner Ice Shelf"). Roberts Inlet can also be distinguished. The large glacier draining from the mainland into Filchner Ice Shelf is Recovery Glacier (in the southeastern corner of the map).

Henry Ice Rise is wedge-shaped with a 135 km N–S extension in its eastern part and a 120 km E–W extension in its southern part; elevations reach above 200 m in the satellite-altimetry-derived map (notice that actual elevations are usually higher in reality, due to effects explained in chapter (C)). Ice surface elevations in the Ronne Ice Shelf between Henry Ice Rise and southern Berkner Island decrease from 100 m (above WGS 84) to 50 m with a fairly even gradient of 10 m in 50 km ($0.02\% = 0.011°$; 100 m to 90 m) to 10 m in 70 km ($0.014\% = 0.008°$; 60 to 50 m) and with a lesser gradient to 40 m.

Part III

Applications

(E) Monitoring Changes in Antarctic Ice Surface Topography: The Example of the Lambert Glacier/Amery Ice Shelf System

(E.1) The Problem of Monitoring Changes

Changes in the mass of the Antarctic Ice Sheet are essential in studying the Earth's changing climate and sea level. The large ice streams and outlet glaciers play a key role, as described in the introduction (chapter (A)).

Maps produced from radar altimeter data, as the maps in this Atlas, provide a good basis for monitoring changes in the large outlet glaciers and ice streams, because

(1) maps show actual elevation, so height/elevation differences may be calculated,

(2) maps have a high grid resolution, which facilitates study of fairly small glaciers (several times 3 km wide),

(3) detailed geostatistical mapping results in maps with fairly accurate slopes (compared to other attempts of mapping from altimetry); this is due to the variography;

(4) the Atlas grids, based on UTM grids with rectangular coordinates with meter units allow to calculate mass gain/loss directly from the digital terrain models (approximately, for the 3 km squares of the grids).

Monitoring outlet glaciers, their extension, advance and retreat and their potential gain or loss of mass from atlas-type maps is also a feasible approach, because altimetry and 3-km-grid mapping both involve relatively low data-storage requirements and computational effort (compared to SAR data, for instance, and their processing).

A monitoring study is presented here for Lambert Glacier / Amery Ice Shelf, (a) to give a demonstration of this important application, and (b) as a study of the most important ice-stream/ice-shelf system in East Antarctica. (For more details, see Herzfeld et al. 1997; Herzfeld 1999.)

Lambert Glacier has a catchment area of about 10 percent of the entire ice sheet. To investigate advance and retreat of an ice-stream/ice-shelf system, it is not useful to look at the shelf-ice edge, because large icebergs break off at irregular time intervals, and the ice edge is difficult to monitor. For an ice-stream/ice-shelf system, the position of the grounding line is directly related to advance and retreat, and to changes in mass. Small changes in ice thickness translate into large changes in grounding line position (for geometrical reasons because of the small surface slope typical of Antarctic ice streams). Therefore the grounding line is a feature suitable to monitor changes – if it can be observed. The geostatistical evaluation technique for the first time provided a means to map the grounding line from satellite data (Herzfeld et al. 1993). The principles are: Based on a model that assumes idealized bedrock and perfect plasticity of ice, Weertman (1974) showed that the surface slope of an ice sheet decreases with the transition from grounded to floating ice. Using this result, a break in slope may be taken as a grounding-line indicator. A transition in slope occurs in the Lambert Glacier map between -7925.000 and -7825.000 N (UTM). In addition, the topography changes here from rougher areas indicative of grounded ice to smoother areas indicative of floating ice. This coincides with the

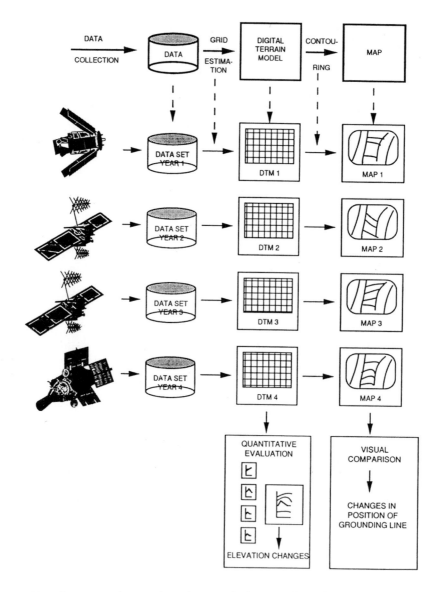

Figure E.2-1. Flowdiagram of mapping, interpolation, and evaluation procedures to monitor changes in ice streams.

location of the grounding line as mentioned in the literature (Mellor and McKinnon 1960; Budd et al. 1982; Zwally et al. 1987; Partington et al. 1987; cf. Herzfeld et al. 1993, 1994).

In this chapter, we use time series of satellite-altimetry-based digital terrain models to study changes in the Lambert Glacier/Amery Ice Shelf system over the time from 1978 to 1989 derived from data of SEASAT and the GEOSAT Exact Repeat Mission (ERM). Lambert Glacier and Amery Ice Shelf form the largest ice-stream/ice-shelf system in East Antarctica, with a drainage

basin of about 870 000 km^2 (Giovinetto and Bentley 1985; McIntyre 1985). Lambert Glacier is the only major Antarctic ice stream whose grounding line lies north of 72° S, well within the coverage of SEASAT and GEOSAT. This makes it the ice stream with the longest record of radar altimeter observations. Field work in the Amery Ice Shelf area conducted by Australian expeditions during 1962–1965 and 1968–1971 (Budd et al. 1982) and in recent years (I. Allison, pers. comm., 1993) included velocity measurements along profiles and determination of the location of the grounding line in a few points. Higham et al. (1995) used a 100 MHz ice radar in a traverse around the hinterland of the Lambert Glacier Basin and discovered that the Lambert Graben structure extends upstream of (lower) Lambert Glacier and is occupied by (upper) Lambert Glacier and Mellor Glacier. Brooks et al. (1983) and Zwally et al. (1987) analyzed SEASAT radar altimeter data, Partington et al. (1987) radar altimeter waveform data from few SEASAT tracks for identification of grounding line positions and crevasse fields. In a comparison of maps derived from geostatistical evaluation of radar altimeter data from SEASAT (1978) and GEOSAT ERM (1987–1989), we observed that the grounding line of Lambert Glacier advanced in this time period (Herzfeld et al. 1993, 1994). This is consistent with an overall increase in surface elevation deduced from crossover analyses (Lingle et al. 1994). These observations raise the questions whether the observed advance of the grounding line occurred gradually, and whether the observed overall elevation increase is common to the lower Lambert Glacier, the grounding zone, and the Amery Ice Shelf. With this goal in mind, and as a first step towards understanding the dynamics of the ice-stream/ice-shelf system, elevation changes are studied separately for grounded ice, floating ice, and the grounding zone.

(E.2) Time Series of Digital Terrain Models and Maps

Time series of digital terrain models are constructed from sets of SEASAT- and GEOSAT-ERM-altimeter data, using the geostatistical method described below. Each point in a time series corresponds to a map of the Lambert Glacier/Amery Ice Shelf area, represented by a DTM, or, in mathematical terms, by a three-dimensional array (easting, northing, elevation). Changes in the ice-stream/ice-shelf system are derived from visual or quantitative comparisons of points in the time series. The procedure is summarized in the flow diagram in Figure E.2-1 and described below.

(E.3) Altimeter Data: Acquisition and Corrections

To produce a time series, we need data from different satellites and different times: SEASAT, GEOSAT, and ERS-1 and -2 altimeter data are utilized. SEASAT was only in operation for a period of about three months in austral late winter (10 July – 9 October, 1978). To avoid seasonal effects such as changes in penetration depth (Ridley and Partington 1988) and precipitation, time windows corresponding in season to the SEASAT operation time have been selected from the GEOSAT ERM data in 1987, 1988, and 1989. Notice that (other than in the Atlas mapping, where GEOSAT GM data are used), here data from the Exact Repeat Mission are used — the tracks of each satellite revolution around the Earth repeat the previous ones exactly, and also those of the SEASAT mission. Track maps for SEASAT and GEOSAT for the study area are given in chapter (B) above (Fig. B.2.1-1 and Fig. B.2.1-2). This reduces errors in the comparison (cf. chapter (B) on Satellite Remote Sensing). The processing is similar to that of the Atlas data: The altimeter-derived surface heights are referenced to Goddard Earth Model (GEM) T2 orbits (Marsh et al. 1989) and corrected for tracking errors (Martin et al. 1983), atmospheric effects and solid Earth tides (Zwally et al. 1983) and water-vapor effects by the ice-sheet-altimetry group at NASA Goddard Space Flight Center, Greenbelt, Maryland, USA. For accurate topographic mapping, the data are also corrected for the effect that the radar signal is first received from the point of closest approach rather than from the point at nadir (slope correction), following a method described in Brenner et al. (1983). For calculation of elevation changes, however, the slope correction is best not applied, because it induces an additional error.

We will present two types of maps:

(1) Maps from corrected, but not slope-corrected data will be presented as the optimal basis to calculate elevation changes, and

(2) maps from corrected and slope-corrected data will be presented to provide a basis for visual comparison with other Lambert Glacier maps.

Surface elevations are given relative to the WGS84 ellipsoid unless stated otherwise (see section (C.4) on the role of the geodetic reference surface). The data, originally referenced to geographical coordinates, are transformed into Universal Transverse Mercator (UTM) coordinates, using 70° eastern longitude as central meridian which is mapped to coordinate 500000 (Snyder 1987). The area is bounded by 66°–75° eastern longitude and 68°–72.1° southern latitude, and is rectangular after transformation into UTM coordinates with dimensions 309 km east–west by 444 km north–south (Herzfeld et al. 1994). The typical spatial distribution of the satellite-altimeter data is shown in a map of groundtracks (Herzfeld et al. 1994, fig. 1); the distribution is characterized by dense measurements along the tracks with gaps of about 25 km between tracks and interruptions in the tracks, which are likely to occur along the glacier margin and in other places of rapid changes in surface gradient, which may lead to loss of the return signal and to retracking problems.

The maps are calculated using kriging with the GEOSAT variogram (Gaussian, nugget effect = 250 m², quasi-sill 593 m^2, and quasi-range 18000 m, see section (C.3)). All DTMs presented have a grid interval of 3 km, which is approximately three times the ice thickness in the grounding zone (ice thickness is 800 m according to Budd et al. 1982). A digital terrain model is calculated for each of the following 10 data sets, 5 time/satellite options and 2 processing options:

(a) SEASAT (1978)

(b) GEOSAT (1987), SEASAT time window

(c) GEOSAT (1988), SEASAT time window

(d) GEOSAT (1989), SEASAT time window

(e) GEOSAT (1987-1989), SEASAT time windows,

with

(1) GEM-T2-referenced, corrected, but not slope-corrected data,

(2) GEM-T2-referenced, corrected, slope-corrected data.

which yields the following maps:

Map Lambert.Seasat.1978.corr (E.3-1)

Map Lambert.Seasat.1978.slcorr (E.3-2)

Map Lambert.Geosat.1987.corr (E.3-3)

Map Lambert.Geosat.1987.slcorr (E.3-4)

Map Lambert.Geosat.1988.corr (E.3-5)

Map Lambert.Geosat.1988.slcorr (E.3-6)

Map Lambert.Geosat.1989.corr (E.3-7)

Map Lambert.Geosat.1989.slcorr (E.3-8)

Map Lambert.Geosat.1987-89.corr (E.3-9)

Map Lambert.Geosat.1987-89.slcorr (E.3-10)

The 1987–89 GEOSAT map is also presented with reference to the GEM-T3 geoid and the OSU91A geoid (see section (C.4.2): Fig. C.4.2-2 and Fig. C.4.2-3).

In Herzfeld et al. (1997), the effects of frequency filtering of the data are also studied. Frequency filtering is found to be disadvantageous.

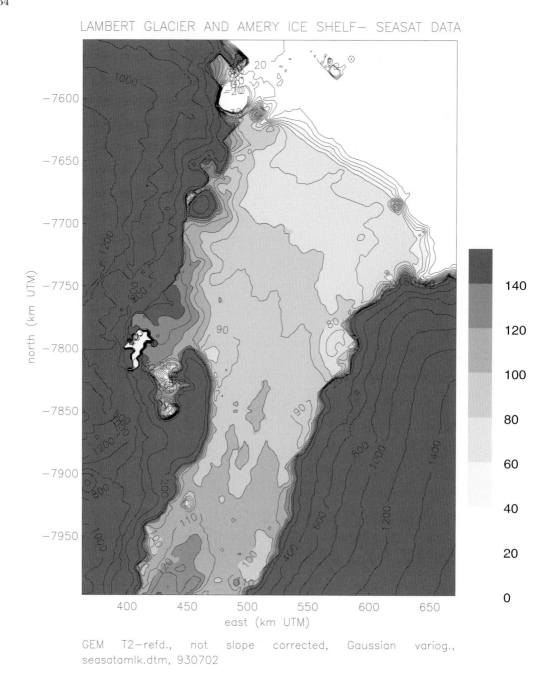

LAMBERT GLACIER AND AMERY ICE SHELF— SEASAT DATA

GEM T2—refd., not slope corrected, Gaussian variog., seasatamlk.dtm, 930702

Figure E.3-1. Map Lambert.Seasat.1978.corr. Lambert Glacier/Amery Ice Shelf area surface elevation used as basis for calculation of elevation changes; not slope-corrected. Scale 1:3.000.000. Elevations in meters above WGS84 ellipsoid. DEM calculated using ordinary kriging with Gaussian variogram model.

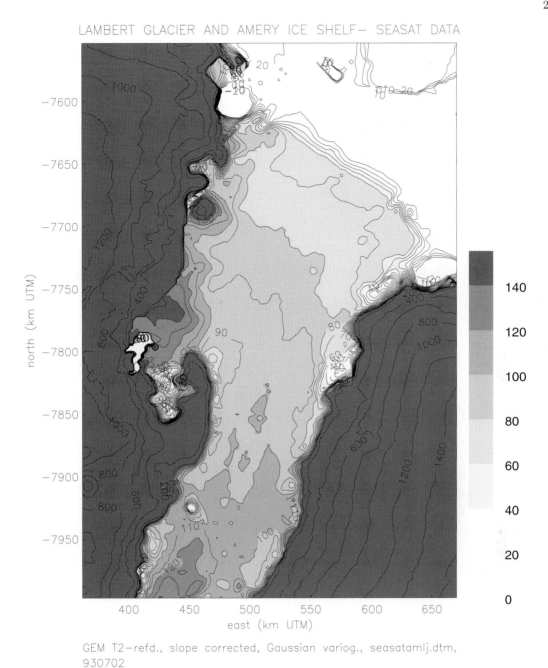

LAMBERT GLACIER AND AMERY ICE SHELF— SEASAT DATA

GEM T2—refd., slope corrected, Gaussian variog., seasatamlj.dtm, 930702

Figure E.3-2. Map Lambert.Seasat.1978.slcorr. Topography of Lambert Glacier and Amery Ice Shelf from 1978 SEASAT radar altimeter data. Scale 1:3.000.000. Elevations in meters above WGS84 ellipsoid, GEM-T2-referenced, slope-corrected. DEM calculated using ordinary kriging with Gaussian variogram model.

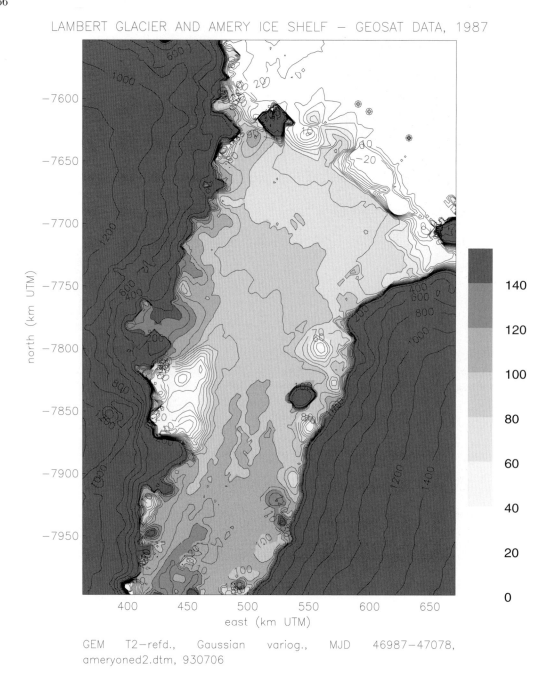

GEM T2—refd., Gaussian variog., MJD 46987—47078, ameryoned2.dtm, 930706

Figure E.3-3. Map Lambert.Geosat.1987.corr. Lambert Glacier/Amery Ice Shelf area surface elevation used as basis for calculation of elevation changes; not slope-corrected. Scale 1:3.000.000. Time frame corresponds in season to SEASAT data. Elevations in meters above WGS84 ellipsoid. DEM calculated using ordinary kriging with Gaussian variogram model.

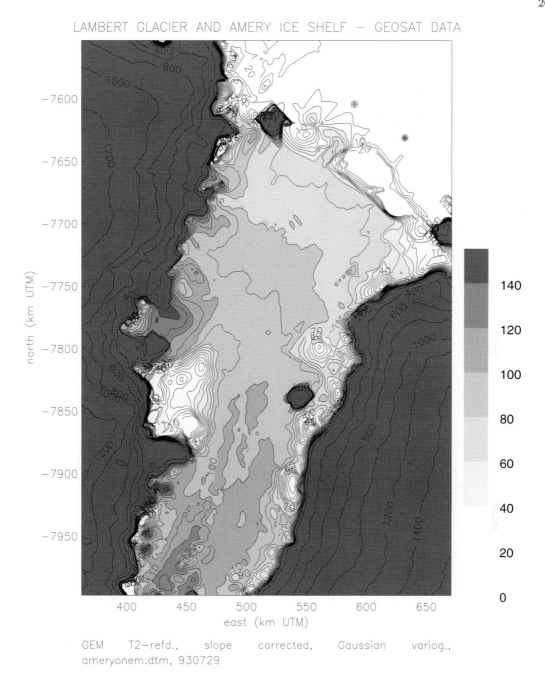

GEM   T2—refd.,   slope   corrected,   Gaussian   variog.,
ameryonem.dtm, 930729

Figure E.3-4. Map Lambert.Geosat.1987.slcorr. Topography of Lambert Glacier and Amery Ice Shelf from 1987 GEOSAT radar altimeter data. Scale 1:3.000.000. Time frame corresponds in season to SEASAT data. Elevations in meters above WGS84 ellipsoid, GEM-T2-referenced, slope-corrected. DEM calculated using ordinary kriging with Gaussian variogram model.

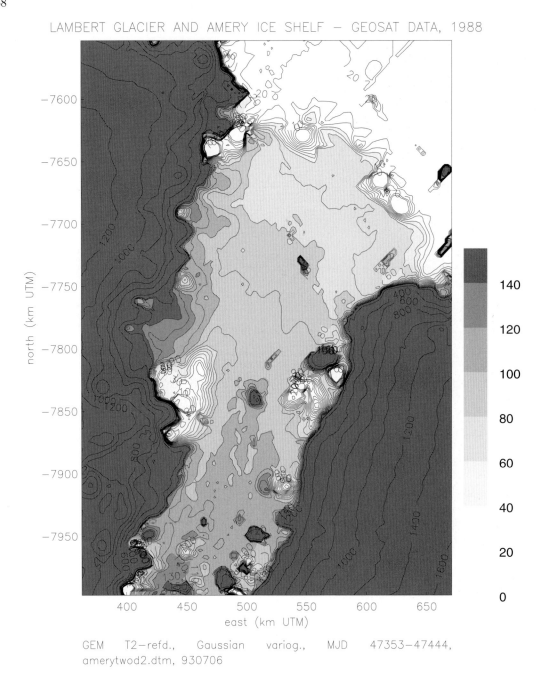

LAMBERT GLACIER AND AMERY ICE SHELF — GEOSAT DATA, 1988

GEM T2—refd., Gaussian variog., MJD 47353—47444, amerytwod2.dtm, 930706

Figure E.3-5. Map Lambert.Geosat.1988.corr. Lambert Glacier/Amery Ice Shelf area surface elevation used as basis for calculation of elevation changes; not slope-corrected. Scale 1:3.000.000. Time frame corresponds in season to SEASAT data. Elevations in meters above WGS84 ellipsoid. DEM calculated using ordinary kriging with Gaussian variogram model.

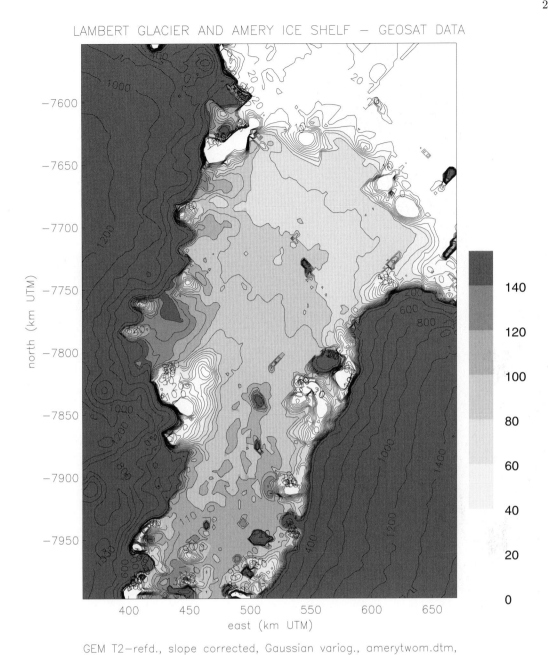

LAMBERT GLACIER AND AMERY ICE SHELF — GEOSAT DATA

GEM T2—refd., slope corrected, Gaussian variog., amerytwom.dtm, 930729

Figure E.3-6. Map Lambert.Geosat.1988.slcorr. Topography of Lambert Glacier and Amery Ice Shelf from 1988 GEOSAT radar altimeter data. Scale 1:3.000.000. Time frame corresponds in season to SEASAT data. Elevations in meters above WGS84 ellipsoid, GEM-T2-referenced, slope-corrected. DEM calculated using ordinary kriging with Gaussian variogram model.

270

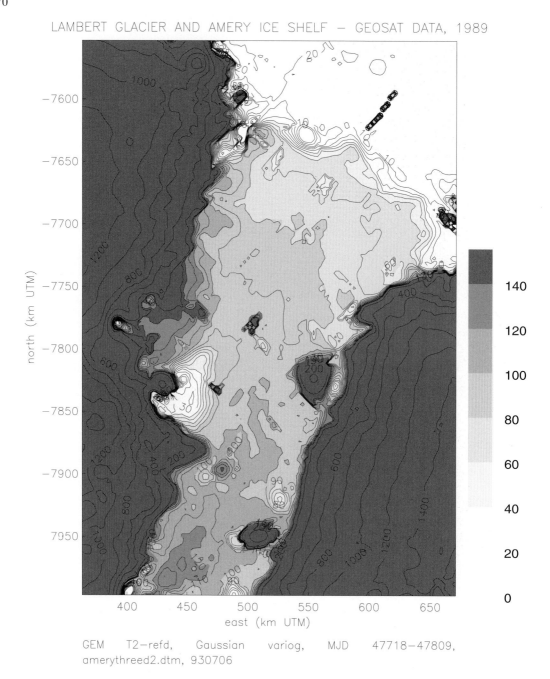

GEM   T2—refd,   Gaussian   variog,   MJD   47718—47809,
amerythreed2.dtm, 930706

Figure E.3-7. Map Lambert.Geosat.1989.corr. Lambert Glacier/Amery Ice Shelf area surface elevation used as basis for calculation of elevation changes; not slope-corrected. Scale 1:3.000.000. Time frame corresponds in season to SEASAT data. Elevations in meters above WGS84 ellipsoid. DEM calculated using ordinary kriging with Gaussian variogram model.

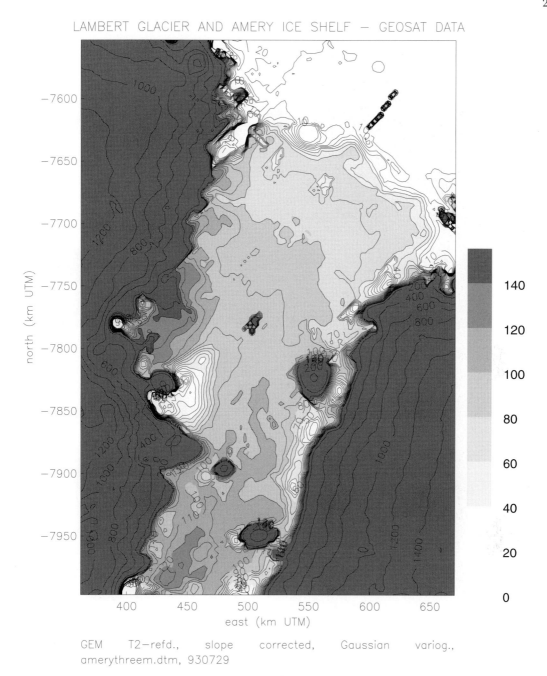

Figure E.3-8. Map Lambert.Geosat.1989.slcorr. Topography of Lambert Glacier and Amery Ice Shelf from 1989 GEOSAT radar altimeter data. Scale 1:3.000.000. Time frame corresponds in season to SEASAT data. Elevations in meters above WGS84 ellipsoid, GEM-T2-referenced, slope-corrected. DEM calculated using ordinary kriging with Gaussian variogram model.

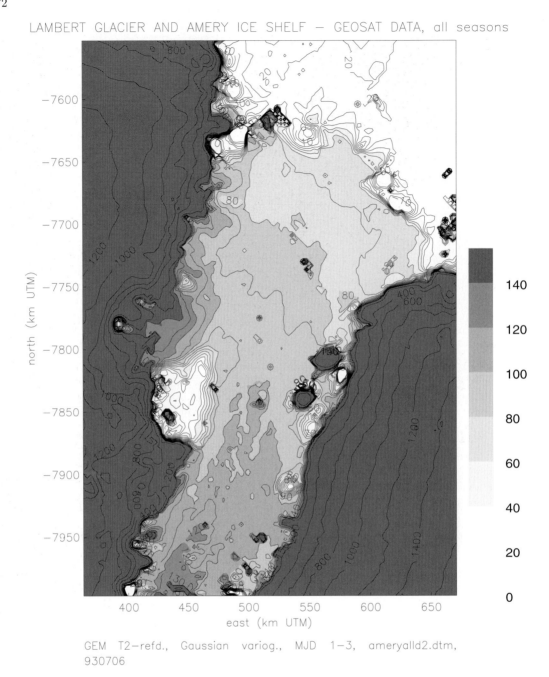

LAMBERT GLACIER AND AMERY ICE SHELF — GEOSAT DATA, all seasons

GEM T2—refd., Gaussian variog., MJD 1—3, ameryalld2.dtm, 930706

Figure E.3-9. Map Lambert.Geosat.1987-1989.corr. Lambert Glacier/Amery Ice Shelf area surface elevation used as basis for calculation of elevation changes; not slope-corrected. Scale 1:3.000.000. Time frame corresponds in season to SEASAT data. Elevations in meters above WGS84 ellipsoid. DEM calculated using ordinary kriging with Gaussian variogram model.

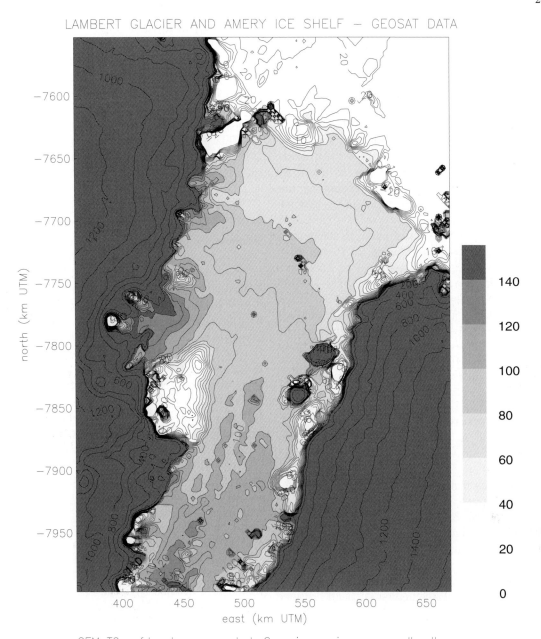

GEM T2—refd., slope corrected, Gaussian variog., ameryallm.dtm, 930729

Figure E.3-10. Map Lambert.Geosat.1987-1989.slcorr. Topography of Lambert Glacier and Amery Ice Shelf from 1987-1989 GEOSAT radar altimeter data. Scale 1:3.000.000. Time frame corresponds in season to SEASAT data. Elevations in meters above WGS84 ellipsoid, GEM-T2-referenced, slope-corrected. DEM calculated using ordinary kriging with Gaussian variogram model.

(E.4) Visual Comparison — Quantitative Comparison

Two approaches are used to investigate changes in the Lambert Glacier/Amery Ice Shelf system: (1) changes in the position of the grounding line (see discussion in chapter on map m69e61-77n67-721 Lambert Glacier), and (2) changes in surface elevation (see flow diagram in Figure E.2-1). It has been established that satellite-altimetry-derived maps, constructed using ordinary kriging as described here, may be used to locate the grounding line (Herzfeld et al. 1993): The location of the grounding line is marked by a characteristic change in surface slope, and by a transition from rougher topography indicative of grounded ice to smoother topography indicative of floating ice. These two criteria have been applied to identify the grounding line on Lambert Glacier/Amery Ice Shelf, and it is most closely approximated by the 100 m contour line (compared to other contour lines at 1 m intervals). The resulting locations are consistent with the few field observations available (see text to map m69e61-77n67-721 Lambert Glacier).

Although changes in surface elevation seem to be the straightforward quantity to monitor, it is actually difficult to get a clear signal out of measurements of surface elevation; elevation changes are relatively small in relation to various geodetic corrections applied in the data processing, and instrument errors. Changes in the apparent location of the grounding line are more reliable indicators of actual changes than elevation changes. Because of the very small angle of the surface slope, changes in elevation on the order of a few decimeters translate into changes in the position of the grounding line on the order of kilometers. A mean advance in the grounding line is visible by eye in the maps. Consequently, changes in position of the grounding line will be derived from maps (left row of arrows in the flow diagram in Fig. E.2-1), using a visual comparison aided by simple measurements with ruler and planimeter. Calculation of elevation changes is based on DTMs with a few necessary mathematical considerations.

An advantage of deriving elevation changes from DTMs rather than directly from differences of averages of along-track elevation data is that each grid node in a DTM is representative of a unit area. An average from along-track data, on the other hand, assigns heavier weights to areas of dense coverage, in the case of Lambert Glacier to the southern part of the map, where the ground tracks converge. This effect is particularly severe in this study, because the map area extends to the southern margin of SEASAT and GEOSAT coverage. The negative effect of weighting may also put DTM-derived elevation changes at an advantage over elevation changes from crossover analysis (e.g. Lingle et al. 1994).

Results from time series of radar-altimeter-data-based maps

Whereas the elevation change data reveal the 1988 season as an outlier in the 1978-1989 trend, each of the three one-season maps for 1987, 1988, 1989 (maps E.3-4 Lambert.Geosat.1987.slcorr, E.3-6 Lambert.Geosat.1988.slcorr, E.3-8 Lambert.Geosat.1989.slcorr) as well as the three-season map for 1987-1989 (map E.3-10 Lambert.Geosat.1987-1989.slcorr) in comparison with map E.3-2 Lambert.Seasat.1978.slcorr shows that the grounding line advanced from SEASAT times to GEOSAT times. The accuracy of the maps (cf. section (C.3.5) on kriging errors) is good enough to warrant an investigation of location changes of larger features, i.e. of features supported by several adjacent 3 km grid nodes in each direction. The area above the 100 m contour is larger and more connected on the Geosat maps ("fingers" extending north rather than a few "islands" north of the connected area above 100 m). At closer scrutiny, the connection of dark-grey areas progresses through the years: The disconnected grey area around UTM-coordinates -7850.000 N/510.000 E on the 1987 GEOSAT map (map E.3-4) may be identified with the smaller area in the same place on the SEASAT map (map E.3-2). The "finger" west and southwest of this grey area has also increased from 1978 to 1987. In 1988 the "finger" breaks up and the "island" connects to the larger area above 100 m, and that process continues to 1989. This observation supports the following hypothesis: The ice grounds on seafloor shoals, and these are formed or capped by sediments that get moved around by the ice, or vice versa, sediment packages form and grow (possibly due to subglacial water flow), pushing the grounding line further seaward. This looks more plausible from the maps than from the graphs of elevation changes and could also serve to explain what seems to be a decrease in elevation in 1988, while an overall advance of the grounding line is observed.

In summary, from a visual comparison of the maps (left side in flow diagram of Fig. E.2-1), it is concluded that the glacier advanced about 10–12 km between SEASAT and GEOSAT ERM times (taken for all or each of 1987, 1988, 1989), and that the apparent interannual variability between 1987, 1988, and 1989 is high relative to the change observed for the 10-to-12-year interval.

(E.5) Calculation of Elevation Changes

The purpose of calculating elevation changes is to answer the following glaciological questions:

(GQ1) How is the general advance of the grounding line between SEASAT and GEOSAT times and overall increase in surface elevation distributed in time?

(GQ2) Is the glacier homogeneously increasing in elevation?

An answer to the first question may give some insight in the variability of the system from one year to another. An answer to the second question may give insight in the dynamics of the system. We should elaborate a bit on the second question. Certain side glaciers of Lambert Glacier have been suspected to be surge-type glaciers.

The first problem (GQ1) initially motivated the construction of Lambert Glacier maps in time series. To approach the second problem (GQ2), the study area has been divided in two different ways: (i) into upper, grounded area (north coordinate -7925.000 to southern margin of outline), grounding zone (-7820.000 to -7925.000), and lower, floating area (northern margin of outline to -7820.000), and (ii) into grounded part (-7875.000 to southern margin) and floating part (northern margin to -7875.000).

We will also answer the following methodological questions:

(MQ1) What type of DTM is best used to study elevation changes? Should data have been slope corrected?

(MQ2) How can robust estimates of elevation changes be derived from the DTMs?

For (MQ1), the two different types of maps – not slope corrected, slope corrected – have been introduced in the previous section. Calculation of robust estimates of average surface elevation from the altimetry-derived DTMs poses a problem (MQ2), because the DTMs are less accurate in areas of high relief such as the mountainous areas on the flanks of the ice stream and ice shelf, and because the overdeepening along the margins of the glacier is larger than in reality due to the snagging effect. Fortunately, the area of interest coincides with the area of high accuracy. Consequently, averages of surface elevation are calculated only from grid nodes inside a polygon that contains the central part of the ice-stream/ice-shelf system.

We have used six different polygon outlines to get an idea of the ambiguity induced by a human drawing an outline on a map; two of those are used in Table E.5-1 and in Figures E.5-1 to E.5-4 for demonstrative purposes. Overlay 2a is characterized by approximately even width throughout length of ice-stream/ice-shelf system in map area. Overlay2c is characterized by narrow width in ice-stream/iceshelf area south of -7800.000 N, widening northwards to include most of thefloating tongue north of -7750.000 N.

Averages have been formed in two ways: arithmetic and statistical averages. To obtain statistical averages, in a first pass, the arithmetic average and the standard deviation of elevation values for all grid nodes inside the polygon outline are calculated; in a second pass, only points with elevation values within three standard deviations of the mean are retained and their values averaged. The combinations of two average types, six outlines, and two types of DTMs yield 24 possible elevation change plots, four of which are presented in methodological order to demonstrate our results (use Fig. E.5-1 vs. E.5-2 for influence of two different outlines, Fig. E.5-2 vs. E.5-3 for difference between arithmetic and statistical averages for the same data, Fig. E.5-3 vs. E.5-4 for influence of slope correction on elevation change estimates).

276

Table E.5-1. Averages of Surface Elevation of Lambert Glacier/Amery Ice Shelf, from Altimetry-Based Grids.

Year	N. margin to S. margin	N. margin to -7875	-7875 to S. margin	N. margin to -7820	-7820 to -7925	-7925 to S. margin
Overlay 2 a, statistical averages						
1978	95.75	88.21	108.03	85.33	99.31	112.59
1987	96.23	88.77	107.88	85.80	100.51	111.83
1988	96.49	89.19	108.41	86.11	101.35	112.42
1989	96.48	88.59	108.50	85.57	100.27	111.65
1987–1989	96.54	88.91	108.49	85.99	100.51	112.60
Overlay 2 a, arithmetic averages						
1978	95.96	88.24	107.90	85.33	99.15	112.34
1987	96.48	88.80	108.36	85.80	100.40	112.37
1988	98.30	90.06	111.05	86.41	102.55	115.95
1989	97.58	89.08	110.73	86.26	101.07	115.17
1987–1989	97.09	89.17	109.34	86.16	100.66	113.82
Overlay 2 c, statistical averages						
1978	91.38	85.30	111.71	83.21	99.43	117.04
1987	91.65	85.73	111.13	83.53	100.68	115.44
1988	91.43	86.04	108.94	83.76	102.47	112.69
1989	91.97	85.64	112.35	83.49	100.82	115.81
1987–1989	91.88	85.89	111.35	83.74	100.98	115.97
Overlay 2 c, arithmetic averages						
1987	91.63	85.30	111.71	83.21	99.43	117.04
1987	91.92	85.66	111.76	83.46	100.86	116.30
1988	92.64	86.87	110.94	83.90	104.16	115.09
1989	92.19	85.88	112.18	83.79	101.53	115.81
1987–1989	92.46	86.08	112.68	83.79	101.38	117.67

In the error maps presented in the section on Errors, the standard error of the kriged elevation maps is mostly below 3 m, with higher values in areas of high topographic relief. Since kriging is a moving-window-averaging operator, the 3 m value is an expression of the number of data in the neigborhood of an estimated grid node, and of the noise in the data (it does not imply that the surface elevation of an area is likely to be 3 m higher or lower than estimated). By averaging over all grid nodes inside a polygon, the standard error of the elevation estimate is reduced by division through the square-root of the number of points. That amounts to an error of 0.03–0.04 m for the entire ice-stream/ice-shelf system, to larger errors for smaller parts, and largest for the smallest and also roughest area, the grounded part south of UTM north coordinate -7925.000 (errors on the order of 0.5 m). For all parts except this southernmost part, the elevation estimates are useful for glaciological interpretation.

The location of the outline obviously influences the actual value of the average. The trend over time (cf. Figs. E.5-1 and E.5-2), however, is the same for the two outlines and other outlines we used, for averages taken over the entire area (column 1 in Table E.5-1), over the Amery Ice Shelf up to the grounding line (average north coordinate -7875.000; column 2 in Table E.5-1), and over the ice shelf excluding the grounding zone (average north coordinate -7820.000; column 4 in Table E.5-1), and for the grounding zone (north coordinates -7820.000 to -7925.000; column 5 in Table E.5-1): The average elevation for any GEOSAT data set is higher than the average for the SEASAT data. For the lower Lambert Glacier, both from the grounding line upward (column 3 in Table E.5-1) and excluding the grounding zone (column 6 in Table E.5-1), the result depends on the outline chosen, and the type of average. Outline 2c is characterized by a narrow (60-100 km wide) section over the grounded part, to avoid difficult areas along the glacier margin, this, on the other hand, includes fewer grid nodes than the wider outline 2a and results in less robust averages.

Both arithmetic averages and statistical averages are suitable; however, arithmetic averages seem a little better (compare Figs. E.5-2 and E.5-3). One would have expected statistical averages to yield lower amplitudes, because statistical aver-

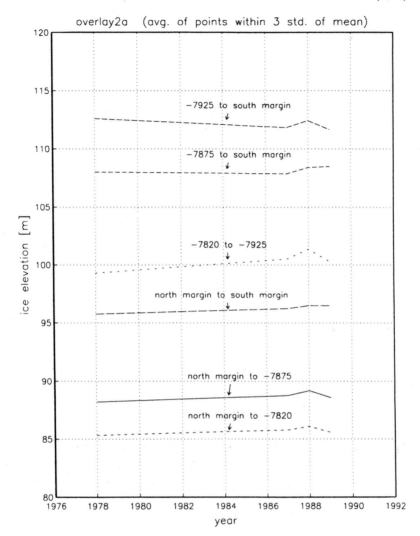

AVERAGE ICE ELEVATIONS OF LAMBERT GLACIER AND AMERY ICE SHELF

GEM T2—REFD. AND NOT SLOPE CORRECTED MAPS (k,d2)

Figure E.5-1. Elevation changes of Lambert Glacier and Amery Ice Shelf from DEMs in section (E.3). Statistical averages of grid values inside overlay 2a. Overlay 2a is characterized by approximately even width throughout length of ice-stream/ice-shelf system in map area. Averages calculated from not-slope-corrected DEMs, see maps E.3-1 (SEASAT 1978), E.3-3 (GEOSAT 1987), E.3-5 (GEOSAT 1988), E.3-7 (GEOSAT 1989)

ages tend to eliminate the error spikes due to the 3-standard-deviations criterion. By a comparison of Figures E.5-3 and E.5-4, it can also be observed that the values of the statistical averages are always lower than the values of the arithmetic averages, which demonstrates that the outliers are high points. The latter observation indicates that a filtering technique, if applied at all, should not be required to preserve the mean (MQ1).

Result: For the Amery Ice Shelf and the grounding zone, calculations indicate an increase in elevation, independently of the averaging algorithm. For the lower Lambert Glacier, the result is not clear, with arithmetic averages yielding mostly decreases, and statistical averages yielding a mixed picture.

Result: A comparison of columns 3 and 6 of Table E.5-1 indicates that a decrease in elevation is

278

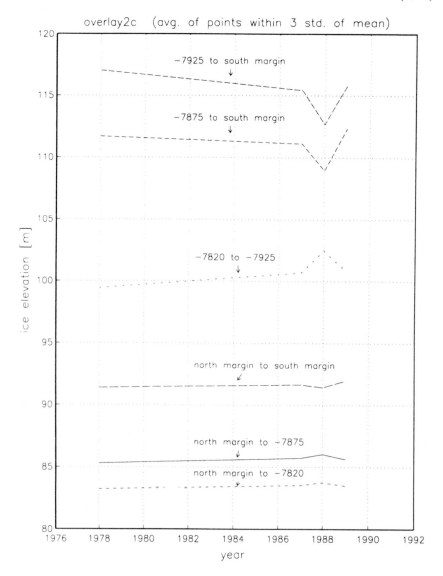

Figure E.5-2. Elevation changes of Lambert Glacier and Amery Ice Shelf from DEMs in section (E.3). Statistical averages of grid values inside overlay 2c. Overlay2c is characterized by narrow width in ice-stream/iceshelf area south of -7800.000 N, widening northwards to include most of thefloating tongue north of -7750.000 N. Averages calculated from not-slope-corrected DEMs, see maps E.3-1 (SEASAT 1978), E.3-3 (GEOSAT 1987), E.3-5 (GEOSAT 1988), E.3-7 (GEOSAT 1989)

most likely for the southernmost part (Lambert Glacier excluding the grounding zone). Whether this observation is indeed showing a stagnation of the average elevation or is caused by the high error levels in this part of the map cannot be determined on the basis of this data (surge wave?).

DTMs calculated from data corrected for geodetical and instrument errors, but not slope-corrected nor frequency-filtered, yield the most accurate calculation of elevation changes (MQ1). Both frequency filtering and slope correction are operations performed over an along-track segment or an area, unlike geodetical and instrument corrections,

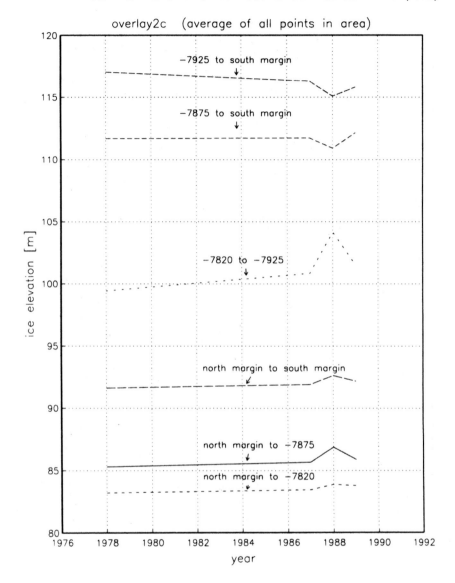

AVERAGE ICE ELEVATIONS OF LAMBERT GLACIER AND AMERY ICE SHELF

GEM T2−REFD. AND NOT SLOPE CORRECTED MAPS (k,d2)

Figure E.5-3. Elevation changes of Lambert Glacier and Amery Ice Shelf from DEMs in section (E.3). Arithmetic averages of grid values inside overlay 2c. Overlay2c is characterized by narrow width in ice-stream/iceshelf area south of -7800.000 N, widening northwards to include most of thefloating tongue north of -7750.000 N. Averages calculated from not-slope-corrected DEMs, see maps E.3-1 (SEASAT 1978), E.3-3 (GEOSAT 1987), E.3-5 (GEOSAT 1988), E.3-7 (GEOSAT 1989)

which are applied pointwise. This is why the former cause a distortion of the data that propagates errors strongly into time-dependent differencing and obliterates the elevation change over time. The slope correction, while necessary for accurate mapping of absolute elevations, induces large er-

rors on the differences between elevations. Theoretically, the slope effect should be approximately the same for two different times, and should therefore cancel out when two DTMs are subtracted from each other. Looking at Figures E.5-3 and E.5-4, this is not the case. Slope correction in-

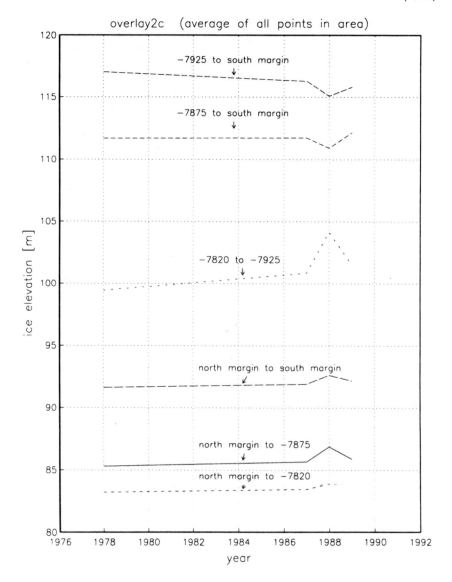

Figure E.5-4. Elevation changes of Lambert Glacier and Amery Ice Shelf from DEMs in section (E.3). Arithmetic averages of grid values inside overlay 2c. Overlay2c is characterized by narrow width in ice-stream/iceshelf area south of -7800.000 N, widening northwards to include most of thefloating tongue north of -7750.000 N. Averages calculated from slope-corrected DEMs, see maps E.3-2 (SEASAT 1978), E.3-4 (GEOSAT 1987), E.3-6 (GEOSAT 1988), E.3-8 (GEOSAT 1989)

fluences in particular the curves for the southern part, where the slope is steepest, and to a much lesser extent those for the floating ice shelf. Frequency filtering with retention of the data mean should be avoided because it neglects topographic and instrument-error sources while retaining the data mean.

In conclusion, average elevation changes derived from DTMs calculated from altimeter data that have not been slope corrected nor frequency filtered (MQ1), and grid nodes inside a polygon outline avoiding the glacier margin (MQ2) are sufficiently accurate for glaciological interpretation, for all subareas of Lambert Glacier/Amery Ice Shelf including the entire ice-stream/ice-shelf system, except for the smallest and roughest subarea, the grounded part excluding the grounding zone. We are now ready turn to the glaciological questions.

To glaciological question (GQ1) – Time distribution of changes. A general increase in elevation for the entire glacier/ice-shelf system is derived from averaging elevations of the DTMs. This is consistent with results from crossover analysis (Lingle et al. 1994) and the advance of the grounding line deduced from DTMs (Herzfeld et al. 1994). The absolute differences between SEASAT and GEOSAT data, however, are lower than the fluctuations between the three GEOSAT years 1987, 1988, and 1989, with 1988 values either too high or too low

to match the trend. These may be due to interannual variation, or more likely, to undetected problems with the altimeter, because apparent interannual elevation changes indicated by the 1988 figures are rather large. With available data, however, we have no means to exclude a sudden change (a surge?).

To glaciological question (GQ2) – Location breakdown. In summary, a significant increase in elevation is observed for the Amery Ice Shelf and the grounding zone (for all outline locations and both averaging algorithms tested), whereas average surface elevations of the rougher and topographically more varied grounded Lambert Glacier may have decreased (but the decrease is not sufficiently supported by the data) (surge indication downstream?).

(E.6) Discussion of Results on Elevation Changes

(E.6.1) Results of the Monitoring Study

Partington et al. (1987) identify the position of several points along the grounding line. Their map, however, does not show any coordinates nor scale, which makes an exact comparison impossible. Their northernmost grounding point is in the same general area (south of Beaver Lake and Jetty Peninsula) as the grounding zone on map E.3-2 Lambert.Seasat.1978.slcorr, the other points are up to about 80 km farther south. Budd et al. (1982) identify a grounding point from a break in surface slope at UTM 485.600 E/-7902.900 S (3 km south of their point T4), this grounding point is located in the southern part of the grounding zone identified on map E.3-2 Lambert.Seasat.1978.slcorr. This comparison might indicate a small advance between the time of the Australian expeditions (1968–1971) and 1978. Brooks et al. (1983) find a break in slope 43 km south of the grounding point of Budd et al. (1982); however, Brooks et al. (1983) remark that their altimeter data processing does not allow to identify the grounding line. According to Partington et al. (1987) most estimates in the literature indicate a net accumulation, except McIntyre (1985). How much of the observed elevation changes and the advance of the grounding line is due to mass

increase cannot be determined from the available data. A more detailed survey of the velocity field of Lambert Glacier/Amery Ice Shelf may contribute to answering related questions.

According to Fricker et al. (2000), the grounding line of Lambert Glacier is located much farther upstream (south of the southern margin of the maps in this study). Even if this is correct, the change in location of the surface characteristics and elements mapped here indicates an advance of Lambert Glacier.

The dynamics of the Amery Ice Shelf and the ice streams that feed into it may be complex. The fast-flowing ice discharges at about 1.2 km/yr into the shelf, only about one third of the discharge is contributed by (lower) Lambert Glacier (Budd et al. 1982). Wellmann (1982) finds geomorphologic evidence that Fisher Glacier, one other tributary of the Amery Ice Shelf, may be surging. Budd (1966) reports a 50 km retreat of the ice shelf front between 1955 and 1965. Consequently, trends in elevation changes need not necessarily coincide for the lower Lambert Glacier and the Amery Ice Shelf and Charybdis Glacier. This same observation is made in our analysis.

Brooks et al. (1983) conclude from a comparison of SEASAT data and data published by Budd et al. (1982) that Lambert Glacier may have retreated between 1968–1971 and 1978, while Charybdis Glacier (another tributary) may have surged. However, a visual comparison between the SEASAT-altimetry-derived map by Brooks et al. (1983) and map E.3-2 Lambert.Seasat.1978.slcorr, using the 20 m-mean offset between the WGS84 ellipsoid and the Geoid (cf. maps E.3-10 Lambert.Geosat.1987-1989.slcorr and E.3-11 Lambert.Geosat.1987-1989.slcorr.gemt3, respectively), shows that the shapes of the contour lines match well in the floating part and that absolute differences are within 5 meters, but differences in morphology exist in the grounded part of Lambert Glacier (except for a general high on the western side of Lambert Glacier, and a low on the eastern side, common to both evaluations). The differences are a consequence of the different approaches to evaluation, Brooks et al. (1983) used an inverse-distance gridding algorithm and contoured by hand, based on elevations from orbits adjusted into a common ocean surface. These results indicate that some of the differences in different studies may be due to different processing algorithms and data references and do not permit a glaciodynamic interpretation.

Reasons for apparent rapid changes are difficult to determine from remote sensing information only.

There is a possibility that a change in the dynamic system of the glaciers feeding into Amery Ice Shelf occurred. Thickening of the lower part of the system and thinning of the upper part, as indicated by the spatial breakdown in our analysis, is typical for surges. The hypothesis of a dynamical event is mentioned here for sake of completeness, as the evidence from altimeter data alone is not sufficient for such a conjecture. The hypothesis is found in the literature (Wellmann 1982; Brooks et al. 1983), but largely without supporting data.

In summary, for the Amery Ice Shelf and the grounding zone, the trend in surface elevation between 1978 and 1987 or 1988 or 1989 is positive, matching the overall increase observed with the same method as well as from crossover analysis (Lingle et al. 1994). For the lower Lambert Glacier, an increase cannot be confirmed. For both elevation changes and changes in position of the grounding line, the apparent interannual variability among the GEOSAT years 1987, 1988, and 1989 identifies the 1988 season as an outlier with respect to the 1978–1989 trend. An advance of the grounding line is generally easier to observe than elevation changes. Interannual variability apparent in the irregular advance of the grounding line may be related to seafloor shoals. The overall advance between 1978 and 1989 is approximately 10–12 km.

(E.6.2) Comparison with Other Maps of Lambert Glacier/Amery Ice Shelf

Atlas maps provide additional points-in-time (1985/86 GEOSAT and 1995 ERS-1 Atlas, maps m69e61-77n67-721 Lambert Glacier, chapter (D)), but they must be used with caution in the monitoring study, because they have not been calculated from the same mission type of data. The 1985/86 GEOSAT map shows a different distribution of ice in particular in the western part of the map than the series of maps used in the monitoring study. More frequent or detailed data would be needed to appropriately address the surge question (see below).

The 1995 ERS-1 maps calculated for the Atlas and for the detail chapter (F) may be used as points-in-time in the monitoring of Lambert Glacier advance/retreat, in particular with reference to a map from 1992 ERS-1 data (Herzfeld et al. 1996). For comparative purposes, a the large 1995 ERS-1 map of Lambert Glacier/Amery Ice Shelf System is shown here (Figure E.6.2-1), next to an additional map constructed from 1997 ERS-2 data for the same area (Figure E.6.2-2).

Using the 100 m contour as a grounding-line proxy, monitoring of advance and retreat of the grounding line and thus of Lambert Glacier are possible. Comparison of 1978 SEASAT and 1987–1989 GEOSAT/ERM data had indicated a 10-km advance of the grounding line (Herzfeld et al. 1994, 1997). Comparison of the 1987–1989 GEOSAT/ERM maps and a 1992 ERS-1 map (Herzfeld et al. 1996) seems to indicate a retreat from 1987–1989 to 1992, however, adjustments of this result may be necessary when a quantitative comparison of ERS-1 specific orbits and other offsets relative to GEOSAT have been calculated. Such offset calculations are not available at present, but are maximally 0.72 m or 0.22 m (as calculated in section (C.1.1) from precision values).

Notice that Lingle et al. (1994) calculated the geographically correlated, systematic orbit bias between SESAT and GEOSAT ERM as $0.083 \pm 0.131m$ (which indicates that the GEOSAT surface is slightly higher), this offset accounts for an apparent northward displacement of the grounding line of about 581 ± 917 meters (which is small compared to the 10-12 km advance (Herzfeld et al.

1994)). Similarly, a GEOSAT-ERS-1 offset of maximally 0.22 m would lead to maximally ≈1.5 km apparent displacement of the grounding line (or any other contour line).

Comparison of the 1992 ERS-1 map (920707 – 921010, same season as for SEASAT and GEOSAT maps) and the 1995 ERS-1 map (950201 – 950801) indicates a definite advance of the grounding line between 1992 and 1995. The advance was about 5 km, calculated from the average advance of the 100-m grounding-line proxy north of UTM coordinate -7925.000. This may be interpreted as a continuation of the trend observed between 1978 and 1987–1989.

Comparison of the 1995 ERS-1 map of Lambert Glacier and the 1997 ERS-2 map may indicate a small further increase in elevation and advance of Lambert Glacier. The northernmost 100 m contour line in the grounding zone in central (lower) Lambert Glacier (between -7800.000 N and -8000.000 N) is located about 10 km further south on the 1997 map than on the 1995 map. However, the shape of the contour line depends somewhat on location of the groundtracks. The area above the 110 m and 120 m contours has increased (110 m contour is about 5 km further north).

An exact assessment of a potentially existing orbit bias between ERS-1 and ERS-2 data from the selected time frames needs to be calculated to determine the accuracy of the conclusion of advance. In the comparison of SEASAT and GEOSAT data, a shift in the contour lines of 10 km was much larger than the effect of a general offset between the satellites. Furthermore, ERS-1 and ERS-2 carry the same type of altimeter, and orbit determination has increased in accuracy (see Table B.2-1). Hence a potential error would most likely induce an error on the location of any contour line that is small compared to the apparent 5 km advance.

To obtain a more conclusive answer, an extended monitoring study would have to be conducted, ensuring that in all maps mission types, months of data, variograms are the same, and orbit accuracies are high. Selection of mission types and months, however, depend on satellite mission design, hence the same type of data may not be available for the same season in any given year.

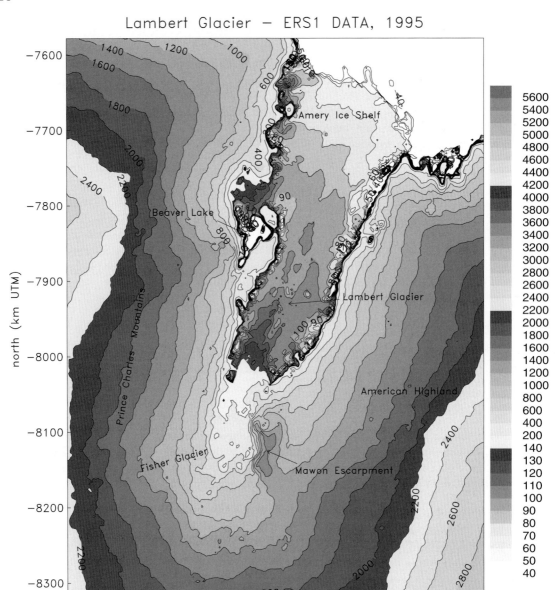

Figure E.6.2-1. Lambert Glacier/Amery Ice Shelf System from 1995 ERS-1 data. Time frame 01.February-01.August.1995. Elevations in meters above WGS84 ellipsoid, GEM-T2-referenced, slope-corrected. DEM calculated using ordinary kriging with Gaussian variogram model.

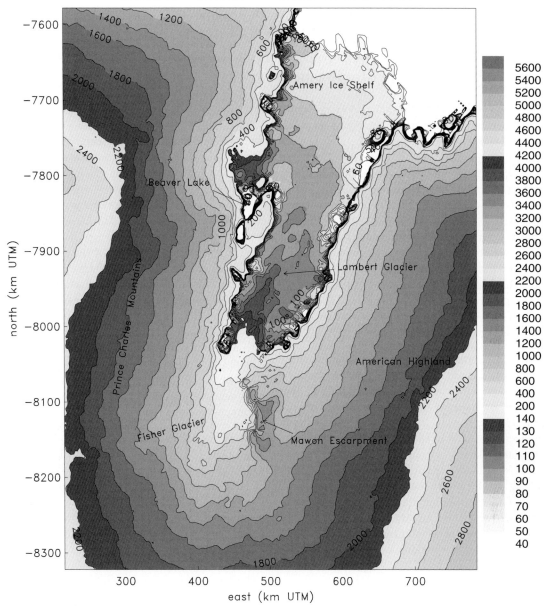

Figure E.6.2-2. Lambert Glacier/Amery Ice Shelf System from 1997 ERS-2 data. Time frame 01.August-31.October.1997. Elevations in meters above WGS84 ellipsoid, GEM-T2-referenced, slope-corrected. DEM calculated using ordinary kriging with Gaussian variogram model.

(E.7) On the Potential Existence of Surge Glaciers in the Lambert Glacier/Amery Ice Shelf System

(E.7.1) Introduction to the Surge Phenomenon and Relationship to Results of the Monitoring Study

A *surge-type glacier* is a glacier that does not move forward with about the same velocity at all times (or, a bit slower in winter than in summer, as usual alpine glaciers). To the contrary, the surge glacier has a long, quiescent phase of continuous movement, during this time, ice builds up in a given part of the glacier. Suddenly, a short surge phase starts, during which the glacier accelerates to about 100 times its normal velocity (the actual factor depends on the glacier), more ice is moved downglacier, and, due to high velocity, the ice surface breaks up in a maze of crevasses. The surge phase typically lasts one to a few years, the quiescent phase 25–100 years, depending on the glacier and the geographic area.

(I) Now, if Lambert Glacier moves with constant velocity throughout time and so do all its tributaries, then an advance of the grounding line indicates an increase in mass in the Lambert Glacier/Amery Ice Shelf system, and likely stems from accumulations in the drainage basin of Lambert Glacier, and/or from mass increases in the higher elevations of the Antarctic inland ice. The time lag for increased mass to show in the advance of the grounding line is many thousands of years, however. So answering the question "Is the Antarctic ice sheet thinning or thickening?" directly requires a very long time lag in observation.

(II) If, on the other hand, Lambert Glacier has surging tributaries, then the dynamics of the Lambert Glacier/Amery Ice Shelf system is very complex, and direct conclusions from ice advance or elevation increase in the system are not possible. For instance, a surge in a tributary would increase the inflow from that glacier rapidly (in a year of observations, in reality even faster), then in the next years the "excess" ice would be moved down the main glacier: A thickness wave would appear to travel down Lambert Glacier, starting from the point of the inflow of the surging tributary. This would lead to otherwise unexplainable ups and downs in the elevation curves. If there were more surging tributaries, matters would get more complicated. This assumes that if part of Lambert Glacier is surging, it would have a cycle with a time scale known from Alaskan glaciers. In a model for Lambert Glacier, however, Budd and McInnes (1978) calculated a surge cycle of 23000 years and a 250 year duration of the surge phase — phenomena clearly on a different time scale.

Since a few tributaries of Lambert Glacier have been suspected to surge, including Fisher Glacier and Charybdis Glacier, the latter possibility is likely. However, it is not possible to detect surges of side glaciers with certainty within the limits of accuracy of the mapping method (altimetry and kriging at 3 km grid resolution). Data of higher reolution and more frequent observations would be needed. But it should be known to the reader that there are sudden changes in nature — and monotonous lines are not the same as error-free lines.

(E.7.2) Discussion of the Surge Hypothesis in the Glaciologic Literature

A discussion of surging in the Lambert Glacier/Amery Ice Shelf system has been going on in the glaciologic community. Two avenues need to be distinguished: In context of potential instabilities of the Antarctic Ice Sheet, it is of interest to investigate whether the entire ice sheet may disintegrate as a consequence of warming, or whether dynamic properties of parts of the ice sheet or the whole ice sheet lead to advance or retreat. Lambert Glacier/Amery Ice Shelf system as a whole could be surging. On the other hand, only some glaciers within the Lambert Glacier/Amery Ice Shelf system may be surge-type glaciers, namely Fisher Glacier (the westerly of the three glaciers that form the main trunk of Lambert Glacier) and Charybdis Glacier (a large glacier that enters NW

Amery Ice Shelf north of Beaver Lake, a small branch of it flows into Beaver Lake), and possibly others.

Budd and McInnes (1978) suggest that the Lambert Glacier/Amery Ice Shelf surges with a periodicity of 23000 years, with surges lasting 250 years and velocities reaching several kilometers per year (velocity now is 1 km a^{-1}). In the same paper, it was suggested that similar behaviour was happening in other Antarctic drainage basins.

From Lambert Glacier and side glaciers joining mostly from the west, the mass flux of Lambert Glacier has been calculated as ≈ 30 Gta^{-1} ($= 33$ km^3a^{-1}) by Allison (1979). This is almost twice as much as Byrd Glacier. But measurements of ice flux on Lambert Glacier at a north coordinate corresponding to that of the head of Beaver Lake have indicated a mass flux of only 11 Gta^{-1} ($= 12$ km^3a^{-1}). How can this discrepancy be explained? One possibility is that it builds up temporarily further inland, and suddenly flows downstream in surge events. Alternatively, the ice may possibly build up inland and not release in a surge event.

In contrast, Robin (1979, 1983) concluded that an apparent imbalance was due to strong basal melting. In a recalculation of the area of the Lambert Glacier drainage basin, McIntyre (1985) determined the size of that basin as 900 000 km^2 rather than 1 090 000 km^2 (which was used in previous calculations) and also assumed a lower accumulation rate, with the result that the input of ice in the Lambert Glacier basin almost matched the flow into Lambert Glacier, hence the dicussed imbalance was no longer existing.

Radok et al. (1987) used a mathematical model to determine which Antarctic drainage basins were prone to surging. As a result, Lambert Glacier and other major glaciers did not appear as likely candidates for surge-type glaciers.

The possibility that Fisher Glacier surges was first put forward by Derbyshire and Peterson (1978) after a study of aerial photographs of the area near Mount Menzies (73° 30' S/61° 50' E, 3335 m), which divides the flow of Fisher Glacier and about 100 km upstream of its junction with Mellor Glacier (and upper Lambert Glacier). Charybdis Glacier has also been suggested to be a surge-type glacier.

Wellmann (1982) also concluded that Fisher Glacier may be a surge-type glacier, he utilized geomorphologic observations deduced from aerial photography. Side moraines of Fisher Glacier show three elevation levels, indicating changes in ice-surface elevations consistent with surges, while other catchments of Lambert Glacier do not have such moraines.

Structural patterns in surging glaciers include typical folded medial moraines (Post 1972) and foliation. Such medial moraines form between a surging and a non-surging glacier during a surge cycle: During the surge, the increased flow of mass down the surging glacier pushes the medial moraine towards the non-surging glacier. After the surge, as the non-surging glacier increases in relative power, the medial moraine gradually moves back to the pre-surge position. Hence the number of folds in medial moraines approximately corresponds to the number of surges.

Based on lack of evidence of ice-surface-structural patterns indicative of surging (from LANDSAT image analysis), Hambrey (1991) concludes that there are no surging glaciers in the Lambert Glacier/Amery Ice Shelf system, (a) in the time of residence of the ice in the system, which he estimates to be 1000 years in the center and several to ten thousand years along the margins of Lambert graben, (b) this still leaves the possibility of really long surge cycles, such as the 23000 year cycle proposed by Budd and McInnes (1978). If Fisher Glacier had surged, then there should be evidence in surface structure. To the contrary, the surface features on Lambert Glacier appear to be indicative of constant-velocity flow. There are no folded nor truncated foliations, nor crevassed areas with clear boundaries to slow-moving ice, which are also typical of surging (see also Herzfeld (1998) and Herzfeld and Mayer (1997) for description of the surface features during the surge of Bering Glacier/Bagley Ice Field system, Alaskas longest glacier and as such most closely related to an Antarctic ice stream).

In contrast, West-Antarctic Ice Stream B (Whillans Ice Stream), which leads into Ross Ice Shelf, shows evidence of surging (Bindschadler et al. 1987; Whillans et al. 1987), and foliations are observed in Ross Ice Shelf.

(F) Detailed Studies of Selected Antarctic Outlet Glaciers and Ice Shelves

(F.0) Introduction

In the glaciologic part of the introduction, the important role that Antarctica plays in Global Change is explained in relationship to temperature change, changes in atmospheric conditions – CO_2 budget – and sea-level rise. There are two main approaches to investigate mass changes in the Antarctic Ice Sheet: a) changes in elevation of the entire ice sheet, and b) changes in the outlet glaciers of the Antarctic Ice Sheet. Changes in elevation can be monitored when comparing maps of different times, for instance, a GEOSAT map and an ERS-1 map in this Atlas, or more time points, as is illustrated in the section on Lambert Glacier. The outlet glaciers, ice streams, and ice shelves are key features, because they are most sensitive *and* change is most evident there. Flux of ice into the ocean occurs across the margin of Antarctica, the ice-ocean boundary.

A major advantage of the geostatistical atlas mapping method over other methods of interpolating and mapping from radar altimeter data is that the topography in areas with complex relief is optimally represented. Optimally means, as good as possible given the constraints of data gaps and technical limitations of radar altimeter observations (see, section (C) on data analysis methods applied in the Antarctic Atlas). Using this method, the margin of Antarctica can be mapped sufficiently accurate for glaciological investigations.

In this section, detailed maps of selected Antarctic glaciers are presented and interpreted. The underlying grid is the same 3-km grid as that of the Atlas maps, but the generally larger scale of the detail maps makes more geographic features visible. The latter demonstrates that the 3-km Atlas grid may be explored locally for its actual information content, which is often higher than that of the Atlas maps. The glaciers presented here have been selected for their glaciologic relevance.

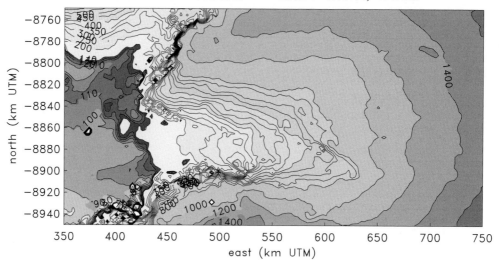

Detail Map 1A. Slessor Glacier. Scale 1:3.500.000

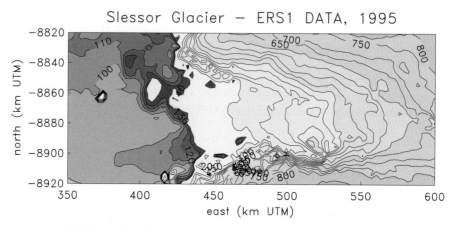

Detail Map 1B. Slessor Glacier. Scale 1:2.500.000

(F.1) Detail Map 1: Slessor Glacier (ERS-1 Data 1995)

Atlas Map: m333e315-351n78-815 Filchner Ice Shelf

Slessor Glacier map area: 1A (350.000-750.000 E/-8750.000- -8950.000 N), scale 1:3.500.000;
1B (350.000-600.000 E/-8820.000- -8920.000 N), scale 1:2.500.000

Slessor Glacier is one of the largest glaciers draining into the Filchner Ice Shelf, in fact it is one of the longest coherent ice streams on Earth. Its large drainage basin is seen on Atlas maps m333e315-351n78-815 Filchner Ice Shelf and m357e339-15n78-815 Western Queen Maud Land (South). It measures about 500 km from Parry Point upstream as determined by flowlines from satellite imagery (Swithinbank 1988, fig. 75, p. B100), plus 350 km from Parry Point to the icefront (Crabtree and Doake 1980), so it is at least 850 km long.

The drainage basin is 575.000 km^2 (McIntyre 199999). Slessor Glacier, at its narrowest point, is 30 km wide. It descends west with a constant slope of 0.32° to a flatter area before joining the Filchner Ice Shelf. Surface elevations, 10–30 m higher than the surrounding ice shelf, mark its protrusion into the ice shelf (at (-8900.000 N/400.000 E); Herzfeld et al. 2000b). Annual discharge is estimated at 34 km^3. No velocity or ice-thickness measurements have been made. The lower drainage basin is best seen in Map 1A (1:3.500.000).

Of the 850 km length of Slessor Glacier, the lower 170 km are mapped at the large scale of 1:2.500.000 (Map 1B). This detail map facilitates study of the longitudinal gradient of Slessor Glacier and of the glacier's continuation in the Filchner Ice Shelf. The gradient is 200 m in 50 km (0.4% = 0.23° between 400 m and 600 m elevation) between 96 km and 146 km from the coast. The center of the basin has a relative constriction at 600–650 m elevation. Between 400 and 350 m elevation is a steeper step with an average slope of 50 m in 6 km (8,3% = 4.8°).

Below 350 m elevation, the slope decreases gradually to 60 m in 2800 m between 200 m and 140 m elevation (above the ellipsoid) (2.14% = 1.2°). At 140 m elevation, the glacier reaches the Filchner Ice Shelf.

The nose at (-8840.000 N/430.000 E) of the topographic high in the north of the detail map is the westsouthwestern end of the Theron Mountains (see Atlas map m333e315-351n78-815 Filchner Ice

Shelf), westsouthwest of that is Parry Point. It is 90 km from Parry Point to Blaiklock Glacier in the Shackleton Range south of Slessor Glacier. Between Blaiklock Glacier and Parry Point, Slessor Glacier enters the Filchner Ice Shelf (if one considers a line Parry Point – Blaiklock Glacier the edge of the Ice Shelf. Notice that this does not mean that the grounding line is there!). The grounding line is the technical edge of the ice shelf.

Using the criterion of a break in surface slope as a determination of the grounding line position, then the most significant break in slope at 350 m elevation would put the grounding line location at (-8800.000 N/526.000 E), 88 km upstream of the line Parry Point-Blaiklock Glacier and of the 140 m elevation contour.

In the south, the glacier is bordered by the Shackleton Range. A 1974 LANDSAT image (Swithinbank 1988, fig. 75, p. B100) of Slessor Glacier and the Shackleton Range shows flow lines indicative of rapid motion on the glacier and on the part that is in the Filchner Ice Shelf.

Swithinbank (1988, p. B101) states that the position of the grounding line is unknown, but that isolated areas of smooth surface undulations downglacier from Parry Point and towards Blaiklock Glacier (visible in a 1974 LANDSAT image, fig. 75, p. B100) appear to be characteristic of glaciers sliding on their beds close to the grounding line, and these areas could be isolated locations of grounding on an otherwise floating ice sheet. Following this argumentation, the grounding "line" would be near the 140 m contour in our map, and likely there is a large grounding zone which extends a bit downward, and possibly a large distance upstream of the line Parry Point-Blaiklock Glacier.

The largest break-in-slope downstream of the 140 m contour is at the 110 m contour, where the slope decreases to 10 m over 40 km (0.025% = 0.014°; 110-100 m), above this it is 10 m over 2 km (0.5% = 0.29°; 140-130 m), 10 m over 6 km (0.16% = 0.095°; 130-120 m), and 10 m over 8 km

(0.125% = 0.072°; 120-110 m); i.e. a less significant break-in-slope is at 130 m. The calculations of breaks in slope leave a few possibilities for the grounding line location, however, combined with the observations on satellite imagery, described in Swithinbank (1988)(see above), the 130 m contour appears the most likely approximate position of the grounding line. The 130 m contour also wanders around in the northern part of the glacier, donwstream of the line Parry Point - Blaiklock Glacier, which matches the location of the lowest patches of likely grounded ice in the satellite image surprisingly well (in particular, if one keeps in mind that the satellite image is from 1974 and the ERS-1 data from 1995). The assumption that there is a grounding zone, which includes the 130 m contour and extends some distance downstream from the line Parry Point – Blaiklock Glacier (40 km from Parry Point), but mostly upstream, is still valid.

Ice surface characteristics in the upper reaches of Slessor Glacier (at 1000–1500 m a.s.l, mapped only in the Alas map m333e315-351n78-815 Filchner Ice Shelf) are typical of those of a large ice stream, with series of escarpments trending generally across the flowline, spaced about 10 km apart and attributed to subglacial ridges (Swithinbank 1988, fig. 74, p. B99). The subglacial topography, as indicated by ice-surface features seen in satellite imagery (Swithinbank 1988, fig. 74, p. B99) indicates that the *Shackleton Range* extends eastward under Slessor Glacier (Marsh 1985, after Swithinbank 1988).

Several small glaciers join Slessor Glacier, both from the Theron Mountain region and from the Shackleton Range, the latter include *Stratton Glacier* and *Blaiklock Glacier* located at the tip of the Filchner Ice Shelf.

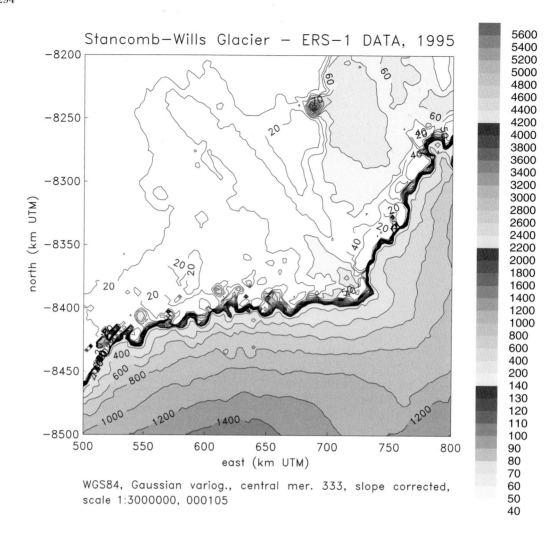

(F.2) Detail Map 2: Stancomb-Wills Glacier (ERS-1 Data 1995)

Atlas Map m333e315-351n71-77 Riiser-Larsen Ice Shelf

Stancomb-Wills Glacier map area: (500.000-800.000 E/-8200.000- -8500.000 N), scale 1:3.000.000

Stancomb-Wills Glacier is located in Queen Maud Land, at the northeastern edge of Coats Land, and drains into the Weddell Sea. Stancomb-Wills Glacier is the largest glacier between Jutulstraumen (m3we11w-5n67-721 Fimbul Ice Shelf) and Slessor Glacier (m333e315-351n78-815 Filchner Ice Shelf). Its drainage basin extends at least to the 2200 m contour line, which is 250 km inland from its terminus (see descriptions of Atlas maps m333e315-351n71-77 Riiser-Larsen Ice Shelf and m357e339-15n75-80 Western Queen Maud Land (North)). From the topographic maps it can be concluded that the glacial drainage basin extends about 500 km farther east into Queen Maud Land, to the 3200 m contour on maps m357e339-15n75-80 Western Queen Maud Land (North) and m357e339-15n71-77 New Schwabenland. According to McIntyre (after Swithinbank, 1988), the drainage basin is only 35.000 km^2, much smaller than the area described here.

The most conspicuous feature of Stancomb-Wills Glacier is its glacier tongue, this is enlarged in the detail map. The glacier tongue is mapped well using satellite radar altimetry and geostatistics, and its length and width can be measured from the elevation maps.

The Stancomb-Wills Glacier tongue was discovered already in January of 1915 by Shackleton (1919), who saw its seaward terminus as an ice-shelf promontory (Stancomb-Wills Promontory) (Swithinbank 1988, p. B97). The ice stream itself was discovered much later, in 1957 by Sir Vivian Fuchs.

The Stancomb-Wills Glacier Tongue extends 230 km to the northwest, it is 30 km wide near its end and up to 50 km wide in various places close to the coast. The elevation of the glacier tongue is about 10 m above surrounding shelf ice on our map, and 30 m in the areas close to the terminus. The fact that the glacier-tongue contour is surrounded by another contour indicates that the ice surrounding the tongue is elevated. Further east is the Riiser-Larsen Ice Shelf with elevations of over 60 m, in comparison, we conclude that the Stancomb-Wills Glacier Tongue extends into sea ice (and not into shelf ice) to the west of the Riiser-Larsen Ice Shelf. A comparison with satellite imagery will show that the ice area surrounding the glacier tongue consists of so-called "low ice shelf", which is a thicker form of sea ice.

Flow lines and crevasses of (part of) Stancomb-Wills Glacier are visible in an annotated LAND-SAT MSS image (from 1981) (Swithinbank 1988, fig. 73; p. B98). The glacier has a wide shear margin to the slow-moving shelf ice of Riiser-Larsen Ice Shelf (to its northeast) and Brunt Ice Shelf (to its southwest). Noticeably, the centerline of the offshore part of the glacier has a smooth low for some part of its length reminiscent of the center of Jakobshavns Isbræ, a fast-moving ice stream in western Greenland (Mayer and Herzfeld 2001). The strain between the glacier tongue and the Brunt Ice Shelf is so large that it cannot be accomodated by deformation anymore, consequently, large icebergs calve from the ice sheet at or near the grounding line and are pulled along by the fast-moving ice stream, which forces them into a counterclockwise rotation (Swithinbank 1988, p. B97). The gaps are filled by so-called "low" ice shelf (sea ice which grows thicker over time). To the contrary, slowly-moving ice shelves usually have a vertical stratification, they consist of a thinning wedge of ice of land origin overlain by a thickening wedge of locally accumulated snow (and sea ice from to the bottom).

Lyddan Island is an ice rise between Stancomb-Wills Glacier and Riiser-Larsen Ice Shelf, it obstructs the westward flow of the ice shelf. Velocities of 1300 m a^{-1} on the right bank of the ice stream (Thomas 1973) and of 4000 m a^{-1} at the ice front (Orheim 1982) have been reported, from these observations and an estimate of 200 m ice thickness, Swithinbank (1988) estimates an ice discharge of 40 km³ a^{-1}.

The lower part of Stancomb-Wills Glacier is much steeper than, for instance, that of Slessor Glacier. The gradient increases from 200 m over 33 km (0.6% = 0.35° at 1000–800 m elevation) to 200 m over 12 km (1.7% = 0.95° at 400–200 m elevation) and 60 m over 4 km (1.5% = 0.86° at 200–140 m elevation).

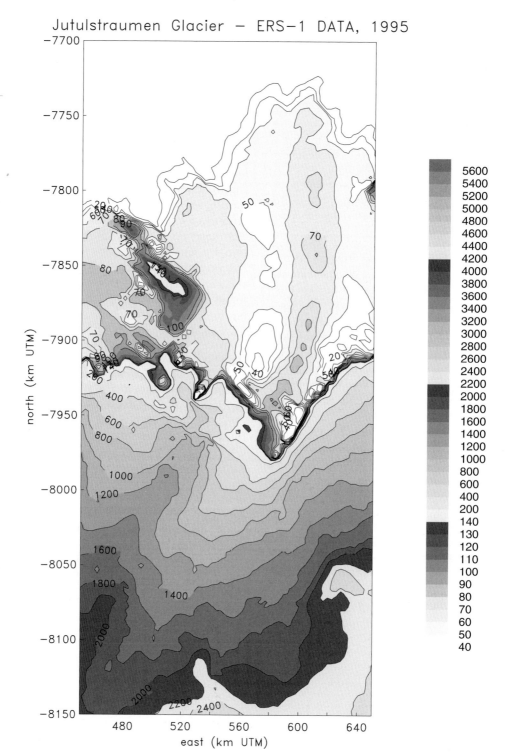

Jutulstraumen Glacier — ERS-1 DATA, 1995

north (km UTM)

east (km UTM)

WGS84, Gaussian variog., central mer. 357, slope
corrected, scale 1:2500000, 000105

(F.3) Detail Map 3: Jutulstraumen Glacier (ERS-1 Data 1995)

Atlas Maps: m3we11w-5n67-721 Fimbul Ice Shelf and m357e339-15n71-77 New Schwabenland

Jutulstraumen Glacier map area: (450.000–650.000 E/-7700.000- -8150.000 N), scale 1:2.500.000

Jutulstraumen Glacier flows north to Fimbul Ice Shelf in Western Queen Maud Land, in Princess Martha Coast. It is a fast-moving glacier whose force lets it extend further seaward than the neighbouring ice shelves (compare ice shelf extensions on map m3we11w-5n67-721 Fimbul Ice Shelf, on which Jutulstraumen Glacier is seen, and on map m9e1-17n67-721 Princess Astrid Coast). The 240 km long tongue seems to actually enlarge Fimbul Ice Shelf northward. Jutulstraumen Glacier has been studied extensively, several references are given in Swithinbank (1988), e.g. Orheim (1979). The geology and gravity field of the Jutulstraumen area have been analyzed by Decleir and van Autenboer (1982).

The tongue and the valley of Jutulstraumen Glacier are mapped here at an enlarged scale of 1:2.500.000.

Jutulstraumen Glacier (the word "glacier" in its name is actually superfluous, as "straumen" means "(ice) stream") is the largest glacier between longitudes 15° E and 20° W. Ice stream is a more precise term, because Jutulstraumen is an outlet glacier of the Antarctic Inland Ice, an ice stream that flows through adjacent slow-moving ice. Jutulstraumen Glacier drains an area of 124.000 km^2 , has a discharge of 12.5 km^3 a^{-1} at 72° 15'S and an ice velocity at the front of 1 km a^{-1}(Swithinbank 1988).

In its lower parts, Jutulstraumen follows a geologic trough that is of structural origin (Decleir and van Autenboer 1982). The bottom of Penck Trough reaches a few hundred meters below sea level. The existence of such a trough enhances the flow volume and velocity in the ice stream, as the trough attracts ice from the surrounding areas gravitationally. Our detail map shows that there are two valleys extending south from the terminus of Jutulstraumen, a western one with a head near east coordinate 480.000 and an eastern one with a head near 600.000 E on the southern map margin.

The western valley has almost the prototype shape of a trough, still at ice surface level, while the eastern valley has a less characteristic morphol-

ogy. The western valley is *Penck Trough*, a geologic trough with Ritscher Hochland to the northwest, in the southern part, and Borg Massif to the west, in the northern part, and the Neumayer Cliffs to the southeast, also in the southern part of the map and the trough. The main subglacial valley extends through Penck Trough as far as 6° W (Swithinbank 1988, p. B92). On a satellite image by IfAG (see Swithinbank 1988, fig. 68, p. B93; IfAG stands for Institut für Angewandte Geodäsie, Frankfurt) it appears that the upper part of the Penck Trough valley has now captured ice that once drained to the coast via Schytt Glacier, which is located west of Borg Massif. Schytt Glacier still exists, but has a small drainage area (see satellite image fig. 69, p. B94, Swithinbank 1988). On our map, the valley of Schytt Glacier is the valley northwest of Penck Trough. The main flow of Jutulstraumen Glacier, which coincides with the flow through Penck Trough, is joined by ice that follows the easterly of the two valleys of lower Jutulstraumen Glacier, i.e. the valley with its head at 600.000 E on the lower map margin. In some maps, the name "Jutulstraumen Glacier" is given to the ice in the easterly branch. We suggest to use "Western Upper Jutulstraumen Glacier" for the branch in Penck Trough and "Eastern Upper Jutulstraumen Glacier" for the branch in the eastern valley.

At the western edge of the terminus of Jutulstraumen Glacier are Passat Nunatak and Boreas Nunatak (220 m) on Giaever Ridge, the eastern edge is Roberts Knoll on Ahlmann Ridge.

Ice velocities of 390 ma^{-1} have been measured at 72° 15'S. Ice discharge at 72° 15'S has been calculated as 12.5 km^{3}a^{-1} (Decleir and van Autenboer 1982). Crevasses near Utkikken Hill, a hill on a land tongue west of Jutulstraumen Glacier, indicate that Jutulstraumen is afloat here (Swithinbank 1988, figs. 69 and 70, p. B92, B95). Orheim (1979) observed a velocity of 1 km a^{-1}at the ice front north of there. In 1967, a large calving event occurred at Jutulstraumen Ice Tongue, one single iceberg was 53 km by 104 km (Swithinbank et al. 1977); the calving event was likely caused by collision of another large, drifting iceberg with the glacier tongue. The time needed for formation of a large ice tongue, such as the one of Jutulstraumen Glacier, is at least 100 years (Decleir and van Autenboer 1982).

The tongue of Jutulstraumen Glacier is at least 40 m higher than the sea ice to its north, and 20 m higher than the surrounding ice of Fimbul Ice Shelf. The center of the tongue appears raised, to above 70 m and about 80 m in several locations, with a trend to decrease in elevation towards the north.

Jutulstraumen Glacier is also part of a study on the influence of variograms on surface features, carried out for the Fimbul maps m3we11w-5n67-721 (see section (C.3.6)). Thus several maps of the glacier tongue are available for comparison.

Features in this map sheet appear more clearly on the GEOSAT map. Ice elevations in the area ap-pear to have decreased from GEOSAT 1985/86 to ERS-1 1995 by possibly 5 m, as deduced from size reductions of the higher-elevation contours on the ice shelves. Most noteworthy, the tongue of Jutul-straumen Glacier has a connected center that is 20 m above the level of the surrounding Fimbul Ice Shelf, in 1985/86 the center elevation is 70 m (above WGS 84) with spots above 80 m, in 1995 the 70 m contour is exceeded only in a few ar-eas. The areas below 40 m and 50 m on the ice shelf may have changed less, so it appears that the offshore part of Jutulstraumen Glacier lost el-evation, and hence mass (since the ice shelves are afloat). These observations need to be adjusted for a general offset between elevations from GEOSAT and ERS-1 observations, such an offset has not been calculated (to our knowledge), however, it is at most 72 cm (or 22 cm, depending on accuracy calculation, see section (C.1.1)), whereas the ob-servations here are on the order of meters.

300

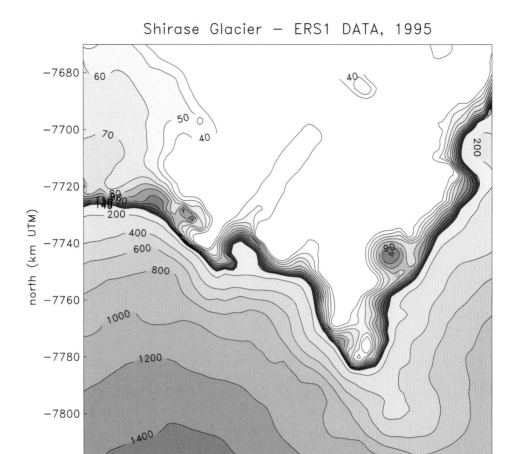

Shirase Glacier — ERS1 DATA, 1995

WGS84, Gaussian variog., central mer. 33, slope corrected, scale
1.1333333, 991312

(F.4) Detail Map 4: Shirase Glacier (ERS-1 Data 1995)

Atlas Map m33e25-41n67-721 Riiser-Larsen Peninsula

Shirase Glacier map area: (600.000-750.000 E/-7670.000- -7820.000 N); scale 1:1.333.333

This map shows glaciers in Lützow-Holm Bay (Princess Ragnhild Coast and Prince Olav Coast, Enderby Land). Shirase Glacier is the large glacier that drains the valley in the southeast corner of the detail map. The most conspicuous feature on this map is a large ice tongue, outlined by an almost rectangular contour. This ice tongue is 48 km long on our map and extends west of Botnneset Penin-

sula, it is also seen on the USGS satellite image map of Antarctica (Ferrigno et al. 1996).

Shirase Glacier is the longest ice stream west of Rayner Glacier, it has a 165.000 km^2 drainage basin that extends 500 km inland (according to Swithinbank 1988). The drainage basin of Shirase Glacier is seen on maps m33e25-41n67-721 Riiser-Larsen Peninsula, m45e37-53n67-721 Prince Olav

Coast, and the southerly adjacent maps m21e3-39n71-77 Sør Rondane Mountains and m45e27-63n71-77 Belgica Mountains. On map m45e37-53n67-721 Prince Olav Coast it is obvious that a regional ice divide separates the drainage basins of Shirase Glacier (west) and Rayner Glacier and Thyer Glacier approximately 500 km farther east.

Shirase Glacier occupies a 9 km wide valley that remains below sea level for 200 km inland from the coast (Swithinbank 1988, p. B79). The terminus is a floating tongue (grounding line 65 km up-glacier in January 1974). The glacier is close to the Japanese Research Station Siowa (Showa), which is located on Ongull Island near the eastern coast of Lützow-Holm Bay, and thus well-studied. Ice-front velocities of 2500 m a^{-1} have been measured by Nakawo et al. (1978), this is the fastest surveyed rate of flow of any Antarctic ice stream, according to Swithinbank (1988, p. B79). Fujii (1981) mea-

sured some points at 2900 m a^{-1}. The glacier front may have advanced 86 km in 37 years between 1936–37 (maps based on photographs) and 1974 (satellite image), but a reinvestigation of the photograph suggested that the glacier tongue may not have advanced. Consequently, estimates of flow velocity, iceberg calving and glacier advance have a wide range of error (Swithinbank 1988, p. B79). Unfortunately, the tongue is not captured in the 1995 ERS-1 data. The overdeepening in the bay near the terminus of Shirase Glacier is likely in part a result of radar altimeter data retracking, but in part a comparison to the satellite image map in Swithinbank (1988, fig. 59, p. B81) shows an elevation low between the floating part of Shirase Glacier and the eastern side of the head of the small bay. (The knob with contour 90 further east is not an artefact.) Lützow-Holm Bay is filled with fast ice.

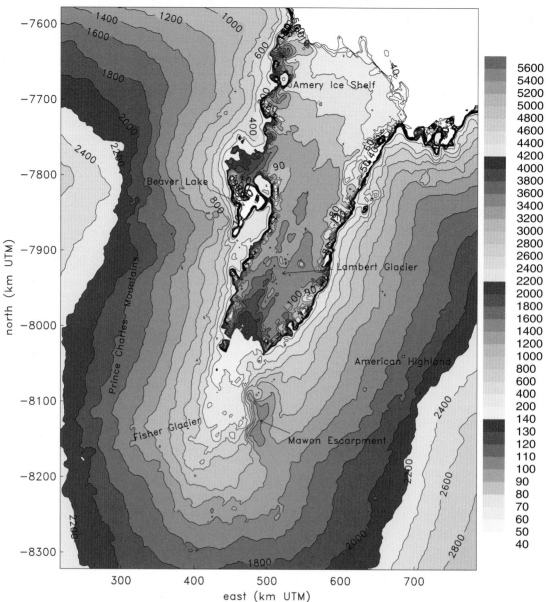

Lambert Glacier – ERS1 DATA, 1995

e59–79n68–75, WGS84, Gaussian variog., central mer. 69, slope corrected, scale 1:5000000, 970730

(F.5) Detail Map 5: Lambert Glacier (ERS-1 Data 1995)

Atlas Maps: m69e61-77n67-721 Lambert Glacier, m69e51-87n71-77 Upper Lambert Glacier

Large Lambert Glacier Map m69e59-79n68-75 ERS-1 1995; Scale 1:5.000.000

Lambert Glacier and Amery Ice Shelf form the largest ice-stream/ice-shelf system in East Antarctica, with a drainage basin that is about 10% of the East Antarctic Ice Sheet (Drewry 1983). Parts of Lambert Glacier / Amery Ice Shelf are seen on maps m69e61-77n67-721 Lambert Glacier, m69e51-87n71-77 Upper Lambert Glacier, m69e51-87n75-80 South of Lambert Glacier, and the drainage basin covers the adjacent maps to the east and west, with ice flowing into Lambert Glacier from as far away as the Gamburtsev Subglacial Mountains (located approximately at (77–79°S/70–80°E; 500 km from the edge of Lambert Glacier), which in turn may extend as far south as below Dome Argus, seen on the southernmost map at the central longitude m69 (m69e51-87n78-815 Dome Argus).

Lambert Glacier has been called the Earth's largest glacier (Allison 1979), because its flowlines can be traced on satellite imagery (LANDSAT) 400 km upstream of its grounding line. The glacier drains an area of 902000 km² (McIntyre; after Swithinbank 1988); the only Antarctic glacier with a larger drainage area is Byrd Glacier in the Ross Ice area.

Since flipping pages to study a large system is inconvenient, we have calculated a large map of Lambert Glacier / Amery Ice Shelf and the surrounding drainage basin for the area m69e59-79n68-75. Technical data for the processing of this map are the following:

The area selected is that of Lambert Glacier/ Amery Ice Shelf, the largest ice-stream/ice-shelf system in East Antarctica. ERS-1 orbital coverage extends to 81.5°S for radar altimeter data, thus it is possible to map the entire Lambert Glacier area including upper Lambert Glacier and other tributaries of (lower) Lambert Glacier (59°–79° E/68°–75°S). Data coordinates were transformed from geographic to Universal Transverse Mercator (UTM) coordinates (cf. Snyder 1987) using 69°E as a central meridian. The nominal map area is 59°–79°E/68°–75°S, the map is rectangular after transformation into the UTM system with dimen-

sions 570 km E–W by 744 km N–+S and UTM coordinates (-8322.000 - -7578.000 N/215.000-785.000 E). The time frame of data in our example is 6 months in 1995 (950201 – 950801). This was part of the Geodetic Phase (April 1994 – May 1996), during which ERS-1 was operated in two 168-day repeat cycles, one offset from the other to increase geographic coverage. The timeframe is the same as for the ERS-1 Atlas.

This large Lambert Glacier map shown here is also calculated with a 3 km grid spacing such that it may be used as a basis for glaciodynamic modeling (see digital version of the drainage basin), following criteria as given in the description of map m69e61-77n67-721 Lambert Glacier. Fewer features are distinguished in the contoured version, just because of the relative size of map to book page.

Lambert Glacier follows a large geologic trough. Ice from a large catchment area drains into the Lambert Glacier trough, accelerating and forming an ice stream. A branch of Lambert Graben extends inland to a distance of 1000 km from the front of Amery Ice Shelf. Lambert Graben is up to 2500 m deep. Mawson Escarpment is a straight-lined escarpment rising to 1000 m above sea level located east of Lambert Glacier. The wall of the graben, measured to the top of Mawson Escarpment, is a total of approximately 3000 m high.

Front thickness (Amery Ice Shelf) is 270 m; ice thickness at the grounding line is ≈1000 m, at a point 600 km upstream of the ice front, the bedrock elevation is still 840 m below sea level, ice thickness is 1950 m, and velocity of Lambert Glacier is 231 m a⁻¹(Swithinbank 1988). In another reference Lambert Glacier near the grounding line is 770 m thick, velocity is 347 m a⁻¹here (quoted after Hambrey 1991, who also quoted thickness at the grounding line as 900 m).

Starting from the head of the glacier, it is obvious from the contours of the surrounding basin that the main flow comes in from the southeast, this matches the observation of the 1958–59 Soviet Antarctic Expedition that ice from the area of the

Gamburtsev Subglacial Mountains (off this map by another 200 km, but covered on maps m69e51-87n75-80 and m69e51-87n78-815) flows northwest to Lambert Glacier. The main ice flow then passes by Mawson Escarpment (here at -8100.000 – -8180.000 N / 480.000 – 500.000 E UTM). The confluence with another major flow/depression is west of here, this is the Mellor Glacier inflow. The small concentric contour lines match some of the nunataks west of Mawson Escarpment, as described earlier (map m69e61-77n71-77 Upper Lambert Glacier).

Lambert Glacier comes in from directly near Mawson Escarpment and has traceable flowlines (on LANDSAT images) for about 100 km farther upstream from Mawson Escarpment; Mellor Glacier flows approximately 20 km farther west, between Mellor and Lambert Glacier lie (from N to S) Cumpston Massif, Mt. Maguire, the Blake Nunataks, and Wilson Bluff; ice of (upper) Lambert Glacier and Mellor Glacier appears to also flow around Mt. Twigg and Mt. Borland; all these are nunataks which stick out of the flowing ice (Swithinbank 1988). The distinction between Mellor Glacier and Lambert Glacier ice does not seem to be unambiguous — on the USGS satellite image map (Ferrigno et al. 1996), all the ice described here is only Lambert Glacier, whereas the name "Mellor Glacier" is attributed to a smaller and subordinate glacier coming in from farther west. So, the main stream of flow into Lambert Glacier is actually termed Lambert Glacier, which makes the most sense. The third branch that joins Mellor Glacier and upper Lambert Glacier to form the main trunk of Lambert Glacier is Fisher Glacier. It is about as wide as Mellor Glacier and upper Lambert Glacier and flows in from the west. Fisher Glacier is important for the flow properties of Lambert Glacier, as it has been hypothesized to surge (see discussion in section (E.7)).

The rugged mass of the Prince Charles Mountains west of Lambert Glacier contrasts the fairly even gradient/slope of the American Highland to the east. Beaver Lake Peninsula is visible (but Beaver Lake is not well mapped, for reasons as discussed earlier for the Atlas map m69e61-77n67-721 Lambert Glacier). The valley north of Beaver Lake continues as a depression up to the 2000 m contour, this is occupied by Charybdis Glacier, several smaller western tributaries are seen. A branch of Charybdis Glacier flows south into Beaver Lake.

One extends west to the 1400 m contour (south of Beaver Lake Peninsula). Some information on the geology of Beaver Lake is given in the description of map m69e61-77n67-721 Lambert Glacier (section D).

The even gradient of the floating ice shelf is well visible, as are the irregular nature of the ice front, the overdeepening along the western and eastern margins of the ice stream due to shear stress (actual part) and to satellite snagging effect (apparent overdeepening, larger than in reality). The inflow of several glaciers from the west (near the front of Amery Ice Shelf) can be traced by the higher surface elevations on Amery Ice Shelf:

1. 50 m higher for 70-km-E-W Glacier (northernmost) at -7640.000 N
2. 40 m higher for 35-km-E-W Glacier at -7660.000 N
3. 40 m higher for 30-km-E-W Glacier at -7700.000 N
4. 30 m higher for 15-km-E-W Glacier
5. 40 m higher for 10-km-E-W Glacier
6. 40 m higher for (80-km-E-W) Charybdis Glacier at -7780.000 N, with large valley.

In between these are rock promontories (mapped as eastward extensions of the 200 m contour). Rock promontories are distinguished from glacier inflows as follows: Inflowing glaciers have contours in steps of 10's increasing from the elevation of the ice shelf, whereas rock promontories are seen in the 200 m contour and have a crowding of the next lower contours. Some inflowing glaciers are also seen on a LANDSAT image (Swithinbank 1988, p. B73, fig. 54), where they are unnamed, as well as several promontories, including Foley Promontory and Ladan Promontory. The advantage of elevation mapping is obvious in this area. The extent of the inflowing glaciers cannot be seen on the LANDSAT image.

Flow units

Hambrey (1991) distinguishes 8 structurally defined flow unit boundaries (labeled 1–8 from W to E) from structural patterns in satellite imagery. The 3 main glaciers that join to form Lambert Glacier are Fisher Glacier (unit 4), Mellor Glacier (unit 5) and (upper) Lambert Glacier (unit 6). Unit (3) consists of ice entering via several glaciers that traverse the Prince Charles Mountains in an area that extends from just south to 300 km south of Beaver Lake. Unit (2) is formed by Charybdis Glacier (which enters north of Beaver Lake, a small branch of Charybdis Glacier flows south into Beaver Lake), and unit (1) is termed "Northwest Amery Ice Sheet", the sources are not identified further. To the east of the main flow, several glaciers and a large ice stream enter from NE of Mawson Escarpment, this is flow unit (7) "Mawson Escarpment Ice Stream", and unit (8) is fed by a large ice stream entering Amery Ice Shelf from the east side less than 100 km from its terminus (unit (8) is termed "NE Amery Ice Sheet").

Ice surface features

Dominant features visible in enhanced LANDSAT iamgery are longitudinal structures (*foliations*), which are surface expressions of three-dimensional structures, generally (but not always) parallel to margins of flow units. The origin of foliation is (most likely) the small and larger-scale isoclinal folding that occurs when the stratified ice from the interior inland ice is compressed into the narrow Lambert Graben (by gravitational forces). The primary flat layering is almost entirely transposed into vertical foliation, oriented parallel to the margins of the flow units. Margins of flow units are medial moraines, which in the case of Lambert, Mellor and Fisher Glaciers consist of very little debris (Hambrey 1991, Hambrey and Dowdeswell 1994). These features have been previously referred to as flow lines (e.g. in Swithinbank 1988).

Fisher and Mellor Glacier have zones of kilometerlong transverse crevasses with uncrevassed areas in between, in upper Lambert Glacier, crevasse fields are larger and occur less regularly, as deduced by Hambrey (1991) from LANDSAT imagery.

After Fisher, Mellor and (upper) Lambert Glaciers have merged, Lambert Glacier passes through a transverse, concave zone of disturbed flow. Likely, there are bedrocks or riegels and the glacier steepens (Hambrey and Dowdeswell 1994). Further downglacier, the crevasses die out, and as Lambert Glacier enters the narrowest part of the Lambert Glacier trough, meltwater streams form on the surface and flow several kilometers (which means that few crevasses exist in this area), and lakes form on the surface. The existence of lakes and meltwater streams is an indicator of a relatively warm climate.

Because lower Lambert Glacier has a very low surface slope but higher velocity, Allison (1979, p. 231) concludes that it moves almost completely by basal sliding. The LANDSAT image (fig. 53, p. B70, Swithinbank 1988) shows an extended area of rumpled ice to the west and south of an observed grounding line position (Morgan and Budd 1975; Budd et al. 1982). The rumpled zone is bounded on one side by the (central) ice stream, on the north by the ice shelf, and in the west by rugged mountain ranges. The thickness gradient is 1.7 m/km (on average between thickness shots of 1000 m and 640 m located 215 km apart). This is low, even for a floating ice shelf. The 640 m point is west of Gillock Island; the 1000 m point is approximately 100 km south of the southern end of Beaver Lake.

On our map, the gradient is approximately 0.02% $= 0.011°$ below the grounding line and 0.032% $= 0.018°$ above the grounding line.

Notes on grounding-zone indicators derived from image data and other data

In his 1991 paper, Hambrey (1991) agrees with the grounding-line position determined by Budd et al. (1982), which is the position that matches our determination from of the grounding zone from satellite radar altimeter data (see section (E)). In a later paper, Hambrey and Dowdeswell (1994) investigate areas located farther upglacier for possible indicators of grounded ice from satellite imagery: Ice surface features visible in LANDSAT imagery near Clemence Massif suggest that the ice may be afloat here. Features include (a) a feature that is flat, 6 km long, located in a depression and could be a lake, (b) two elliptical 3–4 km depressions similar to *ice dolines* found on Antarctic ice shelves (ice dolines have been interpreted as melt lakes that drain periodically through fractures to the sea underneath an ice shelf, Mellor (1960)), and (c) large areas of ice-surface lakes. Not only are the interpretations of the features very vague, all three types of surface features are indicative of a low surface slope only and not necessarily of floating ice. Clearly, interpretation from satellite data is often an indicator rather than a proof. At the time of our determination of the grounding line, radar altimeter data south of 72.1° S were not yet available, and the area identified as the grounding zone is the one that satisfies the break-in-slope and roughness criteria best, in addition to being in the neighbourhood of the position determined by Budd et al. (1982) in the field. LANDSAT data are generally less suited for grounding-line determinations than radar altimeter data, since the criteria are related to elevation and morphology (roughness).

Fricker et al. (2000) conclude that the grounding line is further upstream in Lambert Glacier, based on GPS observations and modeling.

Amery Ice Shelf

It is noteworthy that, while there are meltwater streams and lakes observed on Lambert Glacier, Amery Ice Shelf, located several hundred kilometers farther north, is not an area of ablation. The reason is that accumulation increases towards the coast.

Only a relatively small part of the ice that enters the circum-Antarctic ocean at the front of Amery Ice Shelf stems from the main Lambert Glacier (units Fisher, Mellor, upper Lambert Glaciers).

In addition, forty percent of the ice mass is lost by basal melting near the grounding line and one third of that is replaced by freeze-on of saline (sea) ice (Robin 1983). The relatively high and coastward increasing accumulation leads to an estimated 50 m part in ice thickness that is from snow. As a combination of these effects, less than half of the 270 m total thickness of Amery Ice Shelf at its front is ice from the Lambert Glacier system.

308

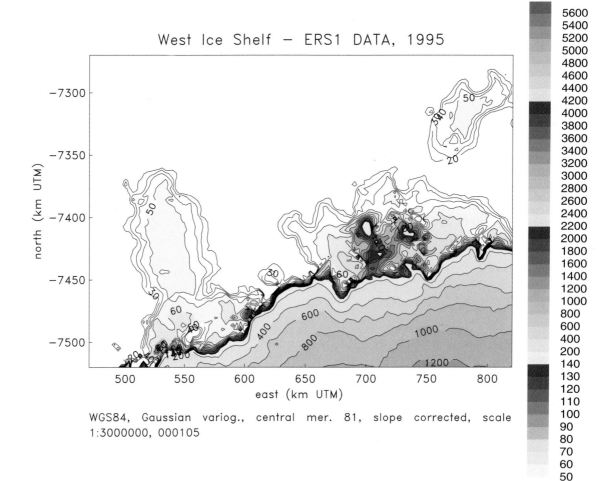

West Ice Shelf — ERS1 DATA, 1995

WGS84, Gaussian variog., central mer. 81, slope corrected, scale 1:3000000, 000105

(F.6) Detail Map 6: West Ice Shelf (ERS-1 Data 1995)

Atlas Map: m81e73-89n63-68 Leopold and Astrid Coast

West Ice Shelf map area: (470.000-820.000 E/-7270.000- -7520.000 N); scale 1:3.000.000

The study of ice shelves provides (1.) an indicator of changes in the mass balance of Antarctic ice, as discharge from outlet glaciers changes the size and volume of an ice shelf, and (2.) an indicator of climatic warming, as ice shelves may warm up (visibly indicated by melt ponds on their surface) and disintegrate in a warming climate. As we shall see, the West Ice Shelf may be an example of the first process. The second process affects northerly ice shelves more than southerly ice shelves — West Ice Shelf is at 66°–68° N (65.8°–67.8° N). Except for

the ice shelves on the Antarctic Peninsula, some of which have already disintegrated, West Ice Shelf is one of the northernmost ice shelves. Shackleton Ice Shelf, about 200 km further east, extends north to 64.5° Southern Latitude.

This detail map shows West Ice Shelf on Leopold and Astrid Coast in East Antarctica, the next ice shelf to the east of Amery Ice Shelf. Leopold and Astrid Coast is bounded by ice shelves on its entire length from Barrier Bay in the west to Philippi

Glacier at Cape Penck in the east; Philippi Glacier (66° 45'S/88° 20'E) is on the eastern edge of the ice shelves.

The extent of West Ice Shelf, the most prominent ice shelf on this coast is marked differently in various sources:

On the USGS satellite image map (Ferrigno et al. 1996), the name West Ice Shelf is given to the entire ice shelf area off Leopold and Astrid Coast, on the National Geographic Atlas map (National Geographic Society 1992, p. 102) the name applies only to the area west of Mikhaylov Island. Philippi Glacier is a 25 km long glacier which flows north to the eastern end of West Ice Shelf (25 km west of Gaussberg), named after a geologist of the German Antarctic Expedition 1901–1903 led by Drygalski. Hence, according to Alberts (1995), West Ice Shelf is the entire ice shelf area.

On our detail map, there are two areas of ice shelf, a western part that extends narrowly into the sea, and a wider eastern part that surrounds Mikhaylov Island and Zavadovski Island. Mikhaylov Island and Zavadovski Island are the western and eastern highlands included in the eastern part of the West Ice Shelf, they are labeled on the Atlas map m81e73-89n63-68 Leopold and Astrid Coast. To the west of the western, narrow shelf is Barrier Bay.

The topography of West Ice Shelf suggests that it is fed by several small glaciers which descend from the rugged and steep coast (there is no satellite image in Swithinbank (1988) to compare with).

The western "tongue" measures 150 km from the coastline to the northwestern ice edge. This tongue has elevations of up to 70 m, located 50 m above the sea level, and 30–40 m above the sea level in the main part of the tongue. The eastern part consists of several areas that extend various distances out into the sea (51 km, 78 km, and 36 km from west to east). The central part around the islands has an elevation of 80 m (60 m above sea level), the ice edge appears to be 30 m high.

Comparing the GEOSAT and the ERS-1 maps, there are notable differences in the seaward extent of West Ice Shelf: The large ice tongue east of Barrier Bay extends a bit farther on the ERS-1 map (1995) than on the GEOSAT map (1985-86), whereas the eastern section (around Mikhaylov Island and Zavadovski Island and east of the latter) appears to have retreated, most drastically in the easternmost part: from 100 km (in 1985-86) to less than 50 km (in 1995). Since Philippi Glacier is located here, this could mean a retreat of Philippi Glacier, or less mass discharge by the glacier, or calving of the ice shelf. However, since the ice shelf appears to have lost extent in the entire eastern area, but has several (about three) "tongues" or areas of locally maximal seaward extent, calving in all those areas seems less likely than decreased glacier flow in the eastern part of Leopold and Astrid Coast and in particular in Philippi Glacier. So, we have an area of an advancing glacier next to retreating glaciers (not an unusual, but an interesting phenomenon).

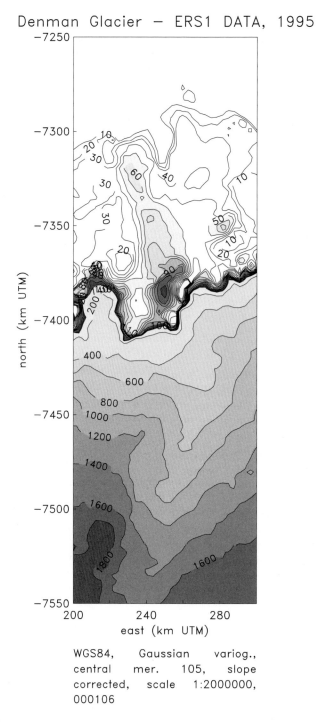

(F.7) Detail Map 7: Denman Glacier (ERS-1 Data 1995)

Atlas Maps: m93e85-101n63-68 Queen Mary Coast and m105e97-113n63-68 Knox Coast

Denman Glacier map area: (200.000-300.000 E/-7250.000- -7550.000 N); scale 1:2.000.000

Denman Glacier is a large glacier on the end of   tica (see also maps: m93e85-101n63-68 Queen
Queen Mary Coast in Wilkes Land, East Antarc-   Mary Coast, m105e97-113n63-68 Knox Coast,

m93e85-101n67-721 Wilkes Land e85-101n67-721, m105e97-113n67-721 Wilkes Land e97-113n67-721).

Denman Glacier occupies a large valley, identifiable on this map and extending at least to 2200 m elevation (see map m93e85-101n67-721 Wilkes Land e85-101n67-721).

From the shape of the Denman Glacier Valley on the detail map, we conclude that the glacier follows a narrow deep trough (approximately 10–20 km wide) at the bottom of a wider valley. The valley is 180 km long from Shackleton Ice Shelf to the 2200 m contour. The drainage basin extends at least to the 3000 m contour, the divide to the west separates Denman Glacier drainage from Lambert Glacier drainage (map m93e75-111n71-77 American Highland, to the east the boundary is Dome Charlie above 3200 m (see map m117e99-135n71-77 Dome Charlie)). Possibly, the flanks of a preexistent geologic trough in the bedrock were eroded in their upper parts to a wider valley, and additional flow comes from the sides of the valley. To the northeast of Denman Glacier are the Bunger Hills (see map m105e97-113n63-68 Knox Coast).

Denman Glacier drains an area of 199.000 km^2 (McIntyre 19999, after Swithinbank 1988, p. B59). At about (66.7° S/99° E), Denman Glacier enters Shackleton Ice Shelf. The glacier's ice can clearly be distinguished from the shelf ice in the ERS-1 data map as an area of 30–40 m higher elevation (distinghished by the contours 40–70 m above WGS 84, whereas the shelf ice is 30–40 m above WGS 84), this elevated area is 30 km wide close to the coast and about 20 km wide near the ice edge. Close to the coast is an area of 120 m elevation (above WGS 84) (the elevated area is too large to be a processing error).

In 1974 LANDSAT MSS (multispectral scanner) imagery (Swithinbank 1988, fig. 47, p. B61), the offshore part of Denman Glacier that cuts through Shackleton Ice Shelf is identified by crevasse patterns, which are: (1.) a braided pattern with blocks of crevasses, (2.) a smoother area in the center, and (3.) a wide shear margin on the western side and a narrow shear margin on the eastern side. Denman Glacier takes a leftward (western) turn, so the western margin is at the inside of the turn, hence there are more compressional forces on the inside. The patterns in area (1.) are similar to patterns observed in the center of Jakobshavns Isbræ (West Greenland), described by Mayer and Herzfeld (2000, 2001), Jakobshavns Isbræ also has a smooth longitudinal center, but symmetrical shear margins on both sides. According to Swithinbank (1988), the braided crevasse patterns are subsectioned by rifts which "represent lines of weakness that, at time of calving, will localize fractures", i.e. they are predecessors of crevasses that determine calving boundaries (later iceberg edges).

Above the grounding line, characteristic flowlines are visible in the LANDSAT image (no crevasse patterns are resolved, only flowline-parallel features). Between the extreme flowline features, Denman Glacier is 15 km wide; the valley is about twice as wide, and ice flows in from elevated areas in the east and west to join the Denman Glacier ice. The ice surface structure allows identification of the grounding line.

About 10 km west of Denman Glacier is Northcliffe Glacier, about 40 km east of Denman Glacier, east of the Obruchev Hills, is Scott Glacier, both are much smaller than Denman Glacier and their flowlines are only distinguishable for a short distance inland: 25 km for Scott Glacier and about 50 km for Northcliffe Glacier). Northcliffe Glacier bifurcates around Davis Island, such that its eastern branch joins Denman Glacier ice just north of the grounding line.

From a comparison between a 1972 LANDSAT image and a 1:1.000.000 scale map of the Australia Division of National Mapping (1969), with 1956 ice positions, Swithinbank (1988) concludes that Denman Glacier advanced 22 km between 1956 and 1972, indicating a minimum rate of advance of 1370 ma^{-1} (more if calving occurred). The Soviet Atlas Antarktiki (Tolstikov 1966, p. 162) gives a maximum rate of movement of 1500 ma^{-1} for Denman Glacier and 1300 ma^{-1} for Scott Glacier.

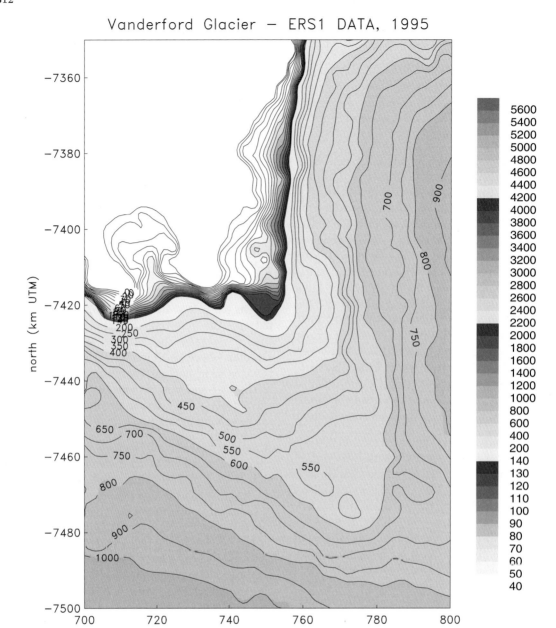

Vanderford Glacier — ERS1 DATA, 1995

WGS84, Gaussian variog., central mer. 105, slope corrected,
scale 1:1000000, 991203

(F.8) Detail Map 8: Vanderford Glacier (ERS-1 Data 1995)

Atlas Maps: m105e97-113n63-68 Knox Coast, m117e109-125n63-68 Sabrina Coast

Vanderford Glacier map area: (700.000-800.000 E/-7350.000- -7500.000 N); scale 1:1.000.000

Vanderford Glacier, mapped here at a scale of 1:1.000.000, is located on Budd Coast, Wilkes Land. It is the largest glacier in Vincennes Bay (650.000 to 750.000 E on the coast). The valley of Vanderford Glacier extends for 120 km southeast around Law Dome, seen here in part in the northeastern corner of the detail map. Totten Glacier is on the other (eastern) side of Law Dome, and its valley reaches southwestward around the Dome (for 130 km), the valleys meet at a surface elevation of 800 m (above the WGS84 ellipsoid).

The Vanderford Glacier Valley has a gradient of about 100 m over 12 km (0.83% = 0.48°) between 400 m and 550 m elevation, and a somewhat larger gradient of 100 m over 6 km (1.7% = 0.95°) between 140 m and 300 m elevation. There are two flatter areas, one at 550 m and one at 350–400 m in the western part of the valley. Ice discharges into Vincennes Bay at three locations: at 720.000 E is a small ice tongue that extends 25 km seaward and is fed by ice in the western part of the valley. At 740.000 E is a small bay filled with glacier ice (note the contours!), and at the head of Vincennes Bay at 750.000 E is glacier ice that fills the head of the bay to 25 km from the coast, it comes from the eastern part of the valley. In the USGS satellite image map (Ferrigno et al. 1996), all this is Vanderford Glacier. It could also be two glaciers, but only one name is given. According to Alberts (1995), Vanderford Glacier is only 8 km wide. It looks as if Vanderford Glacier is the ice in the head of the bay at 750.000 E. The Windmill Islands are north of Vanderford Glacier.

314

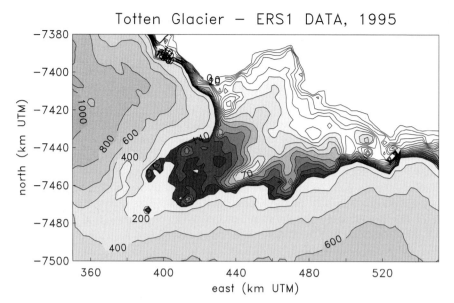

Totten Glacier — ERS1 DATA, 1995

WGS84, Gaussian variog., central mer. 117, slope corrected,
scale 1:2000000, 000106

(F.9) Detail Map 9: Totten Glacier (ERS-1 Data 1995)

Atlas Map: m117e109-125n63-68 Sabrina Coast

Totten Glacier map area: (350.000-551.000 E/-7380.000- -7500.000 N); scale 1:2.000.000

Totten Glacier is located on Sabrina Coast in Northern Wilkes Land, East Antarctica, and southeast of Law Dome. Law Dome is bordered by Budd Coast; the limit between Budd Coast and Sabrina Coast is Cape Waldron at the western margin of Totten Glacier.

To avoid confusion, the following should be noted: *Williamson Glacier* is a small glacier just west of Cape Waldron, and west of Totten Glacier; Fox Glacier is similarly small and 20 km further west of Williamson Glacier (according to Swithinbank 1988, fig. 46, p. B60; and other than indicated on the National Geographic Atlas map (National Geographic Society 1992, p. 102) where Totten Glacier is not named but has the name "Williamson Glacier" printed across it – and ending at the small Williamson Glacier).

The valley of Totten Glacier reaches southwestward around Law Dome from the east for about 130 km, whereas the valley of Vanderford Glacier

reaches southeastward around Law Dome from the west for about 120 km, both valleys meet at a surface elevation of 800 m.

Totten Glacier drains an area of 150.000 km^2 and has a balance discharge of 43 km^{3}a^{-1} (McIntyre 19999, after Swithinbank 1988, p. B57). Radio-echo soundings prove the existence of a subglacial valley. The velocity of ice entering Totten Glacier from the south (where the glacier flows in a west-northwesterly direction) is 280 ma^{-1} (near the 1000 m contour on the Australian elevation map; spot soundings of ice thickness in the same area are 1450, 2000, 1900, 1500, and 1600 m. Ice-front velocities are 850–1200 ma^{-1} (Dolgushin 1966).

While Vanderford Glacier (see detail map 8 (F.8)) does not extend far into the sea, Totten Glacier extends seaward for almost exactly 100 km, filling the bay between Law Dome and Sabrina Coast. The gradient of the offshore ice is very nicely mapped in the detail map. Notice that the gradient is almost constant below the 120 m contour; contours are given at 10 m levels down to 0 m (above WGS 84). The gradient between 80 m and 120 m is 40 m over 20 km (0.2% = 0.11°); the steepest section is between 100 m and 90 m (10 m over 3 km corresponding to 0.3% = 0.19°).

Noticeably, the gradient between 140 m and 120 m is much lower (20 m over 32 km; 0.06% = 0.36°). The gradient of the grounded glacier is 200 m over 32 km (0.6% = 3.6°)(400 m–200 m), that is 10 times as much.

A satellite image map of Totten Glacier (a composite of two LANDSAT 1MSS scenes, both from 26 October 1973) is shown in Swithinbank (1988, fig. 66, p. B60), with contours of an Australia Division of National Mapping map printed on it. On this map, Totten Glacier extends barely 20 km into the sea. Comparison of the LANDSAT map with the GEOSAT map and the ERS-1 map suggests that Totten Glacier advances into the ocean: On the GEOSAT map (1985-86), the glacier extends 85 km into the sea, on the ERS-1 map (1995) 95 km. If one measures "extension into the sea" from a line between the two headlands east (Cape Simonov) and west of the glacier, extent seaward is 20 km (LANDSAT 1973), 30 km (GEOSAT 1985-86), and 40–45 km (ERS-1 1995), so the indication of an advance certainly exists. The advance appears to be about 1 km per year.

On the satellite image in Swithinbank (1988), Totten Glacier has an upper area of relatively smooth ice with subtle flowline-parallel features and a sequence of depressions trending normal to the flowlines and spaced approximately 5 km apart in along-flow direction. More closely spaced and less pronounced depressions indicate a wavy surface morphology further west on the glacier surface. We conclude this is the grounded part of the glacier. Further down, the ice surface is characterized by rifts that trend across the glacier and are spaced increasingly close toward the ice front, where the ice appears crevassed and, further towards the margin, broken into pieces. The boundary between the two patterns probably marks the grounding line. If that position is the grounding line, then it also coincides roughly with the apparent coastline on the ERS-1 map (see detail map), and, furthermore, the seaward extent of the floating part is then 65 km on the 1973 satellite image. This gives the same figures of about 10 km advance between 1973 and 1986 and 10 km between 1986 and 1995.

According to Swithinbank (1988), the grounding line is a small distance upglacier of the depression (seen on p. B59) and thus 110 km from the ice front! The conclusion of an advance is still correct in this case, because the locations of the steepest slopes, indicative of the coastline, are detectable on the satellite image also, and a line connecting those on the east and west sides of the glacier coincides with the boundary of the two patterns (i.e. the same head point for ice-extent measurements will still be used in LANDSAT, GEOSAT and ERS-1 maps).

In summary, a comparison of maps from the three data sets — LANDSAT (1973), GEOSAT (1985-86), and ERS-1 (1995) suggests that Totten Glacier has been advancing at a rate of aproximately 1 km per year.

(F.10) Detail Maps 10: Mertz Glacier, 11: Ninnis Glacier, and 12: Mertz and Ninnis Glaciers (GEOSAT Data 1985-86)

Atlas Maps: m141e133-149n63-68 Adelie Coast, m153e145-161n63-68 Ninnis Glacier Tongue, m153e145-161n67-721 Cook Ice Shelf

Mertz Glacier map area: (350.000-450.000 E/-7438.000- -7530.000 N); scale 1:1.000.000

Ninnis Glacier map area: (495.000-550.000 E/-7535.000- -7590.000 N); scale 1:800.000

Mertz and Ninnis Glaciers map m147e142-148.5n66.5-68.5; scale 1:2.500.000

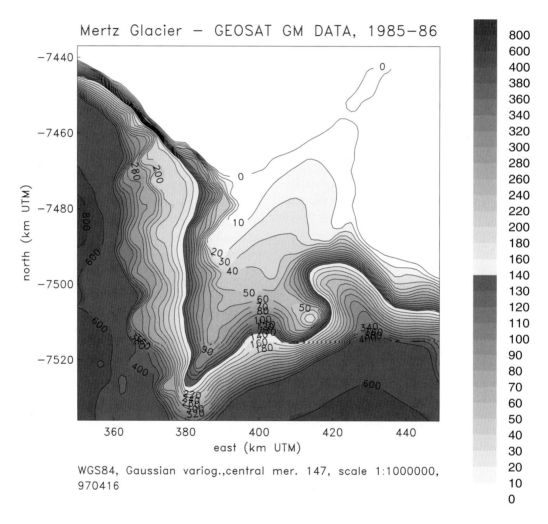

Detail Map 10. Mertz Glacier. Scale 1:1.000.000

Mertz Glacier (67°S/145°E) and Ninnis Glacier (68°S/147.5°E) are two outlet glaciers in East Antarctica with long tongues extending into the southernmost Indian Ocean. These tongues are at interesting, different stages of advance or retreat. Yet, the two glaciers are barely distinguishable in the Atlas maps. The following detail study demonstrates that a lot more information may be avail-

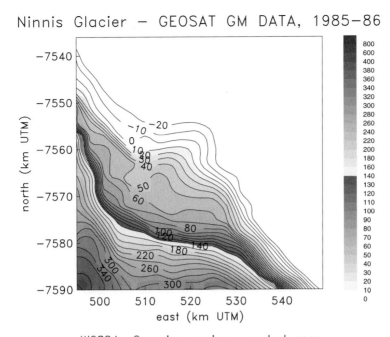

Ninnis Glacier – GEOSAT GM DATA, 1985–86

WGS84, Gaussian variog., central mer.
147, scale 1:800000, 000718

Detail Map 11. Ninnis Glacier. Scale 1:800.000

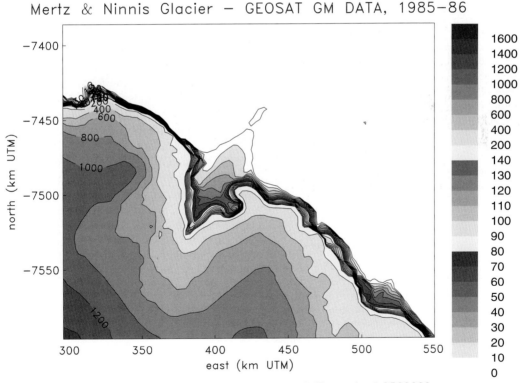

Mertz & Ninnis Glacier – GEOSAT GM DATA, 1985–86

WGS84, Gaussian variog., central mer. 147, scale 1:2500000,
slope corrected, m147e142 1485n665 685, 970507

Detail Map 12. Mertz and Ninnis Glaciers. Scale 1:2.500.000

able — and sometimes hidden! — in the 3 km Atlas grids.

Since both glaciers are located near the margin of the map, a regional topographic map showing both Mertz and Ninnis Glaciers (66.5-68.5° S/142.0-148.5° E) is constructed (detail map 12). To enhance the details in the glaciers and ice tongues, enlarged, individual topographic maps of Mertz Glacier (67.0-68.0° S/143.5-146.0° E) (detail map 10) and Ninnis Glacier (68.0-68.5° S/147.0-148.5° E) (detail map 11) are given. In both maps, isohypses are are drawn for 0, 10, 20, ..., 140m (in steps of 10 m) and for 140, 160, ..., 400 m (in steps of 20 m), and for 400, 600, 800 m (in steps of 200 m). A 20 m spacing is used between 140 m and 400 m to map the transition between the level of the ice cap and the level of the ice tongues.

In the following, we consider accuracy and reliability of the radar altimeter maps, later, we also compare the advantages and disadvantages of radar altimeter and SAR data, while the optimum results can be drawn from combining both data types (Herzfeld and Matassa 1997).

Mertz Glacier is at UTM (360.000-440.000 E/-7525.000- -7440.000 N). Its drainage basin is a broad valley that is distinguishable at the southern map boundary. The sides of the western valley are transected by erosional features. At its head, the glacier is fed by a narrow valley, approximately 4 km wide. The distance from the 80 m contour line at the head to the front is 85 km. In the eastern arm of Mertz Glacier there is a 20 m overdeepening, centered at (415.000 E/-7510.000 N). This is located below the steepest part of the valley walls. Is is not an interpolation error, since the contours are really smooth.

Mertz Glacier is about 27 km wide. The glacier tongue appears to extend about 45 km seaward of the coastline (see discussion below). The grounding line of Mertz Glacier is probably located in the vicinity of the 40-50 m contour. The 60 m contour still exhibits the indentation of the valley that continues subglacially from the valley leading to the head of the glacier, while the 50 m contour does not. The 40 and 50 m contours show the signs of the eastern side valley. At 30 m, the tongue is definitely floating.

The following surface slopes were calculated along local gradients of the Mertz Glacier map: 0.049° (=0.0857%) between the 30 and 0 m contours for the floating tongue; 0.078° (=0.136%) in the direction of the head valley and 0.143° (=0.25%) farther east between the 40 m and 70 m contours on the grounded ice; 0.286° (=0.5%) between the 80 m and 140 m contours in the head valley. The western side of Mertz Glacier valley has a slope of 0.573° (1%) between 80 m and 140 m, and a slope of 0.8275° (=1.4%) between 140 m and 400 m. Above that, the slope increases with increasing elevation. The steepest area in the flanks of Mertz Glacier is located southeast of the overdeepening; it is inclined 2.0045° (=3.5%) on average between 50 and 140 m, and a shallower section (1.2412° = 2.16%) between 140 m and 400 m. In this region, the slope lessens above 400 m elevation.

Ninnis Glacier is at UTM (500.000-540.000 E/-7550.000- -7580.000 N). The glacier tongue appears to extend about 10–15 km seaward from the coastline (the coastline being identified by the steep gradient of contours). Ninnis Glacier lies below a steep cliff, at 90 m above WGS84 and lower. The entire extension from the break in slope at the foot of the cliff to the 0 m contour line is 20 km. Ninnis Glacier is about 35 km wide. It does not really have a tongue anymore (as opposed to 1913), as also noted by Wendler et al. (1996).

The book "Geographic Names of the Antarctic" (Alberts 1995) gives the following information about Mertz and Ninnis Glaciers: Ninnis Glacier is a large, heavily hummocked and crevassed glacier which descends steeply from the high interior plateau to the sea in a broad valley (Alberts 1995, p. 527). Ninnis Glacier Tongue extended seaward for 48 km in 1962 (Alberts 1995, p. 527). Mertz Glacier is a heavily crevassed glacier, 72 km long and, in average, 32 km wide. It reaches the sea at Cape de la Motte and Cape Hurley. Discovered by Douglas Mawson (1911–1914), named after Xavier Mertz, an expedition member who lost his life on a sledge journey far east (7.Jan.1913, see Ninnis). The glacier tongue is given as 72 km long, 40 km wide (no date given; Alberts 1995, p. 487).

Wendler et al. (1996) study the advance and retreat of the Mertz and Ninnis Glacier Tongues and describe their surface features based on satellite Synthetic Aperture Radar (SAR) data from JERS-1 (1993 data). In the sequel, we combine radar altimeter and SAR data to study changes in position, surface morphology, and elevation.

Mertz and Ninnis Glacier Tongues are located close to each other, but their behaviour has been opposite. The glaciers were first charted by Douglas Mawson in 1912–13, which provides a longterm view back in history. Mertz Glacier Tongue has grown both wider and longer since then, compared to 1913, the area of the Mertz Glacier Tongue increased until 1962 to 154.6% of its 1913 area of 3830 km^2, growing mostly wider, and until 1993 to 211.3% (8100 km^2), increasing 26 km in length. The latter corresponds to an advance of 0.9 m per year. Ninnis Glacier Tongue has retreated, in 1913 the tongue had an irregular shape and an area of 6060 km^2, by 1962, it had retreated 90 km in length and shrunk to 56.6% of its 1913 area, by 1993, it had retreated a total of 110 km and shrunk to 35.4% (2150 km^2) (Wendler et al. 1996, from the historic map by Mawson, an Australian map from 1962, and from SAR data).

Surface patterns on Mertz Glacier, seen in SAR data in Wendler et al. (1996) suggest that the Mertz Glacier Tongue is floating for its entire length (the patterns are created in the grounded part of the glacier, in seawater, parts may break off – episodic calving). There is a reduction in strain rate as the tongue becomes afloat. The mechanism for advance are similar to that described by Holdsworth (1974) for Erebus Glacier. The velocity of advance of the glacier tongue is equal to the velocity of the ice stream at the grounding line, plus an increment due to longitudinal strain rates induced by hydrostatic imbalance at the front, minus the calving rate. Rates measured from 1993 and 1994 SAR data are 1.2 km a^{-1}, somewhat larger than 0.9 km a^{-1} averaged from the advance rate 1962–1993 (Wendler et al. 1996). The glacier may have accelerated in the near past, or calving may have occurred, or both.

In contrast, Alaskan tidewater glaciers, typically located in deep fjords, have a cycle of slow advance over timespans on the order of 1000 years and short phases of rapid retreat and disintegration, after which the glacier tongue builds up again (Meier and Post 1987). The advancing process of the terminus involves abrasion of debris from the upglacier side of a frontal moraine (in front of the terminus) and deposition of the debris on the proglacial side ("bulldozer style"). The rate of advance is only 10–100 m a^{-1}.

The surface shows braided patterns similar to those of Jakobshavns Isbræ (Mayer and Herzfeld 2000), a fast-moving ice stream in Greenland. The centerline is visible in the braids, the patterns shift symmetrically to the outside and change to shear margin patterns towards the sides.

The 1993 SAR image of Ninnis Glacier (Wendler et al. 1996) shows different patterns near the centerline, and a large rift across the ice tongue which suggests that the major part of the tongue will soon calve off also. Smaller blocks have already rotated and floated off the margins of Ninnis Glacier Tongue.

Discussion on accuracy and reliability and comparison SAR-RA mapping.

Mapping of atlas type and careful analysis of the spatial structure and distribution of noise levels allows us to extend the limits of use of altimeter data for mapping. Bamber (1994) produced a map of Antarctica (from ERS-1 altimeter data) with 20 km grids, reliable only in areas with a slope of less than 0.65° according to the author. The drainages of Mertz and Ninnis Glaciers are considerably steeper than that. Accuracy depends on topographic relief (see Herzfeld et al. 1993, 1994), with submeter accuracy on ice streams. However, the accuracy of a kriged map in areas of high topographic relief refers to the pointwise error rather than to the error of the "mean" surface elevation mapped. The "pointwise error" is the probable difference between a point on the map (estimated surface elevation) and a radar-altimetry-derived surface elevation; it depends on the averaging process of kriging. Shape and elevation of the surface at the Mertz Glacier drainage basin and on the glacier itself, however, are more accurate than inferred from the noise calculation, which is best conceived from inspection of the maps. The contour lines are smooth on the resolution of the 3 km grid. Smooth contour lines indicate a continuous or differentiable surface function with low error; little islands and edging contour lines indicate a rougher surface function with higher errors. That means the maps are reliable also in areas of steep terrain.

An area with high relief is located on the western side of Mertz Glacier. Notice that the grid spacing of the subarea enlargements is the same 3 km as for the large map, and that a wealth of information only becomes available in the enlargement.

Neither satellite altimetry nor kriging is a tool designed to track the location of an ice cliff. Because of the so-called snagging effect of the altimeter, described in the section on altimetry over ice (C.1.1) (see Thomas et al. 1983; Partington et al. 1987), the location of the ice edge is systematically wrong, differently so in descending and ascending orbits. In the Ice Data Record utilized in the Atlas, the ice edge is in the middle of both errors.

Kriging employs a moving window averaging technique, which results in smoothing the steep edges. However, the ice edge can still be inferred from the location of dense contours. Comparing our maps with the images in Wendler et al. (1996), we note that the results are surprisingly good.

In the satellite-radar-altimetry-derived detail maps, Mertz Glacier Tongue appears to extend about 40 km seaward of the coastline, Ninnis Glacier Tongue about 20 km. On the map in Wendler et al .(1996, fig. 2), Mertz Glacier Tongue extends about 80 km off the (averaged) coastline, Ninnis Glacier Tongue about 20–30 km. One also needs to take into account that the SAR and altimeter data were collected at different times, hence 1993-1985=8 years, assuming 0.9 km or 1.2 km advance per year following Wendler et al.

(1996), the tongue of Mertz Glacier should have been 7.2 km or 9.6 km shorter in 1985 (GEOSAT time) than in 1993 (JERS-1 time).

This indicates that, while SAR data are superior to altimetry-based maps for location of the ice edge, changes in advance and retreat of a glacier can still be monitored from our Atlas maps. Radar altimeter data have the advantage of being available for a longer time span (since the 1978 SEASAT mission). SAR imagery shows surface features in the floating tongues, as interpreted above. The slope and gradient of the glacier surface are lost in the grey-shaded SAR image. Additional information on ice flux can be derived from the surface elevation in the altimetry-based DTMs, calculated for the drainage basins.

In glaciologic summary, the tongue of Ninnis Glacier is retreating, whereas the tongue of Mertz Glacier is advancing, both due to inherent dynamics of the glaciers. This is a good example of glaciers with different dynamic behaviour and advance/retreat behaviour in the same local climate and time.

In a summary of satellite data analysis, a maximal amount of information can be obtained from a combination of SAR and radar altimeter data.

Rennick Glacier — ERS1 DATA, 1995

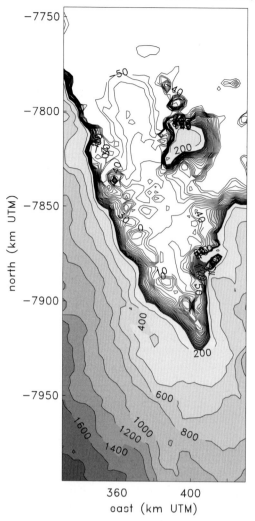

WGS84, Gaussian variog., central mer. 165, slope corrected, scale 1:2000000, 000107

(F.11) Detail Map 13: Rennick Glacier (ERS-1 Data 1995)

Atlas map: m165e157-173n67-721 Pennell Coast

Rennick Glacier map area: (330.000-430.000 E/-7745.000- -7995.000 N); scale 1:2.000.000

Rennick Glacier is located in Victoria Land, it flows northward out of the Transantarctic Mountains and reaches the sea at Oates Coast in Rennick Bay. At 370 km long and 20–30 km wide, Rennick Glacier is one of the largest glaciers of the world. It occupies an almost straight fault-controlled trench (Dow and Neall 1974), the long north-south trending valley shown on maps m165e157-173n67-721 Pennell Coast and m165e147-183n71-77 Victoria Land.

Additional contour lines have been added in the detail map to reveal the surface slope of Rennick Glacier — this was necessary, because the geoid and ellipsoid have an elevation difference near zero level of the WGS84 ellipsoid in this area of Antarctica. The resolution of the underlying DTM is 3 km by 3 km, the same as the Atlas maps.

Offshore and east of the floating tongue is Znamenski Island (70° 14'S/161° 51'E), a high, nearly round (4 km long), ice-covered island. The island was charted by the Soviet Antarctic Expedition in 1958 and named after hydrographer K.J. Znamenski, 1903-1941 (Alberts 1995, p. 833). On a LANDSAT 3 RBV image in Swithinbank (1988, fig. 40, p. B50), the relief of Znamenski Island clearly stands out from the level of Rennick Bay. An RBV (return beam vidicon) image has a very high resolution, revealing glacier flowlines, surface expression of subglacial features and areas of crevasses and individual large crevasses. However, only very few RBV images with correct exposure have been received (Swithinbank 1988, p. B47). On the ERS-1 map, the elevation of the island is over 200 m (its size is too large, a result of retracking altimeter data). The floating tongue of Rennick Glacier extends 40 km beyond Znamenski Island (not a good way of positioning the front because of the inaccurate size of the island), and 85 km north of the eastern shoreline.

The grounding line of a glacier is determined by (a) a break in slope in the ice surface, and (b) a transition from rough ice of the grounded part to smooth ice of the floating part. According to both criteria, the grounding line appears to be crossing the glacier from (-7850.000 N/360.000 E) on its western side to (-7840.000 N/390.000 E) near the apparent southwest end of the island on its eastern side. Another possibility is at (-7900.000 N/370.000 E) to (-7880.000 N/400.000 E) where a side glacier appears to enter from the east.

Swithinbank (1988) places the grounding line across from Sledgers Glacier, using the existence of meltwater hollows and flows as an indication of the break in slope (water collects at the break in slope and flows downward from there, following the small gradient of the floating tongue), hence the glacier would float for 140 km length. Mayewski et al. (1979) concluded from radio-echosoundings that the glacier is afloat for at least 130 km.

Using the second result, on our map, the floating part would be 57x2 totaling 114 km long (according to the first, 64 km). While conclusions from ground surveys are direct geophysical observation and usually better, the measurement from radar altimetry is surprisingly good in comparison.

Ice thickness near the Sledgers Glacier mouth is 800 m, it decreases to 180 m near Znamenski Island (fig. 42 in Swithinbank 1988).

Canham Glacier is the largest glacier entering from the east (approximately 50 km south of Sledgers Glacier).

The western slope of the valley side bordering the glacier and its tongue are steep, reaching 200 m in 2 km between 400 m and 200 m elevation, corresponding to 10% slope, or 5.7 degrees on our map, so the slopes are likely even steeper in reality. To the south, the valley floor ascends gradually (200 m to 400 m over 20 km, 1% slope or 0.6 degrees). The side of the valley south of the coast ascends 1000–1200 m in 12 km, corresponding to 0.0167% or 0.95 degrees, about 1 degree. The satellite images in Swithinbank (1988, fig. 40, 41) show gradually rising ice-covered land to the west, which ascends to the Usarp Mountains 40 km to the west, and a series of mostly north-south trending smaller, rugged hills to the east of the glacier (Bower Mountains). Gressit Glacier comes in from the west, and many small glaciers occupy the mountain valleys to the east.

Swithinbank (1988) attributes about half of the width of Rennick Glacier near its front to ice from glaciers entering from the west, which is indicated by medial flowlines. Mapping of the ice sheet surface inland (Steed and Drewry 1982, after Swithinbank 1988, p. B52) indicates that Rennick Glacier has a small drainage basin and receives little, if any, ice from the interior section of the East Antarctic Ice Sheet. Evidence of flowlines, supraglacial meltwater lakes and a low surface gradient from source to grounding (1 in 85 =0.0117, equal to 1.17% and 0.67°) suggest a low velocity. From the front position seen on two LANDSAT MSS images 8 years apart, Swithinbank (1988, p. B52) concludes the velocity is 190 ma^{-1}, so Rennick Glacier would be one of the slowest glaciers of the region, and ice flux at the front would be 0.6 km^{3}a^{-1}, which is very small for a large glacier.

A larger Rennick Glacier once covered the area, including the Morozumi Range and Gressit Glacier (south of the grounding line position), and likely Rennick Glacier is in a stage of retreat (Mayewski et al. 1979, after Swithinbank 1988, p. B52).

As found in studies of other detail glaciers, the largest amount of information in satellite glaciology is deduced from a combination of satellite images and elevation mapping from satellite. Despite the different technologies and different error sources associated with each technology, the results are surprisingly similar in some cases, and complementary in others.

David Glacier / Drygalski Ice Tongue — ERS—1 DATA, 1995

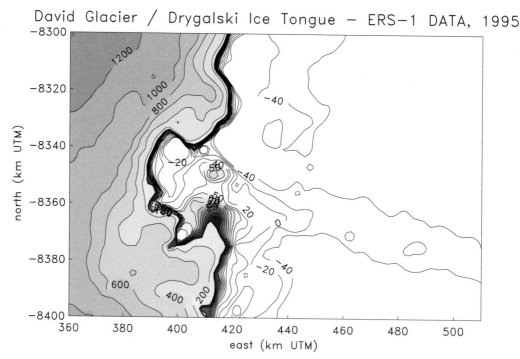

WGS84, Gaussian variog., central mer. 165, slope corrected, scale 1:1333333, 991209

(F.12) Detail Map 14: David Glacier/Drygalski Ice Tongue (ERS-1 Data 1995)

Atlas map: m165e147-183n71-77 Victoria Land

David Glacier/Drygalski Ice Tongue map area: (360.000-510.000 E/-8300.000- -8400.000); scale 1:1.333.333

Drygalski Ice Tongue is the glacier tongue of David Glacier, an over 100 km long glacier that flows from the Antarctic Plateau west of Victoria Land through the Transantarctic Mountains, the Prince Albert Mountains, and reaches the Ross Sea at Scott Coast. The center coordinate of David Glacier is (75° 19'S/162° E). The glacier tongue extends more than 100 km into the Ross Sea (beyond the map margin). David Glacier was discovered in 1908 by the "Northern Party" of Ernest Shackleton's Antarctic Expedition, this group was led by Prof. T.W. Edgeworth David of Sydney University. Drygalski Ice Tongue (center coordinates (75° 24'S/163° 30'E)) was discovered by Robert Scott in 1902 and named for Erich von Drygalski, a contemporary German Antarctic explorer (Alberts

1995, p.201). The capes to the sides of Drygalski Ice Tongue are Cape Philippi and Cape Reynolds.

The drainage basin of David Glacier appears to extend 300 km west on the contour map: David Glacier is the largest glacier in the bowl-shaped area of the Antarctic Plateau west of Victoria Land. Because Priestly and Reeves Glaciers are also outlet glaciers, David Glacier does not drain the entire bowl, but likely a segment delineated by a diagonal line from the northern headland northwest and a diagonal line from the southern headland southwest, the latter being indicated by topography (m165e147-183n71-77 Victoria Land).

According to McIntyre (199999, after Swithinbank 1988) the drainage basin is 224.000 km^2 and should have a balance discharge of 14 km^{3}a^{-1}. According to Steed and Drewry (1982) the drainage basin coincides with a subglacial trench that transects the mountain range and extends inland for more than 300 km. This number matches our conclusions based on ice surface topography. The glacier is 2530 m thick at the ice fall at 900 m surface elevation.

Some flowlines of upper David Glacier are discernable on a LANDSAT 1 MSS image (in fig. 35 in Swithinbank 1988, p.B44) 140 m upstream of the northern headland and 105 km upstream of an area of confluence of several branches, in the lower part of David Glacier. An aerial photograph (Swithinbank 1988, fig. 36, p. B43) of the confluence area shows David Glacier as a heavily crevassed ice stream with distinctive flowlines.

Drygalski Ice Tongue has extensional crevasses and groups of crevasses normal to the flowlines. On its margins there are feathery shear crevasses. These patterns are typical of floating glacier tongues. As the glacier ice leaves the valley, the longitudinal strain rate increases, the glacier thins rapidly (1190 m to 500 m over 40 km) and also spreads out (from 12 km at the mouth to 22 km). Large ice floes can break off near the ice tongue (20 km by 40 km), as the satellite image of 22 February 1973 shows (Swithinbank 1988, fig. 34, p. B43). The landward end of the ice tongue moves at 580 ma^{-1}, the speed 50 km from the coast is 730 ma^{-1} (Holdsworth 1985).

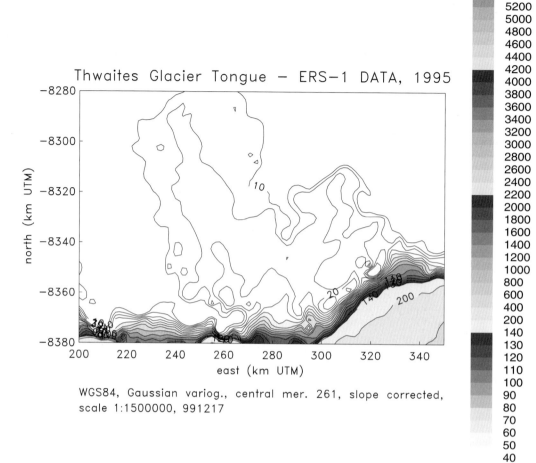

Thwaites Glacier Tongue — ERS-1 DATA, 1995

WGS84, Gaussian variog., central mer. 261, slope corrected, scale 1:1500000, 991217

(F.13) Detail Map 15: Thwaites Glacier (ERS-1 Data 1995)

Atlas Map: m261e243-279n71-77 Walgreen Coast

Thwaites Glacier map area: (200.000-350.000 E/-8280.000- -8380.000 N); scale 1:1.500.000

Thwaites Glacier is located on Walgreen Coast in West Antarctica, about 200 km west of Pine Island Glacier (detail map 16, (F.14)). Thwaites Glacier and Pine Island Glacier are the two largest and fastest ice streams in Ellsworth Land and Marie Byrd Land, flowing at 2900 m a^{-1} (Thwaites Glacier) and 2400 m a^{-1} (Pine Island Glacier; after Ferrigno et al. 1998). According to the literature, they share the property that they are not protected from interactions with the sea by ice shelves (but Pine Island Glacier has a small ice shelf, see

(F.14)). These properties have led to the hypothesis that a possible collapse of the entire West Antarctic Ice Sheet may initiate in the Thwaites and Pine Island Glacier area (the dinsintegration hypothesis is discussed in the Pine Island Glacier section (F.14)).

The ice fronts of Pine Island Glacier and Thwaites Glacier are very different — Pine Island Glacier terminates in a bay, and its ice forms a coverage of this bay. The ice in Pine Island Bay has the characteristics of an ice shelf, as is seen in detail map 16 Pine Island Glacier (F.14). Thwaites Glacier to the contrary, is totally unbounded by shelf ice or coastal confinements, it has a glacier tongue and seaward of that an iceberg tongue, which float loosely into the ocean. The glacier tongue is floating, and the iceberg tongue is grounded on shoals (Holdsworth 1985).

On the Atlas detail map, there is no thick ice associated with either one. The glacier tongue and iceberg tongue extend 100 km into the Amundsen Sea, but do not have a sharp vertical ice edge, and the ice-surface elevation appears to be only 10–15 m above the surrounding sea surface, with a maximum of 20 m in the eastern part (east of the glacier tongue proper).

Hence, Thwaites Glacier extends further out into the sea than any other glacier. According to Lindstrom and Tyler (1985), the Thwaites Glacier Tongue moves 3.5 km a^{-1}. The location of the outline of the glacier tongue matches that in satellite imagery (Swithinbank 1988, p. B126, fig. 93). The disconnected piece may correspond to signals reflected off the iceberg tongue that is drifting away as mentioned in Ferrigno et al. (1998) (see below). The ice is very loosely connected due to lack of constraining forces, and characterized by extensional crevassing throughout. The ice of the iceberg tongue is only connected to that of the glacier tongue in a narrow spot (few kilometers of the total width of 65 km of the iceberg tongue). It appears that Thwaites Iceberg Tongue is broken off from Thwaites Glacier Tongue and rotated about 40° westward, but has otherwise stayed in place, because it is grounded (1973 satellite image, Swithinbank 1988).

Hence, there is clearly a difference in the geometrical properties of offshore ice of Pine Island Glacier and Thwaites Glacier. This observation leads to questioning whether both glaciers can be treated with the same type of disintegration model.

The tongue was discovered in 1946–47 and has been existing since then. It has been hypothesized by Hughes (1977) that Thwaites Glacier surges, because the iceberg tongue increased by 60% between 1965 and 1974, as noted by Southard

and MacDonald (1974) and MacDonald (1976). (A glacier surge is a sudden acceleration of a glacier to up to 100 times its normal velocity. Only some glaciers surge. Surges occur quasiperiodically, with a short surge phase of rapid movement during which the glacier advances, and a long quiescent, normal phase.) Thomas et al. (1979) quote a photogrammetric velocity estimate of 2 km a^{-1} after R. Allen and consider that equilibrium velocity. The glacier drains an area of 121.000 km^2 (McIntyre 19999) and should have a balance discharge of 47 km^3 a^{-1}. In later satellite imagery the iceberg tongue has broken off and is floating away by 1996 (Ferrigno et al. 1998), and the tongue has a shorter length again.

The hypothesis of a possible surge, if correct, would indicate that Thwaites Glacier and Pine Island Glacier have different dynamic properties. Such a difference in the dynamics of the glacier would also result in differences in ice-sea interaction and vice versa.

Another difference between Thwaites Glacier and Pine Island Glacier exists in surface and bed morphology — Pine Island Glacier occupies a large trough or basin but with a low gradient in its midelevation reaches (700–800 m) whereas Thwaites Glacier terminates on a steep section of Walgreen Coast that also has a relatively steep hinterland (800 m elevation 60 km from the coast for Thwaites Glacier, 800 m elevation 200 km from the coast along a longitudinal profile of Pine Island Glacier), as follows from a comparison of the two detail maps. The main part of Thwaites Glacier occupies a deep trench that extends inland well below sea level. The drainage basin extends south to the area of the Hollick-Kenyon Plateau, to 350 km further south (measured to the 2000-m contour). It is not clear, however, how much of the area drains to Thwaites Glacier.

Differences in the geometric characteristics of the floating parts and grounded parts of Pine Island Glacier and Thwaites Glacier as well as of their drainage basins leads to questioning whether both glaciers can be treated with the same type of disintegration model. Indeed, differences in change behaviour have been observed in recent years (as referenced in the introduction).

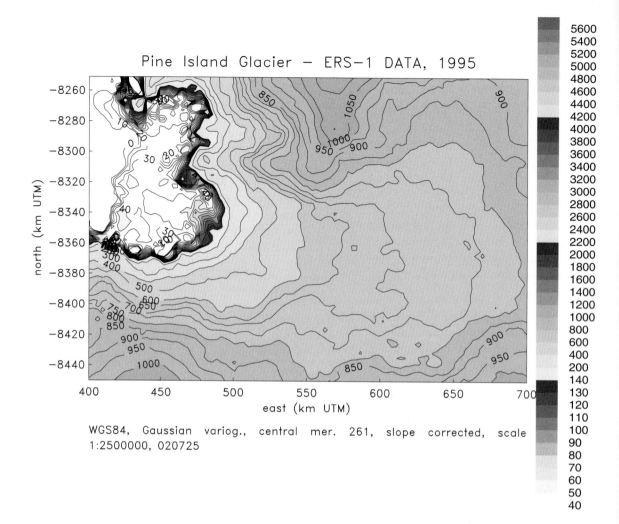

Pine Island Glacier — ERS—1 DATA, 1995

WGS84, Gaussian variog., central mer. 261, slope corrected, scale 1:2500000, 020725

(F.14) Detail Map 16: Pine Island Glacier (ERS-1 Data 1995)

Atlas map: m261e243-279n71-77 Walgreen Coast

Pine Island Glacier map area: (400.000-700.000 E/-8250.000- -8450.000 N); scale 1:2.500.000

Pine Island Glacier is located on Walgreen Coast in West Antarctica, in an area that is rarely visited by expeditions. The glacier reaches the coast in Pine Island Bay south of Canisteo Peninsula.

Pine Island Glacier appears to drain the entire area south of the coastal volcanic ranges that border the northern coast of Ellsworth Land. The high elevation area at (660.000 E/-8180.000 N) indicates the Jones Mountains, the easternmost range of the volcanic province that extends to the

Fosbick Mountains at 145° W (see other maps on the volcanic province, section (D)). Cosgrove Ice Shelf (further north) has a much smaller drainage basin which extends inland about 50 km, whereas the drainage basin of Pine Island Glacier extends 350 km inland (eastward). From the ice divide, that is mapped by the 1200-m contour, ice drains eastward to the Ronne Ice Shelf (Rutford Ice Stream, glacier in Carlson Inlet, Evans Ice Stream, see maps m285e267-303n71-77 Ellsworth Land and m285e267-303n75-80 Zumberge Coast).

In the detail map, the surface topography of this large drainage basin is mapped with 50-m contours (between 600 m and 1000 m), rather than with 200-m contours as in the Atlas maps. The gradient is lowest between 800 m and 700 m of elevation, and steeper above 800 m. Below 700 m, the surface steepens significantly to the coast ($0.013\% = 0.76°$), even in the center of the basin (100-m contours for 200 m–600 m elevation). According to Crabtree and Doake (1982), the drainage basin of Pine Island Glacier is 214.000 km^2 $+/-$ 20.000 km^2 and mass flux at the ice front is 25 $+/-$ 6 gigatons, equal to 28 km^3, per year.

Pine Island and Thwaites Glaciers (see detail map 15 Thwaites Glacier) are the two largest ice streams that drain the northern margin of the West Antarctic Ice Sheet in Ellsworth Land and Marie Byrd Land. Thwaites Glacier is the fastest glacier in this area (2900 m a^{-1}) followed by Pine Island Glacier (2400 m a^{-1}, after Ferrigno et al. 1998). Pine Island Glacier and Thwaites Glacier are glaciologically important, because they play a significant role in scenarios of a possible break-up of the West Antarctic Ice Sheet. In the literature, Pine Island Glacier and Thwaites Glacier are treated together as sharing the rare property of being large glaciers, which do not end in ice shelves. These two properties are important in disintegration models.

Our map shows that Pine Island Glacier extends into Pine Island Bay, and its ice causes an increase of 80 m over the 30-m contour at the outer edge of Pine Island Bay. The entire bay, i.e. Pine Island Bay and the northern bay immediately south of Canisteo Peninsula, appears to be filled with ice with a 30-m step at the margin — so on the map this appears as shelf ice with a typical edge. The northern part is not named in the literature (to our knowledge). In satellite imagery (Swithinbank 1988, Bindschadler 2002, Ferrigno et al. 1998) Pine Island Glacier is also seen to enter a sheltered bay with a small ice shelf. The ice edge location, however, which is clearly seen in the detail map, cannot be seen in any published satellite image, because none of the image subsets extend far enough, hence a comparison is not possible.

In that the ice from Pine Island Glacier fills a sheltered bay, whereas the tongue of Thwaites Glacier extends into the open ocean, the glaciers are different. This raises the question whether it is permissible to treat them similarly in models. The ice of Thwaites Glacier forms the Thwaites Glacier Tongue which is clearly the result of lack of material (ice shelf or coast) that holds the glacier "together".

Grounding line. The location of the 50 m contour (above WGS84) on the Atlas detail map of Pine Island Glacier matches the location of the grounding line determined from ERS-1 radar interferometry (Schmeltz et al. 2002). Criteria for determination of the grounding line from DTMs are (1) break-in-slope criterion: grounded ice above the break in slope, floating ice below the break in slope; (2) surface-roughness criterion: smooth surface of floating ice, rough surface of grounded ice. A break in slope exists at the 50 m contour line. Another less significant break in slope is at the 80 m contour line. From the contours it appears that the grounding line is between the 50 m contour and the 60 m contour. The roughness criterion is also met at the 50 m contour. The 40 m contour approximately outlines the location of the floating tongue of Pine Island Glacier protruding into Pine Island Bay. The grounded Pine Island Glacier extends 30 km into Pine Island Bay (measured from a line connecting points on the north and south side of the glacier margin on the detail map). In Schmeltz et al. (2002) the grounding-line-to-front distance is about 40 km. The front is not determined well in radar altimeter data in general (Partington et al. 1987), so this is a fairly good correspondence. In contrast, the 1400 m sounding that marks the location of the grounding line in Crabtree and Doake (1982) is located much further upglacier (80 km from the front), this location is supported in Swithinbank (1988) in an analysis of a LANDSAT1MSS image (1973).

This comparison is not undertaken to postulate that the grounding line of a relatively small glacier is better determined from radar-altimetry-derived DTMs than from interferograms of SAR data, but to demonstrate that the geostatistical method with a variogram-modeling approach as utilized in the Atlas detail maps is capable of deriving accurate and geophysically useful maps/DTMs in mathematically difficult data situations. (In general, backscatter data and elevation data are complementary in any data analysis as previously established (Herzfeld and Matassa 1997, Bamber and Rignot 2002).) In conclusion, the Atlas detail DTMs have an accuracy that is higher than previously thought possible because of the known effects in altimeter data when crossing a break in slope, crossing an ice shelf edge, or retracking signals returned from areas with topographic relief. Furthermore, geostatistical methods applied to improved data sets (ATM data, GLAS data and RA data) as proposed here may be expected to yield optimal results.

Role of Pine Island Glacier and Thwaites Glacier in instability models of the West Antarctic Ice Sheet.

The hypothesis of disintegration of the West Antarctic Ice Sheet, which raised interest in Pine Island and Thwaites Glaciers because of its catastrophic outlook, was first brought forward by Hughes (1973). Pine Island and Thwaites Glaciers are the fastest flowing glaciers in this part of Antarctica (Thwaites Glacier 2900 m a^{-1} and Pine Island Glacier 2400 m a^{-1}, after Ferrigno et al. 1998). In the literature, Pine Island Glacier and Thwaites Glacier are treated together as sharing the rare property of being large glaciers, which do not end in ice shelves. These two properties are important in disintegration models.

This area would most likely start a collapse of the entire West Antarctic Ice Sheet. The course of events in Hughes' (1973) disintegration model is as follows (after Swithinbank 1988, p. B124): "Surging could produce a basal water layer that would uncouple the ice from its bed and thus draw down the surface level of the ice sheet. In the absence of high bed rock sills to prevent it, the grounding line could migrate inland until ultimately the whole marine portion of the ice sheet is converted into an ice shelf."

As visible in satellite imagery (Swithinbank 1988, Bindschadler 2002, Ferrigno et al. 1998) as well as in the detail map, Pine Island Glacier enters a sheltered bay with a small ice shelf, whereas Thwaites Glacier has a glacier tongue and an iceberg tongue which extend far into the ocean, and the hinterland of the two glaciers is morphologically different (see above). Pine Island Glacier occupies a large trough that extends far inland and has a very low surface gradient, whereas Thwaites Glacier termi-

nates on a steep section of Walgreen Coast. Consequently, the glaciers should be treated differently in models.

The possibility that collapse of the West Antarctic Ice Sheet may lead to rapid sea-level rise in the near future is discussed in Bentley (1997). Assessment of the potential contribution to sea-level rise in the Amundsen sea varies (1.2 m from BEDMAP ice volume above sea level (Lythe et al. 2001) to zero because of absence of significant imbalance (Bentley and Giovinetto 1991)). A summary is also given in Vaughan et al. (2001).

Recent changes. Hughes (1973) and Thomas et al. (1979) suggested that the northern part of the ice sheet could already be collapsing. The exceptionally low ice surface gradient, which is obvious east of Pine Island Bay, might support this hypothesis. In contrast, Crabtree and Doake (1982) modeled the longitudinal profile of Pine Island Glacier using steady-state assumptions and found no evidence of instability. From comparison of 1973 and 1975 satellite images and 1966 aerial photographs, Swithinbank (1988) concluded 10 km retreat or calving.

In recent work it has been reported that Pine Island Glacier is undergoing rapid changes. Retreat of the grounding line of Pine Island Glacier and increase in flow velocity has been observed from interferometric analysis of SAR data (Rignot 1998, 2002; Schmeltz et al. 2002, Rignot et al. 2002), changes in crevassing of the glacier and changes in the ice front are visible in satellite imagery (Bindschadler 2002, Ferrigno et al. 1998), ice surface elevation mapping indicates that the ice in the drainage basin is thinning (Shepherd et al. 2001). Thwaites Glacier exhibits much less activity (Ferrigno et al. 1998, Bamber and Rignot 2002).

Modeling and data analysis work on Pine Island Glacier is ongoing reflecting the importance of the problem. There is still a wide range of possible explanations for glaciodynamic processes that may explain the observed changes and in a more general scenario as well as on a larger scale provide understanding of the physics of ice sheet collapse.

(G) Combination of SAR and Radar Altimeter Data: Lambert Glacier/Amery Ice Shelf

It has been noted previously that a maximum of information can be obtained from a combination of data from two sources, elevation and backscatter data (see for instance, section F.10). The maps of Lambert Glacier / Amery Ice Shelf presented here are examples of such a data combination; they are based on ERS-1/2 radar altimeter data and SAR data from the Canadian RADARSAT (Antarctic Mission). During the Antarctic Mission, the SAR sensor aboard RADARSAT was rotated such that it looked toward the South Pole — hence it covered the area that had been a gap in coverage in altimeter and SAR data from previous satellite missions. There were two Antarctic Missions, data here stem from the first Antarctic Imaging Campaign, which lasted from 9 September to 20 October 1997.

RADARSAT data from the 1997 Antarctic Imaging Campaign are utilized here in the form of a backscatter-data mosaic with 125 m compiled by Jezek et al (1999) at the Byrd Polar Research Center, Ohio State University, Columbus, Ohio, U.S.A., from many individual SAR scenes with 25 m pixel resolution. The RADARSAT mosaic is mapped with polarstereographic projection, the total image has 40141 rows and 48333 columns of pixels, it is stored as an 8-bit image. The mosaic is available via the ftp site of the World Data Center for Glaciology A (National Snow and Ice Data Center, University of Colorado Boulder, Boulder, Colorado, U.S.A., see the website www.nsidc.org).

Unfortunately, RADARSAT does not carry an altimeter. Therefore, ERS-1 and ERS-2 radar altimeter data are used in the compilation of SAR/radar altimeter maps of the Lambert Glacier/Amery Ice Shelf region. First, a combination with 1995 ERS-1 altimeter data is performed,

so it matches the atlas maps in time frame and data base and the large Lambert Glacier/Amery Ice Shelf region map in section F also in area. As for the Atlas maps, the time frame is 1 February – 1 August 1995.

Second, an ERS-2 radar altimeter data time frame was selected to match the RADASAT Antarctic Mission time frame as closely as possible and to contain sufficiently many data points for construction of a DEM (1 August to 31 October 1997). All those ERS-2 radar altimeter data stem from a 35-day-repeat-cycle mission. ERS-1 was not collecting radar altimeter data at this time anymore, the altimeter had decidedly been turned off by ESA, potentially a mistake considering that altimeter data are valubale, but relatively small data sets compared to SAR data. ERS-2 satellite radar altimeter data were processed by the author's group, analogous to ERS-1 data presented in earlier sections of the Atlas. To facilitate comparison, the ERS-1/2 radar altimetry maps are presented here (Figures G-1 and G-2).

The software ENVI (Research Systems Incorporated, Boulder, Colorado, U.S.A.) is employed to load the SAR mosaic and to select the subarea of Lambert Glacier/ Amery Ice Shelf and its neighbourhood. Using a cubic convolution, the reference system of the mosaic subarea is converted to UTM coordinates. The mapped subarea is (215.000-785.000 E/-7578.000- -8322.000 N) in UTM coordinates with respect to central meridian 69° E (same projection as used for the Lambert Glacier/ Amery Ice Shelf region maps in the atlas, section D, and in sections E and F). This corresponds to an area of 570 km east-west and

334

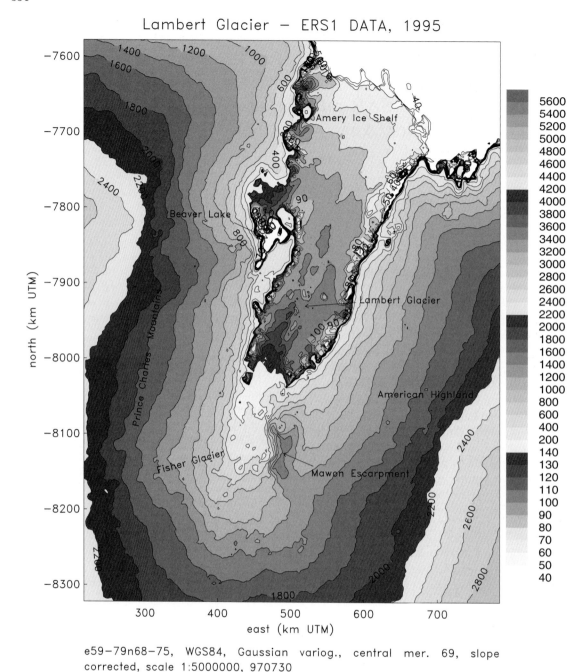

e59−79n68−75, WGS84, Gaussian variog., central mer. 69, slope
corrected, scale 1:5000000, 970730

Figure G-1. Topography of Lambert Glacier/Amery Ice Shelf region from 1995 ERS-1 data.
Time frame 01 February–01 August 1995 of ERS-1 Atlas maps. Elevations in meters above WGS84 ellipsoid,
GEM-T2-referenced, slope-corrected. DEM calculated using ordinary kriging with Gaussian variogram model.

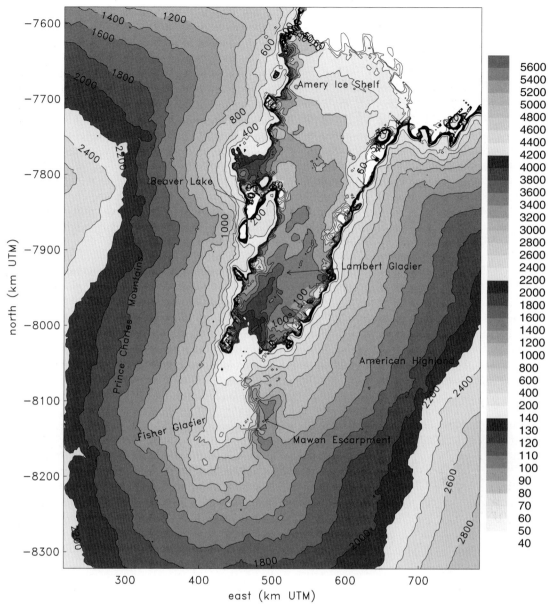

Figure G-2. Topography of Lambert Glacier/Amery Ice Shelf region from 1997 ERS-2 data.
Time frame 01 August–31 October 1997. Elevations in meters above WGS84 ellipsoid, GEM-T2-referenced, slope-corrected. DEM calculated using ordinary kriging with Gaussian variogram model.

744 km north-south and 4560 pixels east-west by 5952 pixels north-south.

However, before the image data set and the altimery-based DEM can be combined, the two data sets need to be coferenced, that is, both data sets need to be jointly geolocated such that geographic feautres in both data sets are associated with the same coordinates. Coreferencing of the RADARSAT SAR data and the gridded ERS-2 altimeter data is a non-trivial task, as the RADARSAT data are distorted. Distortion of image data is a consequence of image data collection, and each correction for location still has a remaining error.

Reference points that can be identified in both data sets can be used to improve the geolocation of the image data and the coreferencing of the two data sets. A total of 30 point were employed in the geolocation. The following 8 points are given as examples for reference points, as they can be identified visually in the maps:

(1) Clemence Massif (72° 11'S/68° 43'E), 1400 m (north of Mawson Escarpment)

(2) Mount Johnston (71° 32'S/67° 24'E), 1770 m (south of Fisher Massif)

(3) Sandilands Nunatak (70° 32'S/67° 27'E) (north of Mount Seaton, Nemesis Glacier)

(4) Single Island (69° 48'S/68° 36'E) (west of Amery Ice Shelf)

(5) Mount Stinear (73° 04'S/66° 24'E), 1950 m (lower Fisher Glacier)

(6) Mount Mather (73° 34'S/61° E) (west of Mount Menzies)

(7) Pattrick Point (73° 28'S/66° 51'E) (between Lambert Glacier and Mellor Glacier)

(8) Grove Mountains (72° 45'S/ 75° E) (east of Mawson Escarpment)

(9) Reinbolt Hills (70° 29'S/72° 30'E) (east of Amery Ice Shelf)

The resultant maps are presented in Figures G-3 and G-4. As a result of the coreferencing, the contour lines of glaciers flowing into the Lambert Glacier system match the flowlines (the surface structures) with astonishing accuracy. Contour lines follow the ups and downs of topography created by surface feautres that are visible in the grey-shade image, as close inspection of the map will easily reveal. The largest differences between contour lines and ice outlines exist at the front of Amery Ice Shelf. Here, altimeter data are not exact, and the front location changes with ice flow and ice-berg calving.

In the combined map, flow units of Lambert Glacier can be matched visually with elevation information.

Surface features of central Lambert Glacier/Amery Ice Shelf are easily identifiable. Dominant features of relatively fast-moving glaciers are longitudinal structures (*foliations*), which are surface expressions of three-dimensional structures, generally (but not always) parallel to margins of flow units. These features are often termed "flow lines" in the glaciologic literature (see definitions in Hambrey 1991 and Swithinbank 1988, summarized in section F.5). These foliations cannot only be traced in Landsat imagery, but also in SAR data.

One can distinguish the flow lines of the ice as it accelerates on its way into the central Lambert Glacier/Amery Ice Shelf area. Lambert Glacier follows a geologic trough, which causes gravitative acceleration of the ice. Hambrey (1991) and Hambrey and Dowdeswell (1994) identify 8 flow units of the Lambert Glacier/Amery Ice Shelf system (see section F.5), more are visible in this map, because the area is larger and hence some of the summarized units can be separated into several smaller glaciers.

Lambert Graben is up to 2500 m deep. Mawson Escarpment is a straight-lined escarpment rising to 1000 m above sea level located east of Lambert Glacier. The wall of the graben, measured to the top of Mawson Escarpment, is a total of approximately 3000 m high. Mawson Escarpment is easily identified to the east of Lambert Glacier.

The main flow of Lambert Glacier can be seen coming in from the southeast, this matches the observation of the 1958–59 Soviet Antarctic Expedition that ice from the area of the Gamburtsev Subglacial Mountains (off this map by another 200 km, but covered on maps m69e51-87n75-80 and m69e51-87n78-815) flows northwest to Lambert Glacier. The main ice flow then passes by Mawson Escarpment (here at -8100.000 – -8180.000 N / 480.000 – 500.000 E UTM). In the SAR data, the flowlines of Lambert Glacier can be traced upstream to the edge of the map (about 200 km from Mawson Escarpment; 100 km in LANDSAT data (Swithinbank, 1988), the difference may be caused by image area selection). Likely, it may be possible to trace Lambert Glacier even farther upstream.

The confluence with Mellor Glacier, which joins Lambert Glacier from the west, is near (-8170.000 N/410.000 E), south of this point Cumpston Massif is located. Cumpston Massif is the most northerly outcrop of a field of nunataks, eg. those south of (-8220.000 N/410.000 E); these include (from N to S) Cumpston Massif, Mt. Maguire, the Blake Nunataks, and Wilson Bluff; ice of (upper) Lambert Glacier and Mellor Glacier appears to also flow around Mt. Twigg and Mt. Borland. Fisher Glacier joins Mellow Glacier from the west, it has a small turn from a north-easterly flow direction to an almost precisely easterly direction near the left map edge at (-8150.000- -8190.000 N), where its northerly branch curves around a nunatak, a flow unit south of that nunatak is also part of Fisher Glacier. Fisher Glacier is important for the flow properties of Lambert Glacier, as it has been hypothesized to surge (see discussion in section E.7). With the help of flow features in SAR data, the individual glaciers are much more easily identified than in just altimeter data.

North of Fisher Glacier, the surface is characterized by undulating structures. Ice flows in many small units into Lambert Glacier from the west. These glaciers traverse the Prince Charles Mountains in an area from just south of Beaver Lake to about 300 km south of Beaver Lake (see also Hambrey and Dowdeswell 1994). The large glacier that comes in from the west-southwest is Charybdis Glacier (at the left map edge near -7900.000 N, flowing diagonally to (-7830.000 N/430.000 E). Whereas in atlas DEMs alone, Charybdis Glacier

can be traced to the 2000 m contour, in the combination map it can be traced all the way to the map edge (corresponding to 2400 m). One branch truns southward into Beaver Lake, in this area, there are too dense contour lines. Beaver Lake peninsula is well-visible as an area of high reflectance in the SAR data and outlined by contours. Just to the north of Charybdis Glacier, another glacier of similar width enters Amery Ice Shelf from the west, this tributary flows almost exactly west-east. Several small, but less distinguishable, glaciers enter northern Amery Ice Shelf from the west. They can be traced better by higher surface elevations marking their inflow than by SAR data:

(1) 50 m higher for 70-km-E-W Glacier (northernmost) at -7640.000 N

(2) 40 m higher for 35-km-E-W Glacier at -7660.000 N

(3) 40 m higher for 30-km-E-W Glacier at -7700.000 N

(4) 30 m higher for 15-km-E-W Glacier;

(5) 40 m higher for 10-km-E-W Glacier

(6) 40 m higher for (80-km-E-W) Charybdis Glacier at -7780.000 N, with large valley.

The rugged mass of the Prince Charles Mountains south of Charybdis Glacier and west of Lambert Glacier contrasts the fairly even gradient/slope of the American Highland to the east.

To the east of the main flow, several glaciers and a large ice stream enter from NE of Mawson Escarpment, these correspond to flow units (7) "Mawson Escarpment Ice Stream" and (8) in Hambrey (1991; see section F.5), the latter is fed by a large ice stream entering Amery Ice Shelf from the east side less than 100 km from its terminus (unit (8) is termed "NE Amery Ice Sheet").

The difference in surface appearance between the grounded and the floating part of the ice-stream/ice-shelf system in the combined map supports the determination fo the grounding as in the atlas maps (see section E.) Two different interpretations of grounding line location are given in Hambrey (1991; close in location to our determination) and Hambrey and Dowdeswell (1994; farther upstream).

338

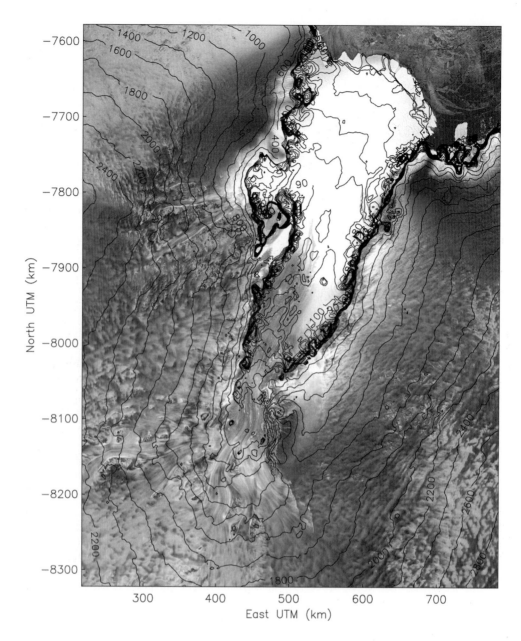

Figure G-3. Topography and morphology of Lambert Glacier/Amery Ice Shelf region from 1995 ERS-1 data and 1997 RADARSAT data. Time frame 01 February–01 August 1995 of ERS-1 Atlas maps. Elevations in meters above WGS84 ellipsoid, GEM-T2-referenced, slope-corrected. DEM calculated using ordinary kriging with Gaussian variogram model. RADARMAP data from first Antarctic mission 9 September–20 October 1997. Mosaic with 125 m pixels.

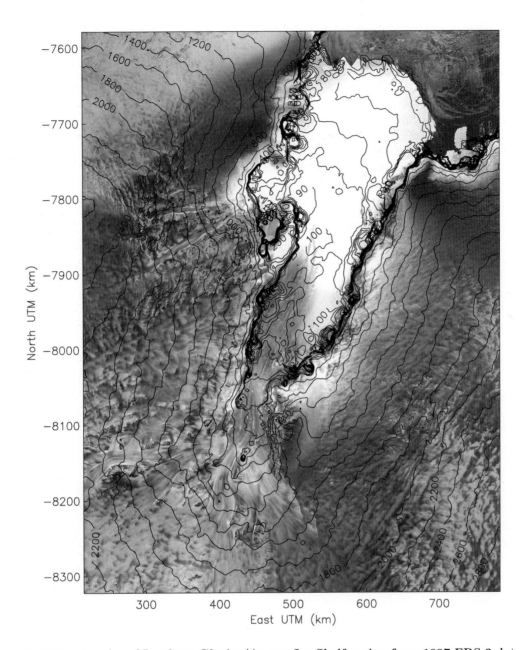

Figure G-4. Topography of Lambert Glacier/Amery Ice Shelf region from 1997 ERS-2 data and 1997 RADARSAT data. Time frame 01 August–31 October 1997 of ERS-2 data. Elevations in meters above WGS84 ellipsoid, GEM-T2-referenced, slope-corrected. DEM calculated using ordinary kriging with Gaussian variogram model. RADARMAP data from first Antarctic mission 9 September–20 October 1997. Mosaic with 125 m pixels.

Comparison of the 1995 ERS-1 map of Lambert Glacier and the 1997 ERS-2 map indicates an apparent small increase in elevation and advance of the glacier, in continuation of the 1978-189 trend (as derived in chapter (E)). An exact assessment of a potentially existing orbit bias between ERS-1 and ERS-2 data from the selected time frames needs to be calculated to determine the accuracy of the conclusion of advance. In the comparison of SEASAT and GEOSAT data, a shift in the contour lines of 10 km was much larger than the effect of a general offset between the satellites. Furthermore, ERS-1 and ERS-2 carry the same type of altimeter, and orbit determination has increased in accuracy (see Table B.2-1). In the areas surrounding the ice-stream/ice-shelf system, contours are largely identical, and small differences should be attributed to differences in ground-track locations and in the search algorithm (Ralf Stosius, pers. comm., used a different search algorithm and kriging implementation in the 1997 map).

The ice front has a different shape in the 1997 RADARSAT data and the 1995 ERS-1 data. Since the icefront (a) changes and (b) is not well-mapped in altimeter data, the advantage of using SAR data is particularly evident here. Several large rifts extend about 25 km into the fron of Amery Ice Shelf, indicating that the ice shelf is prone to calving a few huge icebergs in the near future (as an estimate, on the order of several years to ten years). The question arises whether unsually large volumes of ice will calve, and, given the catastrophic nature of calving events, whether this may be an indicator of climatically induced warming of Lambert Glacier/Amery Ice Shelf, which may lead to partial disintegration of the Amery Ice Shelf, similar to more drastic events observed in iceshelves along the Antarctic Peninsula. On the other hand, occasional calving of large icebergs is common in ice sheets, and connections with warming may be difficult to draw. Several small icebergs are floating in the ocean offshore of the front of Amery Ice Shelf.

In conclusion, combined maps from backscatter and elevation data allow investigation of several variables of glaciers and snow and ice surfaces in general, including surface morphology, ice-flow patterns, ice-fromt location, formation of large crevasses, undulations in slow-moving ice areas, surface gradients and elevation changes — and thus facilitate a broad and integrated approach to the study of the changing cryopshere.

Part IV

References and Appendix

(H) References

Alberts FG (1995) (ed) Geographic names of the Antarctic. U.S. Geological Survey, U.S. Board on Geographic Names, Reston, VA

Alley RB, Bindschadler RA (eds)(2001) The West Antarctic Ice Sheet. Behavior and environment. Antarctic Research Series 77: 296 pp. American Geophysical Union, Washington, D.C.

Alley RB, Blankenship DD, Bentley CR, Rooney ST (1986) Deformation of till beneath ice stream B, West Antarctica. Nature 322: 57–59

Allison I (1979) The mass budget of the Lambert Glacier drainage basin, Antarctica. Journal of Glaciology 22(87): 223–235

Allison I, Frew R, Knight I (1982) Bedrock and ice surface topography of the coastal regions of Antarctica between 48° E and 64° E. Polar Record 21(132): 241–252

Armstrong M (1984) Problems with universal kriging. Mathematical Geology 16(1): 101–108

Australia Division of National Mapping (1969) Australian Antarctic Territory Sheet SQ 47–48, 1:1,000,000 scale. Canberra

Bagnold RA (1941) The physics of blown sand and desert dunes. Methuen, London, 265 pp

Bamber JL (1994) A digital elevation model of the Antarctic Ice Sheet, derived from ERS-1 altimeter data and comparison with terrestrial measurements. Annals of Glaciology 20: 48–53

Bamber J, Rignot E (2002) Unsteady flow inferred for Thwaites Glacier, and comparison with Pine Island Glacier, West Antarctica. Journal of Glaciology 48(161): 237–246

Barrett PJ (1991) The Devonian to Jurassic Beacon Supergroup of the Transantarctic Mountains and correlatives in other parts of Antarctica. In: Tingey RJ (ed) The Geology of Antarctica, Oxford Monographs on Geology and Geophysics 17: 120–152. Clarendon Press, Oxford

Behrendt JC (1962) Summary and discussion of the geophysical and glaciological work in the Filchner Ice Shelf area of Antarctica. University of Wisconsin Geophysical and Polar Research Center Research Report Series no 62-3: 66 pp

Bentley CR (1997) Rapid sea-level rise soon from West Antarctic ice sheet collapse? Science 275: 1077–1078.

Bentley CR, Giovinetto MB (1991) Mass balance of Antarctica and sea level change. In: Weller G, Wilson CL, Severin BAB (eds) International Conference on the Role of the Polar Regions in Global Change: proceedings of a conference held June 11–15, 1990 at the University of Alaska Fairbanks, vol II: 481–488. University of Alaska, Geophysical Institute/Center for Global Change and Arctic System Research, Fairbanks, AK

Bindschadler RA (1991) West Antarctic Ice Sheet Initiative, Science and Implementation Plan, Proceedings of a workshop held at Goddard Space Flight Center, Greenbelt, Maryland, Oct. 16-18, 1990. NASA Conference Publication Preprint, 53 pp

Bindschadler RA (2002) History of lower Pine Island Glacier, West Antarctica, from Landsat imagery. Journal of Glaciology 48(163): 536–544

344

Bindschadler RA Scambos TA (1991) Satellite-image derived velocity field of an Antarctic ice stream. Science 252: 242–246

Bindschadler RA, Stephenson SN, MacAyeal DR, Shabtaie S (1987) Ice dynamics at the mouth of Ice Stream B, Antarctica. Journal of Geophysical Research 92(B9): 8885–8894

Brenner AC, Bindschadler RA, Thomas RH, Zwally HJ (1983) Slope-induced errors in radar altimetry over continental ice sheets. Journal of Geophysical Research 88(C3): 1617–1623

Brenner AC, Koblinsky CJ, Beckley BD (1990) A preliminary estimate of geoid-induced variations in repeat orbit satellite altimeter observations. Journal of Geophysical Research 95: 3033–3040

Briggs JC (1974) Machine contouring using minimum curvature. Geophysics 39(1): 39–48

Brodscholl AL, Herzfeld UC, Sandwell DT (1992) A comparison between satellite gravity data (GEOSAT) and marine gravity data measured in the Weddell Sea, Antarctica. In: Colombo OL (ed) From Mars to Greenland: Charting gravity with space and airborne instruments — Fields, tides, methods, results, IAG Symposium no. 110, Vienna, Aug. 20, 1991. Springer-Verlag, New York, pp 129–138

Brooks RL, Williams RSjr, Ferrigno JG, Krabill WB (1983) Amery ice shelf topography from satellite radar altimetry. In: Oliver RL, James PR, Jago JB (eds) Antarctic earth science. Australian Academy of Science, Canberra, and Cambridge University Press, Cambridge, pp 441–445

Brown GS (1977) The average impulse response of a rough surface and its applications. IEEE Trans. Antennas Propag. AP-25(1): 67–74

Budd WF (1966) The dynamics of the Amery Ice Shelf. Journal of Glaciology 6(45): 335–357

Budd WF, Corry MJ, Jacka TH (1982) Results from the Amery Ice Shelf project. Annals of Glaciology 3: 36–41

Buggisch W, Kleinschmidt G (1999) New evidence for nappe tectonics in the southern Shackleton Range, Antarctica. In: Millar IL, Talarico F (eds) The EUROSHACK Project; results of the European expedition to the Shackleton Range 1994/95. Terra Antartica 6(3–4): 203–210

Byrd RE (1947) Our navy explores Antarctica. National Geographic Magazine 92(4): 429–522

Canadian Space Agency, NASA, RADARSAT International (1994) RADARSAT. ADRO Program Announcement, vol II. RADARSAT System Description. Saint-Hubert

Chelton DB, Ries JC, Haines BJ, Fu L-L, Callahan PS (2001) Satellite altimetry. In: Fu L-L, Cazenave, A (eds) Satellite altimetry and earth sciences. Academic Press, San Diego, pp 1–131

Chiles JP (1977) Géostatistique des phenomenes non-stationaires, Thèse Doct. Ing. Univ. Nancy, I, 152 pp

Clarke GKC (1987) Fast glacier flow: ice streams, surging, and tidewater glaciers. Journal of Geophysical Research 92(B9): 8835–8841

Cornish V (1914) Waves of sand and snow and the eddies which make them. T. Fisher Unwin, London, 383 pp

Crabtree RD, Doake CSM (1980) Flow lines on Antarctic ice shelves. Polar Record 20(124): 31–37

Crabtree RD, Doake CSM (1982) Pine Island Glacier and its drainage basin: results from radio echo-sounding. Annals of Glaciology 3: 65–70

Debenham F (1948) The problem of the Great Ross Barrier. Geographical Journal 112(4–6): 196–218

Decleir H, van Autenboer T (1982) Gravity and magnetic anomalies across Jutulstraumen, a major geologic feature in western Dronning Maud Land. In: Craddock C (ed) Antarctic geoscience. University of Wisconsin Press, Madison, pp 941–948

Delfiner P (1982) The intrinsic model of order k. Notes for a short course at Batelle Seattle Research Center, Seattle

Doake CSM (1982) State of balance of the ice sheet in the Antarctic Peninsula. Annals of Glaciology 3: 77–82

Doake CSM, Frohlich RM, Mantripp DR, Smith AM, Vaughan DG (1987) Geological studies on Rutford Ice Stream, Antarctica. Journal of Geophysical Research 92(B9): 8951–8960

Dolgushin LD (1966) Noviye danniye o skorostyakh dvizheniya lednikov Antarktidy [New data on the rates of movement of Antarctic glaciers]. Informatsionnyy Byulleten' Sovetskoy Antarkticheskoy Ekspeditsii [Information Bulletin of the Soviet Antarctic Expedition] no 56: 17–20

Dow JAS, Neall VE (1974) Geology of the lower Rennick Glacier, northern Victoria Land, Antarctica. New Zealand Journal of Geology and Geophysics 17(3): 659–714

Drewry DJ (ed) (1983) Antarctica: glaciological and geophysical folio. Scott Polar Research Institute, Cambridge, var. pag.

Engelhardt H, Humphrey N, Kamb B, Fahnestock, M (1990) Physical conditions at the base of a fast moving Antarctic ice stream. Science 248(4951): 57–59

Engels J, Grafarend E, Keller W, Martinec Z, Sanso F, Vanicek P (1993) The geoid as an inverse problem to be regularized. In: Anger G, Gorenflo R, Jochmann H, Moritz H, Webers W (eds) Inverse problems: principles and applications in geophysics, technology, and medicine. Proc. Intern. Conf. Potsdam. Mathematical Research 74: 122–166. Akademie Verlag, Berlin

ESA (European Space Agency) (1992a) ERS-1 System. esa SP-1146. ESA Publications Division, Noordwijk

ESA (European Space Agency) (1992b) ESA ERS-1 Product Specification. esa SP-1149. ESA Publications Division, Noordwijk

ESA (European Space Agency) (1993) ERS User Handbook. esa SP-1148. ESA Publications Division, Noordwijk

Feigl KL, Dupré E (1999) RNGCHN: a program to calculate displacement components from dislocations in an elastic half-space with applications for modeling geodetic measurements of crustal deformation. Computers & Geosciences 25(6): 695–704

Ferrigno JG, Mullins JL, Stapleton JA, Chavez PSJr, Velasco MG, Williams RSJr, Delinski GFJr, Lear D'A (1996) Satellite image map of Antarctica. United States Geological Survey Miscellaneous Investigations Series Map I-2560, 1:5 000 000 scale

Ferrigno JG, Williams RSJr, Rosanova CE, Lucchitta BK, Swithinbank C (1998) Analysis of coastal change in Marie Byrd Land and Ellsworth Land, West Antarctica, using Landsat imagery. Annals of Glaciology 27: 33–40

Flöttmann T, Kleinschmidt G (1993) The structure of Oates Land and implications for the structural style of Northern Victoria Land, Antarctica. In: Damaske D, Fritsch J (eds) German Antarctic North Victoria Land expedition 1988/89; GANOVEX V. Geologisches Jahrbuch Reihe E Geophysik 47: 419–436

Fricker HA, Hyland G, Coleman R, Young NW (2000) Digital elevation models for the Lambert Glacier - Amery Ice Shelf system, East Antarctica, from ERS-1 satellite radar altimetry. Journal of Glaciology 46(155): 553–560

Fujii Y (1981) Aerophotographic interpretation of surface features and an estimation of ice discharge at the outlet of the Shirase drainage basin, Antarctica. Nankyoku Shiryo [Antarctic Record] no 72: 1-15

Gandin LS (1965) Objective analysis of meteorological fields. U.S. Dept. Commerce and National Science Foundation, Washington, DC, 242 pp [Original in Russian: Ob"ektivnyi analiz meteorologicheskikh polei, 1963]

Giovinetto MB, Bentley CR (1985) Surface balance in ice drainage systems of Antarctica. Antarctic Journal of the United States 20(4): 6–13

Goldstein RM, Engelhardt H, Kamb B, Frolich RM (1993) Satellite radar interferometry for monitoring ice sheet motion. Application to an Antarctic ice stream. Science 262: 1525–1530

Grafarend E (1975) Geodetic stochastic processes. Methoden und Verfahren der Mathematischen Physik 14: 1–27

Grenfell TC, Cavalieri D, Comiso D, Steffen K (1992) Chapter 14: Considerations for Microwave Remote Sensing. In: Carsey FD (ed) Microwave Remote Sensing of Sea Ice. Geophysical Monograph 68: 291–300. American Geophysical Union, Washington, DC

Hake G (1982) Kartographie I, 6th edn. Sammlung Göschen 2165: 342 pp, Walter de Gruyter, Berlin

Hambrey MJ (1991) Structure and dynamics of the Lambert Glacier-Amery Ice Shelf system: implications for the origin of Prydz Bay sediments.

346

In: Barron J, Larsen B et al. Proceedings of the Ocean Drilling Program Scientific Results 119: 61–75

Hambrey MJ, Dowdeswell JA (1994) Flow regime of the Lambert Glacier-Amery Ice Shelf system, Antarctica: structural evidence from Landsat imagery. Annals of Glaciology 20: 401–406

Hambrey MJ, McKelvey B (2000a) Neogene fjordal sedimentation on the western margin of the Lambert Graben, East Antarctica. Sedimentology 47: 577–607

Hambrey MJ, McKelvey B (2000b) Major Neogene fluctuations of the East Antarctic ice sheet: Stratigraphic evidence from the Lambert Glacier region. Geology 28(10): 887–890

Hansen HE (1946) Antarctica from latitude 68° 40' to 70° 20'S. and from longitude 36° 50' to 40° 20'E. Norges Geografiske Oppmaling, scale 1:250,000, Oslo

Heiskanen WA, Moritz H (1967) Physical geodesy. W.H. Freeman, San Francisco, 364 pp

Helferich S, Kleinschmidt G (1998) The West-East Gondwana boundary in Antarctica; Kirwanveggen, structural continuation of the Shackleton Range? Journal of African Earth Sciences 27(1A): 109–110

Hermichen WD, Kowski P, Wand U (1985) Lake Untersee, a first isotope study of the largest freshwater lake in the interior of East Antarctica. Nature 315(6015): 131–133

Herzfeld UC (1989a) Geostatistical methods for evaluation of SEABEAM bathymetric surveys: Case studies of Wegener Canyon, Antarctica. Marine Geology 88: 83–95

Herzfeld UC (1989b) Variography of submarine morphology: Problems of deregularization, and cartographical implications. Mathematical Geology 21(7): 693–713

Herzfeld UC (1990) Geostatistical software for evaluation of line survey data, applied to radio-echo soundings in glaciology. In: Hanley JT, Merriam DF (eds) Microcomputer applications in geology II. Computers & Geology vol 6:119-136, Pergamon Press, New York

Herzfeld UC (1992) Least squares collocation, geophysical inverse theory, and geostatistics: A bird's eye view. Geophysical Journal International 111(2): 237–249

Herzfeld UC (1996) Inverse theory in the Earth Science — an introductory overview with emphasis on Gandin's method of optimum interpolation. Mathematical Geology 28(2): 137–160

Herzfeld UC (1998) The 1993–1995 surge of Bering Glacier (Alaska) — a photographic documentation of crevasse patterns and environmental changes. Trierer Geographische Studien 17: 211 pp

Herzfeld UC (1999) Geostatistical interpolation and classification of remote sensing data from ice surfaces. International Journal of Remote Sensing 20: 307–327

Herzfeld UC, Brodscholl AL (1994) On the geologic structure of the Explora Escarpment (Weddell Sea, Antarctica) revealed by satellite and shipboard data evaluation. Marine Geophysical Researches 16: 325–345

Herzfeld UC, Higginson CA (1996) Automated geostatistical seafloor classification — principles, parameters, feature vectors, and discrimination criteria. Computers & Geosciences 22(1): 35–52

Herzfeld UC, Holmlund P (1988) Geostatistical analyses of radio-echo data from Scharffenbergbotnen, Dronning Maud Land, East Antarctica. Zeitschrift fur Gletscherkunde und Glazialgeologie 24(2): 95–110

Herzfeld UC, Holmlund P (1990) Geostatistics in glaciology: Implications of a study of Scharffenbergbotnen, Dronning Maud Land, East Antarctica. Annals of Glaciology 14: 107–110

Herzfeld UC, Matassa MS (1997) Mertz and Ninnis Glacier tongues mapped from satellite radar altimeter data. Journal of Glaciology 43(145): 589–591

Herzfeld UC, Matassa MS (1999) An atlas of Antarctica north of 72.1°S from GEOSAT radar altimeter data. International Journal of Remote Sensing 20: 241–258

Herzfeld UC, Mayer H (1997) Surge of Bering Glacier and Bagley Ice Field, Alaska: an update to August 1995 and an interpretation of brittle-deformation patterns. Journal of Glaciology 43(145): 427–434

Herzfeld UC, Mayer H (in press) Seasonal comparison of ice-surface structures in the ablation area of Jakobshavn Isbræ drainage system, West Greenland. Annals of Glaciology 37

Herzfeld UC, Lingle CS, Lee L-h (1993) Geostatistical evaluation of satellite radar altimetry for high resolution mapping of Lambert Glacier, Antarctica. Annals of Glaciology: 17: 77–85

Herzfeld UC, Lingle CS, Lee L-h (1994) Recent advance of the grounding line of Lambert Glacier, Antarctica, deduced from satellite radar altimetry. Annals of Glaciology 20: 43–47

Herzfeld UC, Mayer H, Higginson CA, Matassa M (1996) Geostatistical approaches to interpolation and classification of remote-sensing data from ice surfaces. Proceedings Fourth Circumpolar Symposium on Remote Sensing of Polar Environments, Lyngby, Denmark, 29 April – 1 May 1996, ESA SP-391: 59–63

Herzfeld UC, Lingle CS, Freeman C, Higginson CA, Lambert MP, Lee, L-H, Voronina VA (1997) Monitoring changes of ice streams using time series of satellite-altimetry-based digital terrain models. Mathematical Geology 29(7): 859–890

Herzfeld UC, Matassa MS, Mimler M (1999a) TRANSVIEW: a program for matching universal transverse mercator (UTM) and geographic coordinates Computers & Geosciences 25(7): 765–773

Herzfeld UC, Mayer H, Feller W, Mimler M (1999b) Glacier roughness surveys of Jakobshavns Isbræ Drainage Basin, West Greenland, and morphological characterization. Zeitschrift für Gletscherkunde und Glazialgeologie 35(2): 117-146

Herzfeld UC, Stauber M, Stahl N (2000a) Geostatistical characterization of ice surfaces from ERS-1 and ERS-2 SAR data, Jakobshavn Isbræ, Greenland. Annals of Glaciology 30: 224–234

Herzfeld UC, Stosius R, Schneider M (2000b) Geostatistical methods for mapping Antarctic ice surfaces at continental and regional scales. Annals of Glaciology 30: 76–82

Higham M, Reynolds M, Brocklesby A, Allison I (1995) Ice radar recording, data processing and results from the Lambert Glacier basin traverse. Terra Antarctica 2(1): 23–32

Hofmann-Wellenhof B, Lichtenegger H, Collins J (1993) Global Positioning System: theory and practice, 2nd edn. Springer-Verlag, Wien, 326 pp

Holdsworth G (1974) Erebus Glacier Tongue, McMurdo Sound, Antarctica. Journal of Glaciology 13(67): 27–35

Holdsworth G (1985) Some effects of ocean currents and wave motion on the dynamics of floating glacier tongues. In: Jacobs S (ed) Oceanology of the Antarctic Continental Shelf. Antarctic Research Series 43: 253–271, American Geophysical Union, Washington DC

Hollin JT (1962) On the glacial history of Antarctica. Journal of Glaciology 4(32): 173–195

Holmann H, Rummler H (1972) Alternierende Differentialformen. Bibliographisches Institut B.I.-Wissenschaftsverlag, Zürich, 257 pp

Hughes T (1973) Is the West Antarctic ice sheet disintegrating? Journal of Geophysical Research 78(33): 7884–7910

Hughes T (1977) West Antarctic ice streams. Reviews of Geophysics and Space Physics 15(1): 1–46

Hughes T (1982) On the disintegration of ice shelves: The role of thinning. Annals of Glaciology 3:146–151

Huybrechts P (1993) Glaciological modelling of the late cenozoic East Antarctic Ice Sheet: Stability or dynamism? Gegrafiska Annaler 75A: 221–238

IfAG [Institut für Angewandte Geodäsie] (1982) Neuschwabenland. 1:3,000,000. Frankfurt am Main.

IfAG [Institut für Angewandte Geodäsie] (1996) Filchner-Ronne-Schelfeis. Topographic Map (Satellite Image Map) 1:2 000 000, 2nd edn. Frankfurt am Main.

Jezek KC, Liu H, Zhao Z, Li B (1999) Improving a digital elevation model of Antarctica using radar remotes sensing data and GIS techniques. Polar Geography 23(3): 185–200.

Jonsson S, Holmlund P, Grudd H (1988) Glaciological and geomorphological studies of Scharffenbergbotnen. In: Fütterer DK (ed) Die Expedition ANTARKTIS VI mit FS 'POLARSTERN" 1987/88. Berichte zur Polarforschung 58: pp 186–193, 195

Journel AG, Huijbregts C (1989) Mining geostatistics, 2nd edn. Academic Press, London, 600 pp

Kamb B, Echelmeyer KA (1986) Stress-gradient coupling in glacier flow: IV. Effects of the "T" term. Journal of Glaciology 32(112) 342–349

Kamb B (1987) Glacier surge mechanism based on linked cavity configuration of the basal water conduit system. Journal of Geophysical Research 92(B9): 9083–9100

Kaula WM (1967) Theory of statistical analysis of data distributed over the sphere. Reviews of Geophysics and Space Physics 5: 83–107

Kleinschmidt G, Buggisch W (1993) Plate tectonic implications of the structure of the Shackleton Range, Antarctica. Polarforschung 63(1): 57–62

Kleinschmidt G, Roland NW, Bässler K-H, Kothe J, Olesch M, Schubert W, Braun H-M, Spaeth G, Buggisch W, Peters M, Höfle H-C (1988) Geologische Expedition in die Shackleton Range (GEISHA). In: Fütterer DK (ed) Die Expedition ANTARKTIS VI mit FS 'POLARSTERN" 1987/88. Berichte zur Polarforschung 58: 196–237

Kleinschmidt G, Clarkson PD, Tessensohn F, Grikurov GE, Buggisch W (1995) Geological history and regional implications. In: Thomson JW, Thomson MRA (eds) Geological map of Shackleton Range, Antarctica, scale 1:250000. BAS Geomap Series, pp 57–60. British Antarctic Survey, Cambridge

Korotkevitch ES, Koblents YaP, Kosenko NG (1977) Antarktida. Karta korennogo rel'efa Zemli Enderby [Antarctica. Map of the bedrock relief of Enderby Land]. Scale 1:2,500,000, Arctic and Antarctic Research Institute, Leningrad

Krabill W, Frederick E, Manizade S, Martin C, Sonntag J, Swift R, Thomas R, Wright W, Yungel J (1999) Rapid thinning of parts of the southern Greenland ice sheet. Science 283(5407): 1522–1524.

Krige DG (1951) A statistical approach to some basic mine valuation problems on the Witwatersrand. J. Chem. Min. Soc. S. Afr., 119–139

Krige DG (1966) Two-dimensional weighted moving average trend surfaces for ore evaluation, J. S. Afri. Inst. Min. Metal., Proc. Symp. Stat. Comput. Appl. Ore Valuation, 13–79

Kruchinin YA, Pinter S, Simonov IM (1967) Opredelenie skorosti i napravleniya dvizheniya lednikov v rayone stantsii Novolazarevskoy [Determination of the speed and direction of glacier movements in the area of Novolazarevskaya Station]. Informatsionnyy Byulleten' Sovetskoy Antarkticheskoy Ekspeditsii [Information Bulletin of the Soviet Antarctic Expedition] no 61: 26–31

Laird MG (1991) The Late Proterozoic – Middle Palaeozoic rocks of Antarctica. In: Tingey RJ (ed) The Geology of Antarctica, Oxford Monographs on Geology and Geophysics 17: 74–119. Clarendon Press, Oxford

Lauritzen S (1977) The probabilistic background of some statistical methods in physical geodesy. Publ. 48, Dan. Geod. Inst., Copenhagen, 96 pp

Ledenev VG, Yevdokimov AP (1965) Izmeneniya shel'fovykh lednikov Zapadnogo i Eymeri [Changes in the West and Amery Ice Shelves]. Informatsionnyy Byulleten' Sovetskoy Antarkticheskoy Ekspeditsii [Information Bulletin of the Soviet Antarctic Expedition] no 55: 12–18

LeGrand P, Minster J (1999) Impact of the GOCE gravity mission on ocean circulation estimates. Geophysical Research Letters 26: 1881–1884

LeMasurier WE (1972) Volcanic record of Cenozoic glacial history of Marie Byrd Land. In: Adie RJ (ed) Antarctic geology and geophysics. Universitetsforlaget, Oslo, pp 251–259

LeMasurier WE, Rex DC (1982) Volcanic record of Cenozoic glacial history in Marie Byrd Land and western Ellsworth Land: Revised chronology and evaluation of tectonic factors. In: Craddock C (ed) Antarctic geoscience. University of Wisconsin Press, Madison, pp 725–734

Lerch FJ, Marsh JG, Klosko SM, Williamson, RG (1982) Gravity model improvement for SEASAT. Journal of Geophysical Research 87: 3281–3296

Lindstrom D, Tyler D (1985) Preliminary results of Pine Island and Thwaites Glaciers study. Antarctic Journal of the United States 19(5): 53–55

Lingle CS, Brenner AC, Zwally HJ (1990) Satellite altimetry, semivariograms, and seasonal elevation changes in the ablation zone of West Greenland. Annals of Glaciology 14: 158–163

Lingle CS, Lee, L-h, Zwally H, Seiss TC (1994) Recent elevation increase on Lambert Glacier, Antarctica, from orbit cross-over analysis of satellite-radar altimetry. Annals of Glaciology 20: 26–32

Liu H, Jezek K, Li B (1999) Development of Antarcic digital elevation model by integrating cartographic and remotely sensed data: A GIS-based approach. Journal of Geophysical Research 104(B10): 23199–23213.

Loewe F (1956) Contributions to the glaciology of the Antarctic. Journal of Glaciology 2(19): 657–665

Lythe MB, Vaughan DG, BEDMAP Consortium (2001) BEDMAP: a new ice thickness and subglacial topographic model of Antarctica. Journal of Geophysical Research 106(B6): 11335–11351

MacAyeal DR, Hulbe CL, Scambos TA, Fahnestock MA (2002) Explosive break-up of Larsen Ice Shelf, Antarctica, by a meltwater-triggered iceberg capsize mechanism. Eos Transactions American Geophysical Union 83(47 Suppl.): F301

MacDonald WR (1976) Antarctic cartography. In: Williams RSjr, Carter WD (eds) ERTS-1, a new window on our planet. U.S. Geological Survey Professional Paper 929: 37–43

Madigan CT (1929) Meteorology. Tabulated and reduced records of the Cape Denison station, Adelie Land. Australasian Antarctic Expedition 1911–14, Scientific Reports Series B vol 4: 286 pp. Government Printer, Sydney

Marsh JG et 16 al. (1989) The GEM T2 gravitational model. NASA Technical Memorandum no 100746: 94 pp. Goddard Space Flight Center, Greenbelt, Maryland

Marsh PD (1985) Ice surface and bedrock topography in Coats Land and part of Dronning Maud Land, Antarctica. British Antarctic Survey Bulletin no 68: 19–36

Martin TV, Zwally HJ, Brenner AC, Bindschadler RA (1983) Analysis and retracking of continental ice sheet radar altimeter waveforms. Journal of Geophysical Research 88(C3): 1608–1616

Matheron G (1963) Principles of geostatistics. Economic Geology 58(8): 1246–1266

Matheron G (1971) The theory of regionalized variables and its applications. Cah. Cent. Morph. Math. Fontainebleau 5: 211 pp

Mayer H, Herzfeld UC (2000) Structural glaciology of the fast-moving Jakobshavn Isbræ, Greenland, compared to the surging Bering Glacier, Alaska, U.S.A. Annals of Glaciology 30: 243–249

Mayer H, Herzfeld UC (2001) A structural segmentation, kinematic analysis and dynamic interpretation of Jakobshavns Isbræ, West Greenland. Zeitschrift für Gletscherkunde und Glazialgeologie 37(2): 107–123

Mayewski PA, Attig JWjr, Drewry DJ (1979) Pattern of ice surface lowering for Rennick Glacier, northern Victoria Land, Antarctica. Journal of Glaciology 22(86): 53–65

McIntyre NF (1985) A re-assessment of the mass balance of the Lambert Glacier drainage basin, Antarctica. Journal of Glaciology 31(107): 34–38

McKelvey BC, Hambrey MJ, Harwood DM, Mabin MCG, Webb P-N, Whitehead JM (2001) The Pagodroma Group — a Cenozoic record of the East Antarctic ice sheet in the northern Prince Charles Mountains. Antarctic Science 13(4): 455–468

Meier MF (1984) Contribution of small glaciers to global sea level. Science 226: 1418–1421

Meier MF, Post A (1969) What are glacier surges? Canadian Journal of Earth Science 6: 807–817

Meier MF, Post A (1987) Fast tidewater glaciers. Journal of Geophysical Research 92(B9): 9051–9058

Meier S (1977) Die küstennahe Eisdecke des westlichen Enderby-Landes, Antarktis; Beiträge zu Relief, Bewegung und Massenhaushalt [The near coastal ice sheet of western Enderby Land, Antarctica; observations on topography, movement, and mass balance]. VEB Hermann Haack, Gotha/Leipzig, 104 pp

Meier S (1983) Portrait of an Antarctic outlet glacier. Hydrological Sciences Journal 28(3): 403–416

Mellor M (1960) Antarctic ice terminology: Ice dolines. Polar Record 10(64): 92

Mellor M, McKinnon G (1960) The Amery Ice Shelf and its hinterland. Polar Record 10(64): 30–34

Ménard Y, Fu L-L, Escudier P, Kunstmann G (1999) Jason-1, on the tracks of TOPEX/POSEIDON. Eos Transactions American Geophysical Union 80

Mercer JH (1978) West Antarctic ice sheet and CO_2 greenhouse effect: A threat of disaster. Nature 271(5643): 321–325

Millar IL, Talarico F (eds)(1999) The EUROSHACK Project; results of the European expedition to the Shackleton Range 1994/95. Terra Antarctica 6(3–4)

Minster J-F, Rémy F, Normant E (1993) Constraints on the repetitivity of the orbit of an altimetric satellite: Estimation of the cross-track slope. Journal of Geophysical Research 98: 10410–10419

Moffitt FH, Bouchard H (1975) Surveying. Intext Educational Publishers, New York, 6th edn, 879 pp

Morgan VI, Budd WF (1975) Radio-echosounding of the Lambert Glacier basin. Journal of Glaciology 15(73): 103–112

Moritz H (1980) Advanced physical geodesy. H. Wichmann Verlag, Karlsruhe, 500 pp

Moritz H (1984) Least squares collocation. In: Grafarend EW, Rapp RH (eds) Advances in geodesy. American Geophys. Union, Washington DC, pp 27–36 (from: Rev. Geophys. Space Phys., 1978, 16(3))

Nakawo M, Ageta Y, Yoshimura A (1978) Discharge of ice across the Soya Coast. In: Ishada T (ed) Glaciological studies in Mizuho Plateau, East Antarctica, 1969–1975. Memoirs of National Institute of Polar Research Special Issue no 7: 235–244, Tokyo

National Geographic Society (1992) National Geographic atlas of the world, 6th edn. National Geographic Society, Washington, DC, 138 pp

Nerem RS, Mitchum GT (2001) Sea level change. In: Fu L-L, Cazenave, A (eds) Satellite altimetry and earth sciences. Academic Press, San Diego, pp 329–349

Neuburg HAC, Thiel E, Walker PT, Behrendt JC, Aughenbaugh NB (1959) The Filchner Ice Shelf. Annals of the Association of American Geographers 49(2): 110–119

Nichols RL (1960) Geomorphology of Marguerite Bay area, Palmer Peninsula, Antarctica. Bulletin of the Geological Society of America 71(10): 1421–1450

Nolin AW, Herzfeld UC (2002a) Mapping ice sheet surface roughness with MISR. Third International Workshop on Multiangular Measurements and Models, Steamboat Spr., CO, USA, 10–12 June, 2002 (abstract)

Nolin AW, Herzfeld UC (2002b) Characterization of ice sheet surface roughness from MISR data. International Geosciences Remote Sensing Symposium, Toronto, Canada, 24–29 June, 2002 (abstract)

Nolin AW, Fetterer FM, Scambos TA (2002) Surface roughness characterization of sea ice and ice sheets: case studies with MISR data. IEEE Transactions on Geoscience and Remote Sensing 40(7): 1605–1615

Orheim O (1979) Flow of Antarctic ice shelves between longitudes 29° E and 44° W (abs). Journal of Glaciology 24(90): 484–485

Orheim O (1982) Radio echo-sounding of Riiser-Larsenisen (abs). Annals of Glaciology 3: p 355

Partington KC, Cudlip W, McIntyre NF, King-Hele S (1987) Mapping of the Amery Ice Shelf, Antarctica, surface features by satellite altimetry. Annals of Glaciology 9: 183–188

Partington KC, Ridley JK, Rapley CG, Zwally HJ (1989) Observations of the surface properties of the ice sheets by satellite radar altimetry. Journal of Glaciology 35(120): 267–275

Paterson WSB (1994) The Physics of Glaciers, 3rd edn. Pergamon Press, Oxford, 480 pp

Rapp RH (1992) Computation and accuracy of global geoid undulation models. Sixth International Geodetic Symposium on Satellite Positioning, Columbus, Ohio, March 1992, 8 pp

Rapp RH (1994) The use of potential coefficient models in computing geoid undulations. Course notes for the International School for the Determination and Use of the Geoid, October 1994. International Geoid Service, Politecnico di Milano, Milano, 28 pp

Rapp RH, Pavlis NK (1990) The development and analysis of Geopotential Coefficient models to spherical harmonic degree 360. Journal of Geophysical Research 95: 21885–21911

Raymond CF (1987) How do glaciers surge? A review. Journal of Geophysical Research 92(B9): 9121–9134

Reigber C, Massmann F-H, Flechtner F (1997) The PRARE system and the data analysis procedure. In: Rummel R, Reigber C, Hornik H (eds) Advanced space technology in geodesy — achievements and outlook. Coordination of Space Techniques for Geodesy and Geodynamics Bulletin 14: 73–80, Deutsches Geodätisches Forschungsinstitut, München

Reynolds JM (1981) The distribution of mean annual temperatures in the Antarctic Peninsula. British Antarctic Survey Bulletin no 54: 123–133

Ridley JK, Partington KC (1988) A model of satellite radar altimeter return from ice sheets. International Journal of Remote Sensing 9(4): 601–624

Rignot EJ (1998) Fast recession of a West Antarctic glacier. Science 281(5376): 549–551

Rignot E (2002) Ice-shelf changes in Pine Island Bay, Antarctica, 1947–2000. Journal of Glaciology 48(161): 247–256

Rignot E, Vaughan DG, Schmeltz M, Dupont T, MacAyeal D (2002) Acceleration of Pine Island and Thwaites Glaciers, West Antarctica. Annals of Glaciology 34: 189–194

Robin GdeQ (1979) Formation, flow and disintegration of ice shelves. Journal of Glaciology 24(90): 259–271

Robin GdeQ (1983) Coastal sites, Antarctica. In: Robin GdeQ (ed) The climatic record in polar ice sheets. Cambridge University Press, Cambridge, pp 118–122

Robin GdeQ, Swithinbank CWM, Smith BME (1970) Radio echo exploration of the Antarctic ice sheet. In: Gow AJ, Keeler C, Langway CC, Weeks WF (eds) International Symposium of Antarctic Glaciological Exploration (ISAGE), Hanover, New Hampshire, U.S.A., 3–7 September 1968. International Association of Scientific Hydrology Publication no 86: 97–115

Rose KE (1979) Characteristics of ice flow in Marie Byrd Land, Antarctica. Journal of Glaciology 24(90): 63–75

Rott H, Rack W, Skvarca P, Angelis HD (2002) Northern Larsen ice shelf, Antarctica: further retreat after collapse. Annals of Glaciology 34: 277–282.

Rubin J (1996) Antarctica. A lonely planet travel survival kit. Lonely Planet Publications, Hawthorn, Victoria, Australia, 362 pp

Sandwell DT, Smith WHF (2001) Bathymetric estimation. In: Fu L-L, Cazenave, A (eds) Satellite altimetry and earth sciences. Academic Press, San Diego, pp 441–457

Schmeltz M, Rignot E, MacAyeal D (2002) Tidal flexure along ice-sheet margins: comparison of InSAR with an elastic-plate model. Annals of Glaciology 34: 202–208

Schmidt T, Mellinger G (1966) Bestimmungen von Eisbewegungen am Rand des antarktischen Inlandeises [Determination of ice movement on the margin of the inland ice sheet of Antarctica]. Nationalkomitee für Geodäsie und Geophysik der Deutschen Demokratischen Republik Reihe III no 4: 32 pp

Schutz BE, Tapley BD, Shum CK (1985) Precise SEASAT ephemeris from laser and altimeter data. Adv. Space Res. 5: 155–168

Seelye M, Steffen K, Cavalieri D (1992) Chapter 15: Microwave Remote Sensing of Polynyas. In: Carsey FD (ed) Microwave Remote Sensing of Sea Ice. Geophysical Monograph 68: 303–312. American Geophysical Union, Washington, DC

Shabtaie S, Bentley CR (1982) Tabular icebergs: Implications from geophysical studies of ice shelves. Journal of Glaciology 28(100): 413–430

Shackleton EH (1919) South, the story of Shackleton's last expedition 1914–1917. Heinemann, London, 376 pp

Shepherd A, Wingham DJ, Mansley JAD, Corr HFJ (2001) Inland thinning of Pine Island Glacier, West Antarctica. Science 291(5505): 862–864

Sievers J, Vaughan DG, Bombosch A, Doake CSM, Heidrich B, Mantripp DR, Pozdeev VS, Ritter B, Sandhaeger H, Schenke HW, Swithinbank C, Thiel M, Thyssen F (1993) Topographische Karte (Satellitenbildkarte) 1 : 2000000 Filchner-Ronne-Schelfeis, Antarktis. Institut für Angewandte Geodäsie (IfAG), Frankfurt am Main

Smith AM (1986) Ice rumples on Ronne Ice Shelf, Antarctica. British Antarctic Survey Bulletin no 72: 47–52

Snyder JP (1987) Map projections — a working manual. U.S. Geological Survey Professional Paper no 1395: 383 pp

Southard RB, MacDonald WR (1974) The cartographic and scientific application of ERTS-1 imagery in polar regions. Journal of Research U.S. Geological Survey 2(4): 385–394

Spiegel MR, Schiller J, Srinivasan RA (2000) Theory and Problems of Probability and Statistics, Schaum's Outline Series. McGraw-Hill, New York, 2nd edn, 408 pp

Steed RHN, Drewry DJ (1982) Radio-echosounding investigations of Wilkes Land, Antarctica. In: Craddock C (ed) Antarctic geoscience. University of Wisconsin Press, Madison, pp 969–975

Steffen K, Box JE (2001) Surface climatology of the Greenland ice sheet: Greenland Climate Network 1995–1999. Journal of Geophysical Research 106(D24): 33951–33964

Steffen K, Comiso J, StGermain K, Gloersen P, Key J, Rubinstein I (1992) Chapter 10: The Estimation of Geophysical Parameters Using Passive Microwave Algorithms. In: Carsey FD (ed) Microwave Remote Sensing of Sea Ice. Geophysical Monograph 68: 201–228. American Geophysical Union, Washington, DC

Stirzaker D (1999) Probability and Random Variables, A Beginner's Guide. Cambridge University Press, Cambridge, 368 pp

Stosius R, Herzfeld UC (2004) Geostatistical estimation from radar altimeter data with respect to morphologic units outlined by SAR data – application to Lambert Glacier/Amery Ice Shelf. Annals of Glaciology 39

Swithinbank C (1957a) The morphology of the ice shelves of western Dronning Maud Land. Norwegian-British-Swedish Antarctic Expedition 1949–52 Scientific Results vol 3A: 37 pp. Norsk Polarinstitutt, Oslo

Swithinbank C (1957b) The regime of the ice shelf at Maudheim as shown by stake measurements. Norwegian-British-Swedish Antarctic Expedition 1949–52 Scientific Results vol 3B: 43–75. Norsk Polarinstitutt, Oslo

Swithinbank C (1958) The morphology of the inland ice sheet and nunatak areas of western Dronning Maud Land. Norwegian-British-Swedish Antarctic Expedition 1949–52 Scientific Results vol 3D: 99–117. Norsk Polarinstitutt, Oslo

Swithinbank C (1963) Ice movement of valley glaciers flowing into the Ross Ice Shelf, Antarctica. Science 141(3580): 523–524

Swithinbank C (1966) A year with the Russians in Antarctica. Geographical Journal 132(4): 463–475

Swithinbank C (1968) Radio echo sounding of Antarctic glaciers from light aircraft. In: Ward W (ed) General Assembly of Bern, Commission of Snow and Ice, Reports and discussions. International Association of Scientific Hydrology Publication no 79: 405–414

Swithinbank C (1969) Giant icebergs in the Weddell Sea, 1967–68. Polar Record 14(91): 477–478

Swithinbank C (1988) Antarctica. In: Williams RSjr, Ferrigno, JG (eds) Satellite image atlas of glaciers of the world. United States Geological Survey Professional Paper 1386-B: B1–B278

Swithinbank C, McClain P, Little P (1977) Drift tracks of Antarctic icebergs. Polar Record 18(116): 495–501

Tapley BD, Kim M-C (2001) Applications to geodesy. In: Fu L-L, Cazenave, A (eds) Satellite altimetry and earth sciences. Academic Press, San Diego, pp 371–406

Tessensohn F, Kleinschmidt G, Talarico F, Buggisch W, Brommer A, Henjes-Kunst F, Kroner U, Millar IL, Zeh A (1999) Ross-age amalgamation of East and West Gondwana; evidence from the Shackleton Range, Antarctica. In: Millar IL, Talarico F (eds) The EUROSHACK Project; results of the European expedition to the Shackleton Range 1994/95. Terra Antartica 6(3–4): 317–325

Thomas RH (1973) The dynamics of the Brunt Ice Shelf, Coats Land, Antarctica. British Antarctic Survey Scientific Reports no 79: 45 pp

Thomas RH, Sanderson TJO, Rose KE (1979) Effect of climatic warming on the West Antarctic ice sheet. Nature 277(5695): 355–358

Thomas RH, Martin TV, Zwally HJ (1983) Mapping ice-sheet margins from radar altimetry data. Annals of Glaciology 4: 283–288

Tierney TJ (1975) An externally draining freshwater system in the Vestfold Hills, Antarctica. Polar Record 17(111): 684–686

Tingey RJ (ed)(1991) The Geology of Antarctica. Oxford Monographs on Geology and Geophysics 17: 680 pp. Clarendon Press, Oxford

Tolstikov EI (ed) (1966) Atlas Antarktiki. Glavnoe upravlenie geodezii i kartografii vol 1: 236 pp, Moskva

Torge W (1980) Geodesy. Walter de Gruyter, Berlin, 254 pp

Tscherning CC (1984) Local approximation of the gravity potential by least squares collocation. In: Schwarz KP (ed) Proc. Beijing Internat. Summer School "Local gravity field approximation" 1984, Div. Surveying Engineering Publ. 65003, University of Calgary, Calgary, 277–362

Ulaby FT, Moore RK, Fung AK (1981) Microwave remote sensing, active and passive. Vol I Microwave remote sensing fundamentals and radiometry. Addison-Wesley, Reading, MA, 456 pp

Ulaby FT, Moore RK, Fung AK (1986a) Microwave remote sensing, active and passive. Vol II Radar remote sensing and surface scattering and emission theory. Addison-Wesley, Reading, MA, 608 pp

Ulaby FT, Moore RK, Fung AK (1986b) Microwave remote sensing, active and passive. Vol III From theory to applications. Addison-Wesley, Reading, MA, 1705 pp

Van Autenboer T, Decleir H (1978) Glacier discharge in the Sør Rondane, a contribution to the mass balance of Dronning Maud Land, Antarctica. Zeitschrift für Gletscherkunde und Glazialgeologie 14(1): 1–16

Vaughan DG (1993) Implications of break-up of Wordie Ice Shelf, Antarctica for sea level. Antarctic Science 5(4): 403–408

Vaughan DG, Marshall GJ, Connolley WM, King JC, Mulvaney R (2001) Devil in the detail. Science 293(5536): 56–58

Velicogna I, Wahr J (2002) A method for separating Antarctic postglacial rebound and ice mass balance using future ICESat Geoscience Laser Altimeter System, Gravity Recovery and Climate Experiment, and GPS satellite data. J. Geophys. Res. 107 (B10): 2263, doi:10.1029/2001JB000708

Wackernagel H (1998) Multivariate geostatistics. 2nd edn, Springer-Verlag, Berlin, 291 pp

Wahr J, Molenaar M, Bryan F (1998) Time variability of the Earth's gravity field: hydrological and oceanic effects and their possible detection using GRACE. Journal of Geophysical Research 103: 30205–30229

Wahr J, Wingham D, Bentley CR (2000) A method of combining ICESAT and GRACE satellite data to constrain Antarctic mass balance. J. Geophys. Res. B7(105): 16279-16294

Warrick RA, Le Provost C, Meier MF, Oerlemans J, Woodworth PL (1996) with contributions by Alley RB, Bindschadler RA, Bentley CR, Braithwaite RJ, de Wolde JR, Douglas BC, Dyurgerov M, Flemming NC, Genthon C, Gornitz V, Gregory J, Haeberli W, Huybrechts P, Jóhannesson T, Mikolajewicz U, Raper SCB, Sahagian DL, van de Wal RSW, Wigley TML. Changes in sea level. In: Houghton JT, Meira Filho LG, Callander BA, Harris N, Kattenberg A, Maskell K (eds) Climate change 1995: The science of climate change. Cambridge University Press, Cambridge, pp 359–405

Webers GF, Splettstoesser JF (1982) Geology, paleontology, and bibliography of the Ellsworth Mountains. Antarctic Journal of the United States 17(5): 36–38

354

Weertman J (1974) Stability of the junction of an ice sheet and an ice shelf. Journal of Glaciology 13(67): 3–11

Wellmann P (1982) Surging of Fisher Glacier, Eastern Antarctica: evidence from geomorphology. Journal of Glaciology 28(98): 23–28

Wendler G, Ahlnäs K, Lingle CS (1996) On the Mertz and Ninnis Glaciers, East Antarctica. Journal of Glaciology 42(142): 447–453

Whillans IM, Bolzan J, Shabtaie S (1987) Velocity of Ice Streams B and C, Antarctica. Journal of Geophysical Research 92(B9): 8895–8902

Wrobel BP, Walter H, Friehl M, Hoppe U, Schlüter M, Steineck D (2000) A topographical data set of the glacier region at San Martin, Marguerite Bay, Antarctic Peninsula, generated by digital photogrammetry. Polarforschung 67(1/2) [1997]: 53–63

Zebker HA, Villasenor J (1992) Decorrelation in interferometric radar echoes. IEEE Transactions on Geoscience and Remote Sensing 30: 950–959

Zwally HJ, Brenner AC (2001) Ice sheet dynamics and mass balance. In: Fu L-L, Cazenave, A (eds) Satellite altimetry and earth sciences. Academic Press, San Diego, pp 351–369

Zwally HJ, Bindschadler RA, Brenner AC, Martin TV, Thomas RH (1983) Surface elevation contours of Greenland and Antarctic ice sheets. Journal of Geophysical Research 88(C3): 1589–1596

Zwally HJ, Stephenson SN, Bindschadler RA, Thomas RH (1987) Antarctic ice-shelf boundaries and elevations from satellite radar altimetry. Annals of Glaciology 9: 229–235

(I) Appendix

(I.1) Glaciological Glossary

Blue-ice areas are bare-ice ablation areas. Ablation areas tend to develop downwind of obstacles that break up ice flow. The blue color results from seasonal melting and refreezing.

Calving: One speaks of calving if an iceberg breaks off a glacier that terminates in an ocean, large lake, or fjord, or off an ice sheet, and the iceberg floats away; see also *iceberg*.

Drainage basin: The drainage basin of a glacier, in particular of an outlet glacier of an ice sheet, is the entire area from which ice flows in that glacier.

Equilibrium line: The equilibrium line is the line between the accumulation and the ablation areas of a glacier.

Fast ice is a terminus for sea ice that is frozen together; the term is used for instance to distinguish between shelf ice and internally connected sea ice. The fast ice is much thinner than an ice shelf.

Fast-moving ice: Types of fast-moving ice include (Clarke 1987) continuously fast-moving glaciers, surge-type glaciers, tidewater glaciers, and the West Antarctic ice streams, which form their own class, moving absolutely rather slowly but otherwise sharing properties of fast-moving ice. Fast-moving glaciers are relatively rare but important.

Glacier: A glacier is a body of ice that moves following gravity. In a narrower sense of the word, a *mountain glacier* or *valley glacier* is understood (glaciers in mountain ranges follow valleys). The term "glacier" is also used in a more general sense for an object out of a class of objects that includes glaciers, ice sheets, ice streams, ice shelves, ice rises, ice piedmonts, outlet glaciers and valley glaciers. A glacier has an accumulation area and an ablation area. In the accumulation area, more mass is gained through snowfall, rainfall and condensation than is lost through ablation and evaporation or sublimation. In the ablation area, more mass is lost through ablation (and evaporation) than is gained through accumulation. Ablation needs to occur, however, for the area of mass loss to be termed "ablation area"; see also *mass balance*. The line separating the areas of mass loss and mass gain is the equilibrium line. This model assumes a temperate glacier. It is possible that a glacier is located entirely in a zone where no ablation occurs. A glacier flows downwards, following gravity, in case of a valley glacier it has a narrow tongue. A *piedmont glacier* ends at the foot of a mountain range, where it forms a wide *lobus* rather than a narrow tongue.

Glacier tongue: A glacier tongue consists of floating glacier ice that is still connected (but possibly heavily crevassed) and extends seaward from a glacier; see also *iceberg tongue*. A glacier tongue is the seaward extension of a single land glacier. Most glacier tongues are composed largely of land ice, other processes that contribute to the formation of a glacier tongue are the same ones as outlined for ice shelves; see also *ice shelf*.

Grounding line: The grounding line of a glacier is defined as the transition of grounded ice (ice moving over rock) and floating ice (ice floating on water) of a glacier that extends from land into the ocean or a fjord. Oftentimes, a glacier has a "grounding zone", a transition zone, rather than a clear grounding line.

The position of its grounding line may be used to monitor advance and retreat of a glacier, because small changes in ice thickness translate into large changes in grounding line position (for geometrical reasons because of the small surface slope typical of Antarctic ice streams). The principles are the following: Based on a model that assumes idealized bedrock and perfect plasticity of ice, Weertman (1974) showed that the surface slope of an ice sheet decreases with the transition from grounded to floating ice. Using this result, a break in slope may be taken as a grounding-line indicator.

A second indicator of the grounding-line position is the surface roughness of the glacier and the ice stream: Sliding over the rough glacier bed induces a rougher surface topography than floating on the smooth water surface; this means that grounded glacier ice is rough and floating ice-shelf ice is smooth.

The grounding line is a feature suitable to monitor changes – if it can be observed.

Grounding zone: Use of the term grounding line assumes that a clear line runs more or less straight across the glacier, identifying the boundary between grounded ice on one side and floating ice on the other. On some glaciers the apparent grounding line meanders across the glacier, or patchy areas indicate grounding on seafloor shoals. These features are indicators of a grounding zone rather than a grounding line.

Iceberg: An iceberg is a piece of ice that broke off the floating part of a fjord glacier or an ice sheet; see also *calving*.

Iceberg tongue: An iceberg tongue consists of loosely connected pieces of ice that float off the same land glacier (icebergs); typically, an iceberg tongue is the seaward extension of a glacier tongue (but not every glacier tongue has an iceberg tongue); icebergs calve off along the crevasses visible in the glacier tongue, as the glacier tongue moves seaward; see also *glacier tongue*. An example of a glacier that has both a glacier tongue and an iceberg tongue is Thwaites Glacier (see section F.13).

Ice divide: Between two adjacent drainage basins an ice divide forms; see also *drainage basin*. Imag-ine that at first only a line separates two adjacent drainage basins; from either side of the line, ice flows in opposite directions. An ice divide is hence defined as a line that separates ice of opposing flow direction (more accurately, ice of opposing orientations of direction). In a digital elevation model, an ice divide may be recognized as a line connecting local maxima in elevation; an ice divide in the interior of Antarctica is often a smooth ridge.

Ice doline: An ice doline is a large oval-shaped depression in an ice shelf. Ice dolines have been described as areas where ice caved in; the terms **ice caldera, ice crater, ice volcano**, which are less fitting, have also been used. The term "ice doline" follows Mellor and McKinnon (1960); see Swithinbank (1988, p. B10–11), where an example of an ice doline with a 1 km diameter observed in Princess Astrid Coast is given.

Ice front: The ice front is the (waterward) edge of an ice shelf (which is floating on the water); see also *ice shelf, ice wall*. In Antarctica, ice shelves rise 2–60 m above sea level, hence the ice front is 2–60 m high. According to Drewry (1983), 44% of the Antarctic "coastline" is made up of ice formed by fronts of ice shelves, 39% by ice walls, 13% by ice streams and outlet glaciers (which may have an ice front or ice wall) and only 5% by rock (after Swithinbank 1988 who distinguishes between "ice front" and "ice wall"). The term "ice front" is also used more generally for the forward (in the sense of flow) edge of a glacier, including ice fronts on land.

Ice rise: An ice rise is a dome-shaped ice cap surrounded by shelf ice on the continental shelf, which consists of ice that does not flow in the direction of the surrounding ice but rests on a seafloor shoal on rocks. An ice rise may be considered an ice-covered island, but the bedrock may not necessarily rise above sea level. It is likely that most ice rises originate from localized grounding of an ice shelf (but technically grounded areas are not part of an ice shelf). An ice rise is bounded by the grounding line, or, in case it faces the sea, by an ice wall; see also: *ice shelf, ice rumple, grouding line, grounding zone*.

Ice rumple: Ice rumples form where an ice shelf flows over large seafloor shoals, and the ice thickness and surface elevation increases upstream of the shoal. Ice rumples have characteristic crevasse patterns. Upstream of those patterns, the ice surface rises. — An example of an ice rumple shows concentric bulges transected by radially extending lines of crevasses around a silghtly bulged, but otherwise undisturbed central area (the grounded area); the radially extending crevasse lines open in a direction normal to the lines of extension and hence approximately normal to the bulges. Some waviness disturbs the radial pattern, and some additional straight-line bulges intersect at a small oblique angle (Example of Donald Ice Rumples, Brunt Ice Shelf, Coats Land, Swithinbank 1988, fig. 9; the exact shape of the bulges depends on the shape of the seafloor shoal on which the rumples form). — The difference between an ice rumple and an *ice rise* is that the ice flow on top of a rumple matches that of the surrounding ice shelf, whereas the ice atop an ice rise rests on top of the shoal, and the shelf ice flows around the ice rise or is halted by it; see also *ice rise*. The direction of ice movement can be inferred from crevasse patterns. Some ice rises are more than 100 m high, but ice rumples are only a few tens of meters high.

Ice sheet: An ice sheet is characterized best by the so-called shallow ice approximation of continuum mechanics: Its thickness is much less than its diameter (the shallow ice approximation uses a relationship of 1:100). An ice sheet is a glacier (see *glacier*) that covers a region with complex bedrock topography, which may be understood in the context of shrinking or growing ice sheets: First, imagine a mountain range that contains valley glaciers. If the climate becomes increasingly colder, the valley glaciers will grow, until their ice has filled the valleys, after that, the ice will build up further and form a cap across the then sunken valleys — an ice sheet has formed. Flow is outward, locally still influenced by the location of the then subglacial valleys; see also *ice stream, outlet glacier*. Second, assume a large region that is covered with an ice sheet. In a warming climate, the ice sheet will shrink, and with further warming, the ice sheet will become less connected, until only glaciers in valleys of the bedrock are left. The Earth's largest ice sheets are the Antarctic and Greenland Ice Sheets. But there are also much smaller areas that are covered by ice sheets, examples are an ice sheet on the Kenai Peninsula in southern Alaska and the Northern and Southern Patagonian Inland Ice Sheets.

Ice shelf: Ice shelves are floating ice sheets that fall and rise with ocean tides. Thickness ranges from 10 m to 2000 m (Swithinbank 1988), most are between 100 m and 500 m thick. Ice sheets or glaciers may not terminate on land, but on water, either in fjords, where a glacier may form a floating *glacier tongue*, or on the ocean, where the glacier may form an ice shelf or a glacier tongue or calve off; see also *glacier tongue, iceberg tongue, calving*. An ice shelf consists of ice that originated as ice sheet or glacier ice that formed on land (freshwater ice) with a bottom marine ice layer (ocean ice frozen to the bottom of the ice shelf) and possibly a top of snow or firn that accumulated and transformed on the ice shelf (freshwater ice). An ice shelf moves forward, that is, farther out into the ocean (or water body) by flow induced by the flow of the land glaciers that feed into the ice shelf, and by spreading in response to local snow accumulation. At the (seaward) front of a large ice shelf, no original land ice may be left, however, as during the time it has taken the ice to move from the inner areas to the outer areas of the ice shelf, ice mass may have been lost to bottom melting (into the ocean), and the top layers may have entirely formed from material accumulated after the ice crossed the grounding line (Swithinbank 1988). Since ice has a lower density than water (by 10 percent) and the more, than seawater (whose density varies with composition), about $\frac{1}{8}$ of the thickness of an ice shelf (or an iceberg) sticks out of the water and $\frac{7}{8}$ are under water. Ice shelves can only exist in the cold climates of Antarctica. The only Arctic ice shelf broke off recently (late 2003).

Ice stream: An ice stream is formed by ice within an ice sheet that flows much faster than the surrounding inland ice. Causes of ice-stream formation are multiple, and in case an ice stream is discovered, the causes may not even be known. (1) The ice accelerates (dynamic cause). (2) Ice follows a trough in the bedrock and hence accelerates (geologic cause). (3) Warming or cooling may change ice-stream velocity, it may "switch" an ice stream "on" or "off", as has been hypothesized about some of the West Antarctic Ice Streams (Siple Coast, Ross Sea/ Ross Ice Shelf).

Ice wall: An ice wall is a vertical ice cliff that forms part of the coastline (as does an ice front in the narrower sense of the word; Swithinbank 1988), but, other than at an ice front, an ice sheet is grounded at an ice wall; see also *ice front*. Water depth at an ice wall may be 0–500 m in Antarctica. If the rock basement is at sea level, the ice cliff is termed **strand ice wall**, if it is below sea level, the cliff is termed **neritic ice wall**. From a ship approaching an ice cliff, an ice wall may be hard to distinguish from an ice front, as one cannot see whether the ice shelf is afloat at a given point.

Ice-wall coastline: A typical *ice-wall coastline* may cross an island archipelago. Here, a narrow band of rock is visible at the foot of an ice cliff (except where ice streams cross the coastline). Further advance of the ice wall is prevented by a relatively high marine melting rate, and consequently the ice-wall coastline is likely to remain in a stable position that is mainly controlled by sea level (Hollin 1962). Ice walls can be in shallow water (strand ice wall)(ragged coastline); or submerged (neritic ice wall), the latter rest on rock that is up to several hundred meters below sea level. Typical surface velocities at the ice wall are 10–40 m a^{-1}.

Mass balance: The ice sheet mass balance is the difference between mass input and mass output (or mass loss). Mass input processes are snowfall, rainfall, and condensation. Mass loss from an ice sheet can occur through evaporation, sublimation, surface and bottom melting, water runoff and iceberg discharge. Total mass balance includes surface mass balance processes and ice-flow components.

Melt pond: Melt ponds may form on the surface of an ice sheet or ice shelf or glacier during the melt season. In colder weather, the melt ponds may refreeze, either surficially (which poses a danger to glacier travelers who may break through the surface) or entirely, forming blue-ice areas. Melt ponds and blue-ice areas are easily visible in satellite imagery, because they have a different albedo than the surrounding ice (unless the entire ice area is snow covered). As a large part of Antarctica is not in the ablation zone, new occurrence of melt ponds indicates warming of an ice shelf or an area of the inland ice. Hence melt ponds may be warning signs of impending ice-shelf break-up in the near future, as has been observed in ice shelves that have recently disintegrated along the Antarctic Peninsula.

Neritic ice wall: see *ice wall*.

Outlet glacier: A glacier whose ice starts as part of an ice sheet, but then forms a distinguishable small glacier, as it, for instance, traverses a moun-

tain range near the margin of the ice sheet, is termed an outlet glacier; see also *ice stream*.

Polynia: A polynia is an open water anomaly.

Relief energy: Mountainous terrain is characterized by large changes in elevation over short distances — this is termed "high relief energy" in geography and is a better characterization than absolute elevation. On the altimetry-derived maps, individual mountains are not outlined because of the relatively low resolution of the altimeter compared to the size (area) of a mountain. The relief energy, however, still translates into rough terrain at this scale and type of mapping.

Sea ice is ice that forms by freezing of sea water. There may be recent snow on top of sea ice. Sea ice may last from one season to many years.

Shuga: Shuga is an accumulation of spongy white lumps formed by snow freezing together with crystals of rapidly freezing seawater. Shuga form when cold katabatic winds carry drift snow out from the land over the seawater.

Snow megadunes: "Megadunes" is a term given to surface features observed in the snow and ice surface of Wilkes Land at elevations of roughly 3000 m in the region (126°–132° E/76°-77° S) in LANDSAT 1MSS imagery (Swithinbank 1988, fig. 45, p. B57–58). The features are subparallel, slightly wavy, a bit offset to each other and have a characteristic distance of 2.5 km from one maximum to the next. Swithinbank (1988) attributes the existence of such "megadunes" most likely to the redistribution of newly fallen snow by wind, with lighter bands representing newer snow than darker bands, or to morphologic structures, where slight change in slope may result in change in albedo. Snow megadunes are likely to be physically similar to sand megadunes, as described by Cornish (1914) and Bagnold (1941) for eastern Iran and the Namib desert; the height of sand dunes is several tens of meters, the height of snow dunes 1–2 m, so they may not be noticeable to the surface traveler.

Strand ice wall: see *ice wall*.

Surge-type glacier: A surge-type glacier is a glacier that does not move forward with about the same velocity at all times (or, a bit slower in winter than in summer, as usual alpine glaciers). To the contrary, the surge glacier has a long, quiescent phase of continuous movement, during this time, ice builds up in a given part of the glacier. Suddenly, a short surge phase starts, during which the glacier accelerates to about 100 times its normal velocity (the actual factor depends on the glacier), more ice is moved downglacier, and, due to high velocity, the ice surface breaks up in a maze of crevasses. The surge phase typically lasts one to a few years, the quiescent phase 25–100 years, depending on the glacier and the geographic area.

(I.2) Index of Place Names

George VI Sound	m292e284-300n67-721
	m285e267-303n71-77
Getz Ice Shelf	m237e219-255n71-77
Hays Glacier	m45e37-53n63-68
Heimefront Range	m357e339-359n71-77
Henry Ice Rise	m285e267-303n78-815
	m309e291-327n78-815
Herbert Mountains	m333e315-351n78-815
	m357e339-15n78-815
Hercules Inlet	m285e267-303n78-815
Heritage Range	m285e267-303n78-815
Hillary Coast	m165e147-183n75-80
	m165e147-183n78-815
Hobbs Coast	m237e219-255n71-77
Hollick-Kenyon Plateau	m261e243-279n75-80
	m261e243-279n78-815
Hull Bay	m237e219-255n71-77
Hull Glacier	m213e195-231n71-77
Icestream D	m213e195-231n78-815
Icestream E	m213e195-231n78-815
Ingrid Christensen Coast	m69e61-77n67-721
	m81e73-89n67-721
Jason Peninsula	m297e289-305n63-68
Jelbart Ice Shelf	m357e349-5n67-721
Joinville Island	m297e289-305n63-68
Jutulstraumen Glacier	m357e349-5n67-721
	m357e339-359n71-77
Kemp Coast	m57e49-65n63-68
	m57e49-65n67-721
King Peninsula	m261e243-279n71-77
Kirwan Escarpment	m357e339-359n71-77
Knox Coast	
	m105e97-113n63-68
Korff Ice Rise	m285e267-303n75-80
	m285e267-303n78-815
Kraul Mountains	m333e315-351n71-77
	m357e339-359n71-77
Lambert Glacier	m69e61-77n67-721
	m69e51-87n71-77
Land Glacier	m213e195-231n71-77
Larsen Ice Shelf	m297e289-305n63-68
	m292e284-300n67-721
Lassiter Coast	m285e267-303n71-77
	m309e291-327n71-77
Law Dome	m105e97-113n63-68
	m117e109-125n63-68
LeMay Range	m292e284-300n67-721
Leopold and Astrid Coast	m81e73-89n63-68
	m81e73-89n67-721
Lillie Glacier	m165e157-173n67-721
Luetzow-Holm Bay	m33e25-41n67-721
Luitpold Coast	m333e315-351n75-80
	m333e315-351n78-815

Mac Robertson Land	m57e49-65n67-721
	m69e51-87n71-77
MacKenzie Bay	m69e61-77n67-721
Marguerite Bay	m292e284-300n67-721
Martin Peninsula	m237e219-255n71-77
Matusevich Glacier	m153e145-161n67-721
Maury Bay	m129e121-137n63-68
Mawson Coast	m57e49-65n63-68
	m69e61-77n63-68
	m57e49-65n67-721
	m69e61-77n67-721
Mawson Escarpment	m69e51-87n71-77
Mawson Peninsula	m153e145-161n67-721
McCarthy Inlet	m309e291-327n75-80
	m309e291-327n78-815
Mertz Glacier	m141e133-149n63-68
	m141e133-149n67-721
Mikhaylov Island	m81e73-89n63-68
Minnesota Glacier	m285e267-303n75-80
	m285e267-303n78-815
Moscow University Ice Shelf	m117e109-125n63-68
Mount Lister	m165e147-183n75-80
Mt. Gray	m213e195-231n71-77
	m237e219-255n71-77
Mt. Longhurst	m141e123-159n75-80
	m165e147-183n78-815
Mt. Melbourne	m165e147-183n71-77
Mt. Minto	m165e157-173n67-721
	m165e147-183n71-77
Mt. Northhampton	m165c147-183n71-77
Mulock Glacier	m165e147-183n75-80
Napier Mountains	m57e49-65n63-68
Ninnis Glacier	m141e133-149n67-721
	m153e145-161n67-721
North Highland	m129e121-137n63-68
Oates Coast	m165e157-173n67-721
Ob Bay	m165e157-173n67-721
Paulding Bay	m117e109-125n63-68
Penck Through	m357e339-359n71-77
Pennell Coast	m165e157-173n67-721
Perry Bay	m129e121-137n63-68
Philippi Glacier	m93e85-101n63-68
Pine Island Bay	m261e243-279n71-77
Polar Record Glacier	m69e61-77n67-721
	m81e73-89n67-721
Porpoise Bay	m129e121-137n63-68
Pourquoi Pas Glacier	m129e121-137n63-68
Prestrud Inlet	m213e195-231n75-80
Prince Albert Mountains	m165e147-183n71-77
	m165e147-183n75-80
Prince Charles Mountains	m57e49-65n67-721
	m69e61-77n67-721
	m69e51-87n71-77

(I.3) Antarctic Expeditions

(I.3.1) Early Seagoing Expeditions

The goal of early expeditions to southern oceans was to answer the question of existence of a southerly "Antarctic" continent. The following list summarizes some important steps of discovery and claims to fame. A short narrative is found in Rubin (1996).

17 January 1773 James Cook (England), first crossing of Antarctic circle, three world circumnavigations 1768-1771, 1772-1773, 1776-1779, discovery of islands (no Antarctic mainland discovered)

27 Jan 1820 Fabian von Bellingshausen and Mikhail Lazarev (Russia), first sighting of Antarctic mainland near (69° 21'S/2° 14'W), mapping of Princess Martha Coast and Princess Ragnhild Coast

1819-1822 James Weddell (England), seagoing expeditions, 1822 discovery of South Orkney islands, reached 74° 15' S in Weddell Sea, but no part of the Antarctic continent

30 Jan 1820 Bransfield (British expedition), sighting of Antarctic Peninsula

16 Nov 1820 Nathanial Palmer (USA), sealing expedition, sighted Trinity Island, Palmer Land, Antarctic Peninsula. Expedition member Powel went ashore at Coronation Island.

1820, 1821 Two sealers landed on the Antarctic Peninsula, first reported landings on the Antarctic continent, John Davis (Nantucket, USA) on 17 Feb 1821, John McFarlane (London, England) at an unreported date in summer 1820/21.

1780–1892 Circa 1100 sealing ships to Antarctica (possibly earlier unreported landings), only 25 exploratory expeditions

1895 Claim to first landing in Antarctica, Leonard Kristensen (Captain) and expedition member Carsten Borchgrevink and ship's boy Alexander Tunzleman

1899 First overwintering expedition on Antarctic continent, led by Carsten Borchgrevink (near Cape Adare)

(I.3.2) Expeditions to the Antarctic Continent

Table I.3.2-1. List of Selected Expeditions to the Antarctic Continent. List follows Alberts (1995).

1838–42	United States Exploring Expedition (Charles Wilkes)
1897–99	Belgian Antarctic Expedition (Adrien de Gerlache)
1898–1900	British Antarctic Expedition (Carstens E. Borchgrevink)
1901–03	German Antarctic Expedition (Erich von Drygalski)
1901–04	British National Antarctic Expedition (Robert F. Scott)
1901–04	Swedish Antarctic Expedition (Otto Nordenskjöld)
1902–04	Scottish National Antarctic Expedition (William S. Bruce)
1903–05	French Antarctic Expedition (Jean B. Charcot)
1907–09	British Antarctic Expedition (Ernest H. Shackleton)
1908–10	French Antarctic Expedition (Jean B. Charcot)
1910–13	British Antarctic Expedition (Robert F. Scott)
1911–12	German Antarctic Expedition (Wilhelm Filchner)
1911–14	Australian Antarctic Expedition (Douglas Mawson)
1928–30	Byrd Antarctic Expedition (Richard E Byrd)
1929–31	British-Australian-New Zealand Antarctic Research Expedition (BANZARE; Douglas Mawson)
1933–35	Byrd Antarctic Expedition (Richard E Byrd)
1934–37	British Graham Land Expedition (John Rymill)
1938–39	German Antarctic Expedition (Alfred Ritscher)
1939–41	United States Antarctic Service (Richard E. Byrd)
1946–47	United States Navy Operation Highjump (Richard E. Byrd)
1947–48	Ronne Antarctic Research Expedition (Finn Ronne)
1947–48	United States Navy Operation Windmill (Gerald L. Ketchum)
1947–	Australian National Antarctic Research Expedition (ANARE; various leaders)
1948–	French Antarctic Expedition (various leaders)
1949–52	Norwegian-British-Swedish Antarctic Expedition (John Giaever)
1955–58	Commonwealth Trans-Antarctic Expedition
1955–	Soviet Antarctic Expedition (various leaders)
1955–	United States Navy Operation Deep Freeze (various leaders)
1956–	Japanese Antarctic Research Expedition (various leaders)
1956–	Norwegian Antarctic Expedition (various leaders)
1957–58	Belgian Antarctic Expedition (Gaston de Gerlache)
1957–	New Zealand Geological Survey Antarctic Expedition (various leaders)
1958–	Victoria University of Wellington Antarctic Expedition (various leaders)
1962–63	New Zealand Federated Mountain Clubs Antarctic Expedition

(I.3.3) Antarctic Expeditions after the International Geophysical Year

The time of the International Geophysical Year (1 July 1957 – 31 December 1958) was selected to coincide with a time of extraordinary sunspot activity. The most important results of the International Geophysical Year (IGY), however, concern scientific discoveries of Antarctica. During the IGY, 20 countries built research stations in Antarctica.

Up to the time of the IGY, only a few expeditions had explored Antarctica. Since the IGY, many countries have been conducting Antarctic Research Programs and visiting Antarctica every year, in both summer and overwintering programs.

Printing and Binding: Stürtz AG, Würzburg